TOWARDS A NEW ASIAN ORDER

TOWARDS A NEW ASIAN ORDER

Foreword by
N.S. Sisodia

Editors
Ali Ahmed
Jagannath P. Panda
Prashant K. Singh

Institute for Defence Studies and Analyses, New Delhi

SHIPRA

Disclaimer: Views expressed are those of the contributors and do not necessarily reflect those of the Institute for Defence Studies and Anlyses or the Government of India.

ISBN: 978-81-7541-615-4

First Published in India in 2012

Towards a New Asian Order

SHIPRA PUBLICATIONS
H.O.: LG 18-19, Pankaj Central Market, Patparganj,
I.P. Ext., Delhi-110092, Tel: 91-11-2223 5152, 2223 6152
Br.Off.: 4617/18, Ansari Road, Darya Ganj,
New Delhi-110002, Tel: 6594 6453, 9810522367
E-mail: info@shiprapublications.com
www.shiprapublications.com

Foreword

The volume 'Towards a New Asian Order' is an ambitious undertaking by the contributors to conjure up an Asian order for the Asian Century, now in its second decade. The authors of the papers in this volume have attempted to conceptualise a future of peace and stability in Asia at a level beyond the sub-regional. This ambitious foray into defining such an order takes regionalism and thinking on integration further than its usual confines so far to Asia's sub-regions. The new order which has been conceptualised subsumes all Asian powers – China, India, US, Japan, South East Asia, Iran, to name but a few.

The endeavour is to visualise a non-confrontational future for Asia, even if a competitive one. Pre-existing spaces for cooperation made possible by strategic cultures informed by Asian values have been furthered over the last quarter century by forces of globalisation. This mellows commentary on the instability said to accompany the simultaneous and rapid 'rise' of powers such as China and India, even as the off-shore 'balancer', the US, remains engaged. It is not so much 'balance of power' that will keep Asia on even keel, but the lack of need to resort to power balancing. This volume helps initiate thinking along such lines, in order that alternative futures emerge to some prognostications that Asia's future will imitate Europe's past.

The sense of optimism of the volume however is not at the expense of the practical and realistic. How Asian states engage with respective strategic challenges and how Asian societies confront socio-political dilemmas will determine Asia's trajectory. Traditional security concerns have not gone away, even as non-traditional ones loom as equally compelling. However, what emerges from the papers is that traditional tools are inadequate, forged as these have been over the past two centuries of considerably different context. New ways of looking to the future enables new ways of helping cope with it. The volume identifies the significant one, namely that an Asian security architecture can help deliver on the desirable public goods, prosperity and security. This has global implications in that Asian rivalries, being non-existent or moderated sufficiently, do not then get to disrupt security complexes elsewhere.

India clearly has a stake in such a future. It will help keep it from diverting its power and attention to power games. The energy thus disengaged can be productively employed for bettering the lives of its people. South Asian prosperity will be an outcome. Better indices on this subcontinent will substantially alter Asia's cumulative credentials. That two thirds of the seven billion people on the planet are better off has implications for the overall state of humanity as such.

It is with this vision the conference at which these papers were presented was organised. These papers have been further refined by a peer review by India's strategic community in attendance at the conference this February. We believe that the ideas that emerged will not only inform the agenda on how to shape the Asian century but will remain influential over the middle term as the many and profound recommendations are taken up in earnest for implementation.

New Delhi
November 2011

N.S. Sisodia
Director-General, IDSA

Contents

A New Asian Order for the Asian Century*

A.K. Antony

I have great pleasure in welcoming you all to the 13th Asian Security Conference. We all have gathered here on the occasion of Id-e-Milad, the birthday of the Prophet Mohammad. I extend my warm greetings to all of you, particularly to all our Muslim brethren. The Prophet's message of peace is today all the more relevant for all nations and societies, as they grapple with violence and terrorism, besides many other nefarious activities.

Today, we all miss the presence of Shri K Subrahmanyam. He would surely have been amidst us, had destiny not willed it otherwise. His strategic vision has helped the IDSA achieve its present status. I am happy to announce that a K. Subrahmanyam Chair is being established at IDSA in his memory. We hope that the outstanding scholars selected to occupy the Chair will carry forward Shri Subrahmanyam's outstanding work.

The theme of this 13th Asian Security Conference—'Towards a New Asian Order' is extremely relevant in today's context. The 21st century has for long been termed—'Asian Century'. As we enter the second decade of the 21st century, the building of a new Asian order is crucial, if Asia has to emerge as a world leader. This will be possible only in an atmosphere of peace and harmony and not under the shadow of guns. The contours of the Asian order will have a significant bearing on the world order.

The eminent gathering at this Conference includes distinguished speakers from the world over. Their experiences, thoughts and

---

* Based on the Presidential Address delivered at the 13th Asian Security Conference, Institute for Defence Studies and Analyses, New Delhi, February 16, 2011.

perspectives on Asia will go a long way towards setting the agenda for a stable, equitable and prosperous Asia.

For the Asian century to become a living reality, firstly, it is the responsibility of all Asian states to create and sustain such an order. Secondly, the traditional Asian ethos based on the principles of Panchsheel has to be at its core—i.e. respect for territorial integrity and sovereignty, non-aggression, non-interference in each other's internal affairs, equality and mutuality and peaceful co-existence. Sustained peace and security will lead to a more lasting economic prosperity, social well-being and equity with justice for the people of Asia. There will be contentious issues along the way, but they will have to be resolved through mutual understanding, dialogue at bilateral and multilateral levels and within a pan-Asian framework. However, if nations permit themselves to be swayed by the burden of historical factors, we will be putting the peace and security of the region and the world in grave peril.

India has always supported the resolution of issues through peaceful means and dialogue between nations. Soon after India's independence, New Delhi's avowed objective was to "provide a cultural and intellectual revival and social progress in Asia." Over the years, Asian nations have achieved this—individually, as well as collectively.

The last century witnessed two world wars. However, any large-scale conflict in nuclear age will have unforeseen consequences. The stability of the Asian region is crucial for world peace, as several nations of the region have become seething cauldrons of violence and unrest. Transition of power must invariably be through peaceful and non-violent means.

India's location on the map of Asia gives it a central and strategic importance. Nations such as India, China and Japan have historical and cultural linkages. Given such a unique interconnectivity between these and other Asian nations too, security-related challenges and their solutions require a unified approach from the entire region. Some of the prominent challenges that face the Asian region include—terrorism, nuclear proliferation, piracy and smuggling of arms. The leadership of Asian regions—individually, as well as collectively—needs to work upon building a common identity and an efficient security architecture. If the European Union can do it successfully, there is no reason why the Asian region can't better it.

This thirteenth Asian Security Conference must generate the fine print for a common, yet overarching security architecture. Nations of the region have to cooperate on a much larger scale. Some existing institutions like the East Asia Summit and SAARC can and must be broad-based, developed and fine-tuned. Such organisations can really go a long way towards ensuring lasting peace, harmony and stability that is rooted in mutual economic growth and sensitivity for each other's security concerns.

The twin processes of globalisation and liberalisation have made the probability of full-blown wars a remote possibility. Peace and stability have become irreversibly interlinked with economic growth. However, despite the economic interdependence, several Asian nations continue to be plagued by internal disturbances and militancy. Most often the expression of discontent by the people actually reflects their burning desire for good governance and all-round development.

The concept of security is rapidly expanding and evolving. Security today involves the measuring of economic growth in terms of the general standards of living of the poor and the underprivileged. Climate change and the direction in which the global economy is headed, form the core of any discourse on fresh security issues. I believe that climate change and the direction of the global economy also feature as issues to be addressed by this conference.

At times, there is a misplaced scepticism on the space available for the rise and growth of Asian powers. I have no doubt that provided we have the will, creation of space for all Asian nations to grow will never be a constraint.

Friends, India's democratic fabric empowers its citizens by giving them a direct stake in governance through universal adult franchise. Political transition has always been peaceful, non-violent and by the people's will. Successive governments have sought to bring about a transformation in India's diverse and pluralistic society within this democratic framework.

Our relations with other nations are dictated by our security imperatives. We have never wanted to further our relationship with a particular nation at the cost of relations with any other country. In our neighbourhood, we have always been in favour of resolving issues through dialogue and constant engagement.

As two of the largest nations of the region, India and China need to enhance cooperation levels on common issues such as climate change and trade. Recently, the 13th round of discussions at the level of special representatives on the border issue with China were concluded. The complicated border issue can be settled only through dialogue and discussions. We are not unduly concerned about China's military modernisation, but we need to carry out a comprehensive review of our defence preparedness and remain vigilant at all times.

The recent foreign secretary level talks with Pakistan demonstrate a desire on the part of both nations to carry forward the dialogue process.

However, any meaningful progress can only be made only when Pakistan takes concrete steps to address India's concerns on 26/11 Mumbai terror attack.

Other nations in our immediate neighbourhood are experiencing unrest. India has always desired peaceful relations with all its neighbours—Nepal, Sri Lanka, Bangladesh, Bhutan and Myanmar. We know fully well that security of one nation is closely linked and dependent on the security of its neighbouring countries.

India has always played a stabilising role in the Asian region, despite itself being plagued by terrorism. We have always been at the forefront of relief and rebuilding initiatives—be it the reconstruction efforts in Afghanistan, relief and rehabilitation in the wake of tsunami in 2005, or anti-piracy operations. India has also voluntarily abided by nuclear 'No-first use' and unconditional negative security guarantees for non-nuclear states.

We firmly believe that nations of the Asian region have close historical, cultural and civilisational linkages. These values must be constructively utilised to build an Asian order that reflects our unique Asian traditions.

Introduction: Super Asia

The Thirteenth Asian Security Conference 2011 tackled an ambitious topic, 'Towards a New Asian Order'. The onset of the 'Asian Century' requires fresh thinking on regional architectures in Asia. Will this be able to deliver on the promise of the future? What needs to be done, and can be done, to enable this? Given the conflictual history attending the rise of powers in Europe, the timely development of pan-Asian frameworks could help in constructing a different future. The conference aimed to examine efficacy of existing structures and probe Asian thinking on making the future benign for Asia and the world. The intent was to conceptualise a new pan Asian order by surveying existing capacities against possibilities. A pan-Asian framework is to necessary to visualize for addressing security challenges and for facilitating cooperation.

Conference Themes

It has been long apparent that the 21st century belongs to Asia. Given that history has moved its arc lights towards Asia, building a New Asian Order would be critical to this century now already into its second decade. The contours of the Asian order will have considerable impact on the world order. It therefore becomes necessary to deliberate upon how this Asian order should look and be shaped. The topic was designed to challenge thinkers from around the world to share their perspectives about Asia as a whole. The intent was to enable their thoughts to help enlighten the way towards achieving such a stable, secure, equitable, prosperous and just order in Asia. That they come from Asia's various sub-regions and from states interested in Asia's trajectory, helped in diverse insights informing the New Asian Order.

Asia is the site of great geopolitical flux. History of the last century bears witness that jostling of rising powers is fraught with uncertainty. Ensuing conflict can potentially set back the clock of progress. To repeat history in the nuclear age is unthinkable. Given the strides made by Asia since overthrowing the colonial yoke, there is much to lose in treading down the path of strategic tryst alone. While forging power equations may be necessary to stability of a regional order, managing these through mutual cooperation and mechanisms created for the purpose is prudent. It was intended that the conference deliberations would throw up ideas on this score.

Asia is more than its sub-regions. Take for instance, India's strategic location in Asia. While being central to South Asia, it is on the flanks of West, Central and South East Asia. Likewise, China, though seemingly an East Asian state, has a presence across Asia due to shared borders with Central, South and South East Asia. Even states such as Japan, that are apparently on one flank, help link Asia with the Pacific. Japan's trade and energy related interest in the Indian Ocean makes for its wider interest in Asia as a whole. The Gulf states on the other flank similarly have concerns, such as relating to security that are extra regional. The US, being a global player, has pan-Asian presence and interests evidenced by its multipronged partnerships that exist across Asia. Similarly, Russia is equally an Asian power with a trans-Asian presence.

Given this interconnectedness of Asia, one subject to constant enhancement in the age of communication and globalisation, it is but natural that security and associated problems and their solutions require a continental as against merely a regional approach. The challenges of nuclear proliferation, terrorism, managing the commons and the economic order, militate against thinking in sub-regional compartments. While sub-regions are basic building blocks, an over-arching Asian identity and architecture needs to evolve. Thinking and action must however not be confined by geography. Here on, Asia needs to be thought of holistically.

The series of Asian Security Conferences, now in its thirteenth edition, has been designed to progress such thinking. The conference had at inception dwelt on the security challenges facing Asia and in subsequent editions those confronting its sub-regions. This time round it has set itself the ambitious task of conceptualising the security architecture for Asia going beyond mere balancing but to envisaging cooperative security. For this, strategic wisdom needs to move beyond its traditional confines of realist thinking to include insights from interdependence theory and from new disciplines as peace studies.

Clearly, innovative ideas on construction and shaping of an Asian Order need to be built on pre-existing foundations. These are already present in highly developed sub-regional arrangements such as the last East Asia Summit having as its theme 'Towards an Asian Community: From Vision to Action'. Other regional efforts are not absent, but merely 'under construction' to eventually fulfill their potential, such as the one in South Asia, SAARC. The example of Europe exists as useful precedence in two senses. One is in the integration they are attempting today. The second is that they have got this far after traversing two great wars and the Cold War. Even as their destination is a laudable standard, their path also has its lessons. The pitfalls they encountered can be better avoided by Asia following as it is in Europe's wake.

Next, even as fresh insights from latest International Relations and Strategic Studies theory must inform reflection, it must not be at the cost of the remarkable cultural trove that is the legacy of Asia. India for its part has been the beneficiary over the millennia of these intellectual breezes across frontiers. Sun T'su and Chanakya continue to shape strategic thinking today. Equally salient are other cadences in indigenous Asian thinking. For those in South Asia, many protagonists for peace come to mind beginning with Lord Buddha. In the larger Asian pantheon number the prophets and sages who have inspired the world's leading religions—Hinduism, Judaism, Christianity, Islam, Buddhism, Confucianism and Shintoism, among others. Issues of conflict avoidance, confidence building and security cooperation need being informed by Asian philosophy.

Third is the major finding of the two preceding conferences, that the concept of security has vastly expanded of late. The referents of security not only include the state as hitherto, but also populations; and not only people as inhabit states and regions today but also those generations as yet unborn. In other words, security encompasses human security and has to be reckoned with as trans-generational. The security decisions taken today cannot merely be in relation to the immediate causes of friction or long standing differences between states, but must take into account consequences over the long term and in their echoes and ripple effects into the future. This negates unilateral self-interested choices by states and is instead in favour of a security conceptualised as a shared bank balance. The more the contribution to the 'kitty', the more there is to be shared around and the more that can be bequeathed to the future. This is true in particular in respect of the fresh security issues such as climate change, managing the commons and joint direction of the global economy.

Lastly, scepticism on the space available in Asia for the rise of Asian powers needs being dispelled. The prime minister has articulated India's position that sufficient space exists for the rise of all of Asia, together. Space for cooperation is not necessarily inevitable, but would require constant effort. The forces of globalisation are in favour of this, yet conscious steering is required in face of retarding modes of thought and forces. The relationships Asian states and peoples share with varying degrees of proximity with extra-regional actors, such as Europe and the USA, need to be harnessed with this end in view.

India aims to create a security environment that not only facilitates its economic, social and political aims, but in doing so enhances the security of neighbours and the wider region. India treasures its inherent democratic values. Its democracy provides a direct stake, a sense of participation and control for the common citizen. India has been attempting a rapid transformation of its society under the democratic framework. What security means is a conducive environment for furthering this. It can only be possible when neighbours are assured that India's rise is benign. This way the security dilemma of both is moderated. The conference was to arrive at such ideas not only for India but for other security complexes in Asia.

India's endeavours are in conjunction with its neighbours near and far arranged in four concentric arcs: in South Asia, in a wider arc comprising Southern Asia, with nations in the next circle encompassing all of Asia and lastly non-Asian states. India is forging multi-vectored partnerships with all powers so as to create a multipolar world. It has contributed to global confabulations, such as on trade, economy, climate change, nuclear security and changes in the UN system, in keeping with its growing strength and deepening interests and with a sense of self-regard as a major power. President Obama in his visit to New Delhi noted this stating 'India is not simply emerging: India has already emerged.' India's credentials as the largest democracy, its economic strides even during the recent downturn, status as an IT superpower and potential in terms of expanding infrastructure, defence sector acquisition expenditure, free trade agreements as the one with ASEAN and youthful demographics have helped earn the endorsement. India's hectic diplomatic schedule over the turn of the decade is a pointer to India's 'arrival'. Not only did principals of the P5 states visit India, but India was a major presence at the G20, at Cancun and at the East Asia Summit. India won the non-permanent member seat in the Security Council with 187 of 190 votes.

India's approach to issues with its neighbours is of engagement. The thirteenth round of discussions at the level of special representatives on the border issue was completed with China. With Pakistan, the peace process that was interrupted by Mumbai 26/11 has been restarted. Its relations with its smaller neighbours enable asymmetric advantage for each of them. In respect of AfPak, India has been constructive in its position being supportive of both Afghan led development and reintegration efforts. The logic behind India's foreign and security policy and posture is that India's sense of security is dependent on the perception of security of its neighbours.

India's growing stature and its right to higher aspiration owes to its strategic doctrine of deterrence. Its policy of strategic restraint, repeatedly tested, has helped maintain regional stability in the larger global interest against terrorism. India's defence budget has consistently been in the region of 2-2.5 per cent of the GDP. It has historically employed force only as permissible in self-defence. Use of force in internal security has been restrained and has several lessons for those similarly beset countering insurgency in the region and elsewhere. Its participation in peacekeeping is renowned. It has been to the forefront in lending a helping hand, whether it is against natural disasters as the Tsunami in mid-last decade or in situations of armed conflict such as obtains in close vicinity in Afghanistan. It is increasing its reach for measuring up to its weight. A demonstration of this is in the anti-piracy operations in the Indian Ocean and several joint military exercises it conducts with states ranging from Oman to Singapore and US to China. Though heading towards a nuclear 'triad', India abides by nuclear No First Use (NFU) and unconditional negative security guarantees for non-nuclear states.

India's efforts are in sync with its civilizational values, the ideals of its freedom struggle and norms of interstate behaviour. Since these values are present and shared across Asia, it is possible to build an Asian Order with Asian traditions as guide. These are not incompatible with a 'flat world'. Indeed the emphasis on peace through mutual tolerance, acceptance and accommodation makes them an abiding cultural resource against the militarisation of disputes and the idea of inevitability of conflict. The conference was designed to rediscover the Asian Way so as to create, in the words of a participant, a 'Super Asia'.

Contents of the Volume

The Conference was held in the immediate aftermath of the passing away of Mr. Subrahmanyam, the doyen of India's strategic community. The Conference was designed to address both traditional and non-traditional security issues in their political, social and economic dimensions. It does so in five subthemes: The Geopolitics of Asia; Traditional Security Challenges; Globalization and the Rise of Asia; Non-traditional Security Challenges; and Managing the Challenges. The key questions that were addressed were:

- What are the implications of rising powers in Asia for the powers themselves, for external powers and for the region?
- How can the fruits of the economic miracles in Asia be preserved?

- What are the variants if any to the traditional security paradigm?
- Can cooperative security deliver on non-traditional security challenges?
- Is it premature to think of 'One Asia'?

This volume containing some of the papers presented at the conference, duly updated, covers a wide array of themes that are closely related to Asian regionalism. It discusses Asian regionalism on the larger level of its comparison with European regionalism and investigates overarching trends of Asian regionalism. Then, it focuses on China and India-China relations as key factors in shaping Asian regional order. The contributors to this volume provide perspectives on regionalism in the sub-regions of East Asia, South Asia and West Asia. Another salient feature of this volume is that it contains good number of contributions on Non-traditional threats in Asian context.

It has been a long debate whether Asia is a region? Besides, the question whether there is any similarity of pattern between Asia's present and Europe's past has also been posed. William R. Thompson avers that the Asian region is a super-region in making. According to him, Asian International Relations resembles in some ways with those of an older Europe. However, applying historical patterns into the future is not "as straight forward a process as it seems". It requires a great amount of historical knowledge and deft use of this knowledge. Learning from European experience has its own utility for Asian regionalism. However, exact imitation and repetition is not possible and advisable. Over the years Asian regionalism has developed its own merits. Zhang Zhenjiang compares the process of East Asian regionalism with the West European regionalism in 1950. Zhang underscores the differences between the two processes, and distinct characteristics of East Asian regionalism. In this context, he underlines openness of East Asian regionalism as a unique feature.

However, does Asia really lack regionalism and institutional frameworks for it? Quantitatively, 'No', but qualitatively, 'Yes', according to Niklas L.P. Swanstrom. Swanstrom raises a pertinent point about Asian regionalism arguing that Asia does not lack cooperative structures. Instead, the real problem is lack of depth and substantiality in the existing structures. He argues that intra-Asian trade has increased, but there is no meaningful region-wide initiative to sustain economic stability in the region. He brings to light that the cost of intra-regional trade is relatively much higher in Asia; particularly, he points out, in SAARC. He holds political sensitivity and lack of mutual trust responsible for the lack of meaningful cooperative structures in Asia.

Any study on emerging Asian order is incomplete without probing the nature of relations of India and China: the two Asian giants who cover more than on-third population of the world. Their relations present interesting facts. These two countries share a long but disputed boundary. They are actively looking for economic opportunities but, at the same time, they vie for influence and trade. The successful emergence of an Asian order greatly depends on how much these two countries cooperate with each other in this direction. Vincent Wang Weichang underscores the importance of the role of India-China relations in the realization of "Asian Century". He argues that "whether Asia will finally arrive after five centuries of Western dominance" depends on how the two simultaneously rising countries manage their relations. He underlines that the bilateral relationship of the two countries is marked by contradictory tendencies of challenges and promises, contrasts and convergence to such an extent that the future of this relationship is condemned neither to rivalry nor to hospitality. According to him, evolving structure of elite identities and their preferences will essentially shape the course of this relationship.

In the years after Indo-US defence deal in 2005, the shadow of US has loomed large over the Sino-Indian relations. It has been argued in this period that China is not comfortable with the growing Indo-US relations. It perceives that anti-China agenda is behind forging this relationship. In this context, Robert Art approaches this whole question of India-China-US triangle from a different angle. He basically explores by drawing some historical parallels what would be the implications of Sino-American relations for India. His core argument is that India has more leverage over America in highly conflictual (but not war) scenario of Sino-US relations whereas it is likely to have less leverage in a cooperative scenario between them. However, the two scenarios will definitely have impact on Indo-US relations in their own ways. China's strategic behaviour has caused concerns in many of its neighbouring countries in recent years. Its assertive behaviour about its territorial claims has raised questions about its claim of peaceful rise.

Masako Ikegami and Mumin Chen take its intentions with a pinch of salt in their respective articles. Masako Ikegami presents very strong views about China's grand strategy. According to her, China is following a policy of 'stealth imperialism' which is akin to Japan's pre-war Manchukuo strategy. In the support of her hypothesis, she cites the examples of recent Chinese assertion in the Korean peninsula and its firm support to North Korea against South Korea in Cheonan and Yeonpyeong incidents. She also points out recent Chinese territorial assertions in the maritime disputes in South and East China Sea. For her, this assertive behaviour on part of China is an example of its growing military, particularly naval, muscle.

She further argues that Japan's Manchukuo policy in 1930s which was defined by "a large investment in economic infrastructure for extracting natural resources, military interventions for protecting economic interests and socio-political absorption by means of a puppet government" is an apt historical model to understand China's present behaviour. She claims that this framework indicates that "China is a new imperial power, notwithstanding its rhetoric of "peaceful rise".

Again, Mumin Chen argues that China's strategic intentions and rivalries have to be factored in any calculation about Asian security. He studies China's strategic behaviour towards Japan, India and Taiwan in his paper. He argues that although China is showing somewhat reconciliatory attitude towards Taiwan but it becomes very assertive when it comes to India and Japan. He underscores how China becomes very assertive dealing with a neighbour when the neighbour is capable enough to challenge China's strategic interest. Furthermore, he points out that in the case of Taiwan and India, China has displayed an interesting trend of behaving assertively and showing benevolence in words.

Alexander Lukin, Arshin Adib-Moghaddam, Joseph Chinyong Liow and Paul Salem's contributions widen the geographical horizons of this volume. There may be a debate whether Russia is a European power. For Alexander Lukin, it is very much an Asian power. Alexander Lukin states that Russia is an integral part of East Asia unlike the US. Its Siberian and Far Eastern regions make it an East Asian power. He argues that Russia's East Asia policy comes out of Russia's "New Diplomacy". He underlines "greater integration into the world economy", "emphasis on multilateral approaches in dealing with issues and problems", "recognition of distinct interests" and "a pragmatic pursuit of economic and strategic objectives" as four main characteristics of Russia's new foreign policy and diplomacy. These characteristics very much shape Russia's East Asia policy.

The role of the nature of US-Iran relations in shaping regionalism in West Asian sub-region of Asia should not be underestimated. Therefore, Arshin Adib-Moghaddam chooses to analyze their foreign policy behaviour in a comparative manner and highlights some crucial differences. Arshin Adib-Moghaddam focuses on four binaries around which the foreign policy cultures of Iran and US revolve. They are "unipolarity versus multipolarity, America's pro-Israeli policies versus Iran's subversion of Israeli power within the region and beyond, America's efforts to contain populist Islamic movement versus Iran's support for them, and the United State's opposition to populist lefitist movements especially in Latin America and Iran's close cooperation with them." Arshin's analysis alludes that these critical binaries between Iran and US impact regionalism in West Asia.

Paul Salem argues that global stability and cooperation depend on regional stability and cooperation and makes a case for a sub-regional cooperative framework in West Asia. According to him, continued tensions in the region have blocked the emergence of such a framework in spite of flourishing political and economic exchange among the sub-regional countries. He more directly discusses the role of US in shaping regionalism in West Asian sub-regionalism. He accuses the US policy of playing one regional power against other of not allowing a successful cooperative framework to emerge in the sub-region.

In the sub-regions of Asia, it is basically East Asia that has set some standards of regionalism and way ahead of regionalism in other regions. However, regionalism has to go a long way in East Asia also. Joseph Chinyong Liow highlights the overarching trend of institutional minimalism in East Asian regionalism. He is critical of this trend and argues that institutional minimalism sacrifices institutional efficacy. For East Asian Summit, he says that its creation is a welcome development. However, like other existing institutions, it will also need to grapple with the issue of functional clarity, and clear and specific mandates to succeed. Moreover, he argues, it has also to confront challenges "arising out of three unresolved tensions in East Asian regionalism: between inclusive and exclusive regionalism, between process and membership, and finally, between ASEAN centrality and major power leadership."

As has been mentioned that this volume on Asian Security gives coverage to non-traditional security, it covers role of technology, energy security, disaster management and issues relating to water. Tai Ming Cheung argues that technology and technological innovations are a source of "power, influence and change in international relations"; however, they have not been studied properly. His paper highlights the role technological developments have played in shaping the relationship between economics and security in Northeast Asia. In this region, Japan and South Korea are "technological front-runners", "while China is catching up them", he argues.

Xia Liping identifies energy security as one major non-traditional security issue. He points out that it is difficult for any single country to realize its energy security on its own. He argues that although "some progress in bilateral and multilateral dialogue, cooperation and coordination" has been made, greater efforts are required to materialize a meaningful energy security cooperation mechanism.

Yaacov Vertzberger reminds of the fragility of human existence by evoking memories of the December 2004 Indian Ocean earthquake and tsunami. He argues that these catastrophes are a reminder about need of collective human efforts on the face of fury of nature. He avers that construction of a disaster management regime is a need of hour as it is beyond the capabilities and resources of individual countries to stand up and face these catastrophes on their own.

The issues relating to water is discussed again in India-China context. Syed Iqbal Hasnain and Uttam Sinha's contributions highlight the emerging criticality of water in India-China relations and the role and importance of water originating from China's Tibet in shaping regional concerns. Syed Iqbal Hasnain places Himalayan-Tibetan Glacier melt in a geopolitical context and discusses the pace and impacts of Himalayan glacier melt. Furthermore, he examines the Chinese claim that dams and reservoirs on Tibetan Plateau will, in fact, reduce floods in the Brahmaputra. According to him, this claim does not hold scrutiny. In fact, flooding may become worse as the damming of the river will slow down the river flow which will, in turn, create massive silting. However, according to him, India does not have will power to take up this issue with China.

Uttam Sinha underscores the criticality of the Himalayan mountain system, the origin of ten major Asian rivers. He argues that the impact of global warming on the Himalayan mountain system will be felt the region-wide. On a geo-political level, he accuses China of being unilateral in its water development approach. He emphasizes the need of collective initiative on part of lower riparian states to draw China into a water dialogue.

Any discussion on South Asian regionalism warrants a discussion on various dynamics of Indo-Pak relations. Sumit Ganguly examines the US-India relations under the Obama presidency in the Af-Pak context. Prakash Menon discusses the role of emotions and passions in Islamic world in general and Pakistan in particular. He argues that systematic, national and personal factors have brought about changes in US-India relations during the Obama administration, distinct from earlier administrations. He holds that Pakistani military's ties with Afghan Taliban, need to be clear about America's own commitment level (both military and civilian) in Afghanistan and Indo-Pakistani competition in Afghanistan are the three main challenges the Obama administration has to face in the remainder years of its tenure.

Prakash Menon basically focuses on analyzing psychological aspects of India-Pakistan relations. He uses the constructs of "passion", "imagination" and "emotion" in his paper. On a larger level, he argues for "being sensitive to the feeling of humiliation that grips some parts of Asia especially the Islamic world". He urges Asian and global leaders to handle their relations with the countries which have this feeling of humiliation sensitively.

Lastly, the question arises what is the take of India on Asian regionalism? What informs its security perceptions in Asian context? Arvind Gupta identifies the rise of China as a major new security challenge for India. According to him, managing the relationship with China will remain a daunting task for India. However, he is very much conscious that new security challenges for India go beyond China. The issues like terrorism, maritime security and instability in parts of Asia are also security concerns for India. He argues that India has to generate diplomatic capacities to handle these challenges more intensely.

Finally, it is hoped that this volume on Asian order will provide readers some useful insights about the issues and themes involved in emerging Asian order.

Conclusion

Asia, comprising as it does the regions of West Asia, Central Asia, South Asia, South East Asia and East Asia, is at the centre stage of global events and at the fulcrum of the global power shift. A number of issue require attention: implications of the rise of China and India; enduring conflicts in West Asia; the continuing military entanglements of the US; access and control of resources of Central Asia; expansion of security regimes in Asia-Pacific. The maturing strictures against aggression, nuclear backdrop to conflicts between major powers, lethality and costs of modern weapons, and the spread of globalisation have made military power usable mostly for purposes of deterrence. Threats exist less from direct military confrontations between states than from non-state actors acting either autonomously or under state sponsorship. The traditional security domain has changed. Asia, therefore, needs to take note and reshape its military profile accordingly. Can one visualise a non-realist future for Asia, one based on growing interdependence? The participants have set the terms of the debate in this direction.

The liberal-capitalist model of consumption has its perils in being unsustainable, unless tempered by Asian traditions of abnegation. The two continent-sized states, China and India, are justifiably catering to their growing populations. However, they have to face up to the limits of growth. Competition for access to resources and markets in Asia and elsewhere has the potential to disrupt relationships. The recent economic shocks indicate inherent fragility of the global economic order. Non-traditional security thinking is displacing traditional modes given that these challenges are both ubiquitous and more demanding. They have the potential to both mitigate and aggravate existing fissures, depending on how they are approached. Given population densities, social disparities, expanding urban clusters and the fragility of an environment under assault from economic growth, the future could be very bleak. The opportunity for turning it around through enlightened moderation of aspirations, application of technology, creating responsive policing organisations and by democratisation, is in the here and now.

Visualising a benign future helps in the creation of structures to bring about more cooperation in future. The 'Asian' input, relying on a cultural legacy of centuries, needs expansion. The continent currently views itself in multiple 'regions' that have regional structures of varying efficacy to integrate and mellow nationalisms. It is an idea only geographically. In such a circumstance, would broaching a pan Asian concept be in order? This volume answers the question resoundingly in the affirmative.

Editors

Section One

IN THEORY

1

Patterns of Conflict and Stability in the Asian Region from a Long-term Perspective

William R. Thompson

What is the future of Asian geopolitical stability? Will it be as tranquil as the currently integrating European region? Will it be as placid as the South American region now seems to be, at least in interstate terms? Will it be as chaotic as some of the less stable, sub-Saharan African regions—such as east, central, or west Africa? Or, will it resemble the extreme violence of the pre-1945 European region which led to regular incidents of major power warfare in addition to the two world wars and contributed mightily to a post-1945 global Cold War. This question of future regional stability/instability is interesting not so much as an exercise in crystal ball gazing but as a test of our theoretical understanding of regional dynamics. Do we know why some regions are stable and others are not? If the answer is yes, we should be able to make some informed predictions about the future of Asia. If the answer is no, we should ask why not and proceed to remedy what would constitute a significant gap in our knowledge.

Unfortunately, the answer to the question of whether we understand regional dynamics sufficiently to make strong predictions must be negative. We are not without attempts, however, to unravel these dynamics. There are a variety of frameworks for specifying the critical variables leading to regional war and peace but none have yet demonstrated substantial breakthroughs in explanatory power subject to a wide analytical consensus.[1] No doubt, they are on the right track but more work needs to be done before we can rely on their interpretations of regional order. Another approach is to predict more of the same. Whatever level of order, other things being equal, that has prevailed in the immediate past could be forecast to characterise the immediate future. That approach could conceivably work for the short-run but in the longer run—in this case more than a decade or so—there is not much reason to place a great deal of confidence in such predictions.

A third approach involves making use of regional geopolitical history.[2] Friedberg (1993-94) does this by projecting the possibility of a turbulent European past on the Asian future. Kang (2010) does it as well by examining the 1368-1841 history of East Asia and arguing that interstate warfare was unusually scarce in this selected time period due to Chinese centrality in the regional hierarchy and other the states' acknowledgment and acceptance of this fact of regional structure. Wang (2010) focuses on 960-1279 and 1368-1644 China and concludes that China tends towards belligerence when it is strong and accommodation when it is relatively weak. After nearly 500 pages of some 4,000 years of East Asian historical review, Cohen (2000:478) reaches the same conclusion by advancing a minimalist model for the future of Asia when he stipulates that "[China] has been aggressive when it was strong and defensive when it was weak." However, he weakens the model's efficacy in the next paragraph by noting that there are important checks on Chinese behaviour in the 21st century (the United States and Japan) and Chinese leaders could well conclude that peace was the best avenue to advance Chinese economic interests.

Geopolitical history is an attractive laboratory for concocting future predictions but there are problems with this approach, just as there are with other approaches. One problem is that we need to exercise care when comparing two regions. One region's past might offer useful clues to another region's future if the two regions have or had similar geopolitical structures. But if they do not, the comparative-based forecast may not be all that helpful.

A second problem concerns the selection of a period of geopolitical history as the base for forecasting. Regions may have histories extending back thousands of years. Should we ignore some years and highlight others? If, for example, we wanted to forecast the future of Europe based on history, must we go back to Rome or can we profitably restrict our focus on the past 500 years of European international relations? If we want to forecast Asian relations, can we make do with the past several hundred years or must we go back to the emergence of something resembling a region in eastern Eurasia? The answer to this question is that it depends on whether one needs the longer time frame to interpret the shorter time span. It may be that the dynamics of the shorter time span are sufficient to establish a framework for the future. So much may take place in a relatively short time period that the longer patterns are rendered obsolete. But it may be necessary to rely on the longer pattern to make sense of the shorter span.

A third problem is that things have an awkward way of changing. Cohen's view, for instance places a premium on the novelties of contemporary Asia at the expense of the longer pattern concerning the waxing and waning of Chinese foreign policy. He might have taken the opposite approach and privileged the longer Chinese foreign policy pattern over the newer developments of US and Japanese competition and checks but did not. It is certainly not unreasonable to hedge one's analytical bets in the face of changes in the structural environment. It may not be all that satisfying for prediction purposes; yet, it is a problem that we need to keep in mind. Change happens. We need to be alert to the fallacy of placing too much faith in discerned historical patterns and assuming their unaltered continuity into the future.

Each of these problems deserves some further elaboration and consideration and will be treated accordingly in the discussion to follow. The position that is taken is that these problems are not insurmountable. It is possible to study regional dynamics and to use other regions' geopolitical histories to study the dynamics. But we have to be careful in how we do it. The Asian region, encompassing East, South East, South, and Central Asia and probably the south Pacific, is a super-region in formation. In that respect, it does not have much of a history as a super-region. Yet it is possible to use older Asian and European dynamics for indicators to evaluate questions of future regional stability. The European past need not be Asia's destiny but it is not inconceivable that Asian international relations will come to resemble those of an older Europe in some ways. The question is in which ways.

The Problem with the European Past as a Model for Other Regions

Friedberg (1993/94) makes a case for the past of Europe possibly serving as a model for the future of Asia. In some respects, the logic could be said to be predicated on the premise that contemporary Europe is vastly different from its past and that contemporary Asia more closely resembles Europe's past than its present. Europe is culturally homogenous, democratic, affluent, and not especially characterised by salient nationalism. European states are moving towards de-emphasising national boundaries. Territorial disputes are a thing of the past. Populations, goods, and capital move throughout the region in a highly interdependent fashion. These interactions are increasingly regulated by regional laws and institutions. While European states possess and have access to highly lethal weaponry, there is little expectation of interstate conflict and even less anticipation of anything to be gained by military belligerency.

Europe was once a very different region. As in the case of Asia, it once encompassed nationalistic liberal and non-liberal political systems, uneven economic growth patterns, and numerous territorial disputes. Regional law and institutions were scarce. Economic interdependence was uneven. Lethal weaponry was available and at least some national decision-makers did entertain the possibility of gains from military coercion and expansion. Early modern and modern Europe was transformed into post-modern Europe in the second half of the 20th century. Since contemporary Asia more closely resembles the earlier Europe, it is possible that a post-modern Asia is a possibility somewhere down the road. But, in the present, the immediate prospects for contemporary Asia could look like the extreme turbulence of Europe in the first half of the 20th century.

Friedberg's argument cannot be dismissed as simply a case of mis-applying European history to a different region's possible future. His comparison is based on the argument that regions possess various characteristics that either encourage or discourage conflict. The balance of factors, that he highlights for older Europe, encouraged conflict. Many of those same factors seem to be present in Asia and one can easily arrive at the conclusion that the factors promoting conflict in the region are stronger currently than the factors constraining it. In other words, this is a respectable theory of war and peace and quite suitable for regional applications.

Yet there are two inter-related problems with adopting European transformations as the expected or conceivable template for other regions. One is that the European experience may prove to be quite unique due to the processes that led to the transformation of old Europe into post-modern Europe. The second problem is that analysts do not always take those processes into consideration when they are comparing regions. Friedberg's comparison, for example, emphasises attributes (regime type, wealth, homogeneity, dispute propensity) that changed but not the dynamics that transformed those attributes.

The distinctiveness of Europe is traceable to its history over the past 500 years. During that time, the region entered into increasingly intense conflicts that ultimately exhausted the survivors.[3] Exhaustion did not lead automatically to the contemporary Europe as we now know it, but it greatly facilitated its development. The reason the region was able to sustain this upward spiral for so long can be linked to its unique geopolitical structure. Attempts to unify regions are not uncommon and early modern/modern Europe was no exception. However, the geopolitical structure of the region worked against successful conquest on a regional scale. In the case of Europe, the feature of the geopolitical structure that is most important is the tendency for aspiring regional hegemons to be defeated by a coalition of Western maritime powers (the Netherlands and England/Britain) and eastern land powers (at various times, Austria or Russia/the Soviet Union). This geopolitical structure did not always exist—it only gradually emerged. It came to be termed the balance of power. It was not guaranteed to thwart powerful hegemons because it usually came into play late, required intensive warfare, and was always subject to intra-coalition disagreements. But the three advantages the balancing coalition possessed was that it could mobilise both land and sea power (when the aspiring hegemon was usually restricted to land power), it could mobilise more extra-regional resources than could the aspiring hegemon, and it could force the hegemon-to-be to fight on multiple fronts simultaneously. Add in some probability of various strategic errors on the part of hegemonic aspirant and the balancing mechanism worked adequately to stave off European unification for a number of centuries.

Ever since, we have tended to assume that all regions play this same game but it may turn out that the classical balance of power mechanism was unique to Europe thanks to the distinctive duality of maritime and land power that made it possible.[4] On the down side, the hegemonic thwarting mechanism did not prevent the escalation in weapon lethality which, in turn, was due to a combination of regional competitiveness and technological innovation. Of course, the balancing also only worked for a while, even in Europe, because the Western sea power flank eventually had to be substantially reinforced and then substituted entirely by an extra-regional power. But in the process, aspirations for regional hegemony ultimately were extinguished. Regional unification has since been pursued in more pacific ways.

Even so, the switch from coercive to pacific regional unification was neither easy nor fully premeditated. The Franco-German rivalry, unlike most others, had not been terminated in 1945. Regional integration was to some extent a fall back plan after France failed to persuade its alliance partners to deal with the German threat to French security once and for all. The fall-back plan involved encouraging German commitment to regional decision-making and institutions. This coincided nicely with the need for German decision-makers to escape their post war pariah status. Post-World War II Franco-German relations are not the whole of the European regional integration story but they were a critical ingredient in the early years.[5]

So, we need to be cautious about imputing an older European future for Asia—not because the regions are incomparable or because the attributes are not really similar—because regional circumstances are not isomorphic. So far, Asia has not developed the geopolitical structure of maritime-land power duality that was so significant to the European regional experience. Nor has it experienced the upward spiral of intensive conflict that was the signature of European international relations in the early modern/modern era. If the Asian structure and experience is different, it is quite conceivable that its future trajectory is likely to be different as well.

That said the argument here is not that European developments of the 1494-1945 era are entirely irrelevant to the Asian future. One of the perennial issues in pre-1945 Europe was what is sometimes called the Western question. Which state would exercise predominance in the region? Some 450 years of European international relations history can be reduced to the perceived efforts of France, Spain, and Germany to unify Europe.[6] It is not difficult to add a fourth state, Russia/the Soviet Union, to this list and extend the sequence through the Cold War. The European historical relevance for Asia's future probably lies in the competition for regional predominance by states ascending in relative strength in a multi-polar setting. But even here there are application problems. This is a subject to which this essay will return at a later point.

The Problem with Relying on Asia's Own History as a Predictive Guide

If we need to be cautious about extrapolating possible futures from other regions, can we safely base speculation about future trajectories on the region's own long-term patterns? Unfortunately, there are problems here as well. One problem is that the contemporary Asian region is increasingly bringing together a number of different sub-regions which have not shared precisely the same regional experiences in the past. Maximally, one could argue that Asia tends to encompass (or is tending toward encompassing) East Asia, South East Asia, South Asia, Central Asia, Siberian Russia, and the South Pacific. Within these nested sub-regions, East Asia has long been central but South Asia and the South Pacific have tended to be much more autonomous in the past as separate regions. In this respect, a new region is forming with variable path dependencies brought into the mix. Two path dependencies include the tendency for China to be predominant at least for some of the time. The other is the maritime-continental geographical division observed in South East Asia which has merged more recently with Cold War propensities in East Asia to differentiate between Japan and Taiwan and the mainland. The common denominator is distance and geographical separation from China. The irony here is that that these relatively new developments (a mainland-rimland division) could lead to the creation of something like the older European dualistic geopolitical structure in Asia.

Nevertheless, East Asia has long been salient in whatever definition of Asia is embraced. Kang (2010), in this respect, argues that an appreciation for not how East Asia resembles an older Europe but rather how it differs provides the best handle on speculating about stability in Asia. Kang focuses on what he calls early modern East Asia (1368-1841). The period is delimited by the restoration of an indigenous Chinese regime (although it includes the non-Chinese Manchu or Qing regime) to the period of extensive European penetration of China after the first Opium War. If one zeroes in on this period, the relative absence of inter-state warfare is pronounced. Clashes between Chinese forces and nomads or pirates are not uncommon. What is uncommon is war among China, Korea, Vietnam or Japan. China invaded Vietnam in the early 15th century and Japan invaded Korea and fought China in the late 16th century. Only two interstate wars in nearly 500 years adds-up to an unusually long peace for Kang.

In addition, Kang suggests, that the history of East Asia marks it as distinctive and unlike most other regions. The region, dominated by China, was stable because there were accepted territorial limits to Chinese ambitions and because the other states acknowledged China's predominance as legitimate, as symbolised by the tribute system. There were exceptions to this stability but the exceptions prove the rule in this case. The implication is that the latest rise of China might very well be expected to re-create the same stable order as before. Kang realises that the regional circumstances are not the same as before and that, that could cause new problems. But, given the past history of East Asia, he thinks it is conceivable that a stable regional order might still re-emerge after a century and a half of turbulence stimulated by Western imperialism and a breakdown of the old order. Alternatively put, he is privileging the 1368-1841 history over the post-1841 history.

Kang is certainly right to label the early modern East Asian experience as distinctive. It is hard to say just how unique it is because we are not equally familiar with all regional historical trajectories. The North American region after about 1870 comes to mind as structurally similar in terms of hegemonic stability but less long-lived so far (and without the Confucian cultural overlay). The fact that other examples do not leap to one's mind readily is probably telling, however, as to its distinctiveness. Yet there is a difference between the 1494-1945 half-millennia in Europe and the 1368-1841 half-millennia in East Asia. The nearly 500 years in Europe marked a departure from what had occurred before in Europe.

Prior to 1494, there had been a Roman unification and a long period in which few actors (other than perhaps Charlemagne) cared much about regional hegemony. What happened in East Asia between 1368 and 1841 was not so much a departure as it was part of a longer sequence of alternation between strong and weak Chinas. Moreover, the strong China of the early modern era was in part a product of earlier iterations of strong Chinas.

Wang (2010), in marked contrast, finds support for a variation of the Johnston (1995) thesis that when China possessed superiority in capability, it was more likely to pursue a ruthless, expansionist grand strategy.[7] When China was relatively weak, its posture was more likely to be less belligerent and more accommodating.

Although the analyses of both Kang and Wang include the nearly 300 year Ming dynasty as a major segment of their data base, they arrive at much different conclusions. Of course, they are also examining different dependent variables—interstate war for Kang and the use of force for Wang—but it is really a matter of comparative historical context on which they differ most. Kang lacks a comparative context because he isolates a specific period of time for analysis. Wang does have an explicit comparative frame even if choosing the besieged Song dynasty as a contrast to the vengeful Ming dynasty would seem to almost guarantee the type of analytical outcome that is generated.[8] What they have in common then is an equally selective approach to Chinese foreign policy and, by implication, East Asian regional tendencies. Being selective is not a sin but we need to be cautious in embracing the resulting generalisations.

Perhaps the best guide to the long-term sequence in East Asia is Ledyard's (1983) Yin Yang cycle. During the first two millennia of international relations in what became China, dynastic hegemonies alternated with periods of multi-polar conflict among multiple states. Yang phases are characterised by an aggressive and expansive Chinese state focused initially on pressure from the northern steppes from nomads that eventually was displaced by Russian pressures. Only gradually did external maritime pressure build up on the western coast. When China was strong and unified, it could manage the nomadic pressures in the north. When it was not, the outsiders could overwhelm Chinese defences which were no longer subject to robust central management.

In the Yin phases, China either fragmented into smaller states and/or was subjected to non-Chinese rulers. With some exceptions, Chinese strategies in these periods emphasised defensive preparations and survival against increasing pressures until they succumbed. In the Yin phases, external pressures triumph. In the Yang phases, they are pushed back.

Throughout China remains the East Asian region's critical mass. Sometimes it is able to resist successfully external forces. Other times, the external forces overwhelm Chinese defences.

The point of injecting this long-term interpretation is that China will no doubt be one of the central players, if not *the* central player, in 21st century Asia. It is ascending in relative economic and military strength. If there will be a regional hegemon in Asia, China is now the most likely candidate. The question then is what kind of behaviour should we anticipate for a rising China? Yet it cannot be a monadic question. Whatever China and its historical record represents it has not operated in a vacuum. Ledyard is right to stress not only Chinese behavioural regularities but also their variable context of northern and western pressures.

The nomadic threat was not a permanent feature of the East Asian strategic context. Once in play, it became a dominant feature of the East Asian landscape through roughly the 18th century CE and the development of effective guns that could trump nomadic military advantages (see, for instance, Chase, 2003). Yet as the nomadic threat eroded, the Russian movement into Siberia began gaining momentum. One type of northern pressure replaced another. Initially, China could repel Russian intrusions but gradually that changed as well.

The Western pressure on East Asia was slow to develop. China resisted Western incursions until almost the mid-19th century and then succumbed fairly quickly to complicated and uneven forms of subordination by various external powers including, but not limited, to Britain, France, Russia, and Japan. European developments caused Britain to reduce its Asian position very early in the 20th century. But the ascent of Japan in this era pitted Japan increasingly in opposition to both Russia on the mainland and the United States in the Pacific. Japan gradually occupied much of East Asia further intensifying its Russian competition over control of Manchuria and Korea and its rivalry with the US. World War II ended these competitions and Japan's bid for military-political hegemony in East Asia. The defeat of Japan created an opportunity for the re-assertion of Chinese autonomy and the conclusion of the Chinese civil war and the victory of the communist party over the Goumindang.

East Asia then became characterised by a shifting triangular competition among the China, the Soviet Union, and the United States. The Chinese-Soviet alignment lasted little longer than a decade and was followed by increasing Sino-Soviet conflict and persistent US attempts to contain Chinese influence in both East and South East Asia. A period of short-lived Sino-American alignment predicated on mutual animosity toward the Soviet Union and the US desire to exit from Vietnam followed.

But the logic of the alignment could hardly survive diminishing Cold War tensions and the collapse of the Soviet Union. Some level of co-operation between China and the Soviet Union/Russia made more sense, especially since now China was the stronger member of the dyad (in marked contrast to 1949-1950) and both states needed to concentrate on economic development.

Both China and Russia also agreed on opposing the US presence in the various parts of larger Asia. Circumstances had also worked to break down some of the barriers to interaction among the Asian subsystems. South East and East Asia had become more closely meshed thanks to Chinese involvement in two of the three Indochina wars and the subordination of Vietnam's own efforts to predominate in one part of Southeast Asia. Central Asia was made formally independent of Russia but that made the area more open to competition among Russia, China, and the United States, among others, for influence and for access to its resources and strategic location. South Asia had become linked more closely to East Asia due in part to Chinese linkages to Pakistan and Indian linkages to the Soviet Union and in part to an overlapping Sino-Indian rivalry initially over boundaries but increasingly over position and leadership in the larger Asian space.

Thus, the larger Asia is characterised by multiple subsystems and multiple contenders for the central position in the region. In this respect, Asia has come to look more like the older Europe. Older Europe had its subsystems too (Scandinavian, western, central, and south eastern) and multiple great powers vying for relative position. In both new Asia and older Europe, one large land power was/is perceived to possess some threat of military-political expansion. In older Europe, these land powers were thwarted by a series of increasingly total wars within the context of a distinctive and possibly unique geopolitical context. The threats posed by older Europe's largest land power were blocked by coalitions combining flanking land and sea power.[9] Only gradually has some level of regional autonomy been restored—some respectable portion of which is due to the collapse of the Soviet Union. In its post-modern phase, however, the old Western question seems unlikely to be revived in a political-military way.

In Asia, the "Eastern question" has an ancient history as we have seen but its context has undergone considerable change.[10] Core Chinese territorial expansion has been intermittent with contemporary boundaries being (re)gained and lost in different phases. Yet, as Kang (2010) stresses, there are historical limits to Chinese territorial expansion. Aside from important maritime disputes in the East and South China Seas and disputed territory along the border with India, the 21st century is not so much one of Chinese territorial expansion as it is of regional hegemony and subordination to Chinese preferences as the 800 pound regional gorilla that cannot be ignored.

The Eastern question has never worked the way the Western question played out in early modern and modern Europe. The threat environments historically have been much different. Eastern Eurasia had nomads and western Eurasia did not (at least after the 10th century CE). Western Eurasia had sea powers and land powers while eastern Eurasia, for the most part did not have, or has not had, indigenous sea powers with different geopolitical orientations than land powers until fairly recently.[11] This orientation distinction needs to be emphasised. It's not simply a matter of whether or not a state possesses naval fleets. What is distinctive about older Europe is that it had a series of maritime powers (Portugal, the Netherlands, and Britain) that were more interested in pursuing commercial advantages outside the region than they were in exploiting opportunities for territorial expansion within the home region. When large European land powers threatened to become regionally hegemonic, he home bases of these maritime powers were also threatened (most clearly in the Dutch and British cases) leading to attempts to organise resisting coalitions funded by commercial wealth and aided by naval superiority and access to extra-regional resources. While we think we understand balancing processes as universal propensities, it seems quite possible that successful balancing processes have been contingent on this geopolitical duality (land versus sea powers with contrasting strategic orientations).

If we need to be cautious in using European regional history as a template for possible future international relations in the Asian region because old Europe does not resemble future Europe in some respects, it does not mean that we cannot draw on European history in other respects. The past 500 years of European international relations do possess one highly useful attribute. It is not the only a region with a history of iterations of hegemonic combat but it is the most contemporary one. It has also been studied extensively. Consequently, we know a fair amount about the structural changes it experienced and the behavioural processes that followed. We know enough (or think that we do) to be able to specify what to look for as regional conflict escalates towards periods of intensive conflict.

Of course, it is quite possible that Asia will escape the traditional expectations about struggles over regional hegemony.[12] A new and unprecedented era of peace and prosperity may be just around the corner. But since such an era would be unprecedented, it would also be rather hard to forecast what it might look like—other than falling back on the contemporary European example. But, since there are reasons to think that the European pathway to an increasingly integrated and peaceful region may not be replicable elsewhere, unless other regions experience and survive similar spiralling conflict intensity, a pacific Asian future could be a distinctively unique phenomenon.

Applying the European Historical Dynamic to Asia

Putting aside, then, the pacific and highly cooperative alternative, what can we extract from the older European experience that might be of use in determining the course of Asian developments? There are at least five clusters of structures and processes: 1) challengers and leaders, 2) coalitions and bipolarisation, 3) increases in economic competitiveness, 4) multiple rivalries "ripening," and 5) relatively weak Kantian constraints.[13] They need not all be equally useful nor do they all necessarily have the legs required to move from one end of Eurasia to the other. Yet all five seem to have an applicability to the Asian region. Some, perhaps all, are already in play.

To make full use of these changes, however, one important assumption must be clarified. There are many perspectives on how international relations work. Perhaps most prevalent is what might be called an undifferentiated landscape in which large and small states interact around the globe in one planetary-wide international system. A different view sees the world as roughly and often vaguely partitioned into regions, in which some regions are more important than others. But, overall, most states are primarily interested in what happens in their own backyard or the regional neighbourhood. A third view combines the two in a selective fashion. There is a global system in which a few actors are particularly salient because they have interests in and the appropriate capabilities to act over long distances. Trade looms large as a global policy concern because important segments of it must travel long distances to reach consumers. So, too, do order and stability which serve to optimise industrial production and commerce. The states that are most interested in these questions tend to have the most complex economies and sophisticated military capabilities to protect them. Collectively, we have called these states superpowers, great powers, and major powers. They are the elite states of the world system but even they tend not to be equal in clout.

Regional subsystems abound with different types of structures. Some have no resident major powers. Others have a number of them. Still others are characterised by considerable penetration by outside major powers that, in turn, become competitors for local/regional power and influence.

All have hierarchies with one or more states established or contending for position at the top of the power gradient. States at the top of the regional power gradient are regional leaders or hegemons. States near the top may aspire to be regional leaders. A few regional leaders may also aspire to global leadership.

Challengers and Leaders

Thus, there is a global hierarchy and multiple regional hierarchies. States that are high on the global hierarchy may have little to do with some regions while in others their persistent penetration makes them part of regional hierarchies. All hierarchies have some variable potential for challenges emanating from states seeking to move up the hierarchy and targeted on states already high in the hierarchy. Given our present interest in the Asian region, two types of challenges are of special interest. One involves challenges to the position of the global system leader and other involves challenges to the position of the regional subsystem leader or, if there is no established leadership, contention for regional leadership.

In the European past, these two types of challenges tended to become fused because they involved the same challenger operating at two levels simultaneously. As long as Europe was the most central region in the world system, a European regional hegemon threatened global leaders (Netherlands and Britain), located in or on the fringe of Europe, directly. As a consequence, the Netherlands sought to organise coalitions to oppose regional hegemons intent on subordinating the Dutch; and Britain historically reacted to any attempt to seize control of coasts in the vicinity of the English Channel. But there was also a more indirect threat. Control over the European region meant that the European regional leader would possess the capabilities to compete with the global leaders in other parts of the world. Global leaders, as a consequence, tended to react strongly to these threats.

Simultaneously, regional contenders competed for the position of regional leader or hegemon. More recently, some of the Cold War was very much concerned with US protection of western Europe from the perceived threat of Soviet expansion.

If we put aside the place proper names, what do these European patterns suggest for Asia? To the extent that Asia has considerable potential to replace Europe as the most salient region in the world system, control of this region would constitute a quite significant threat to the global system—in much the same way European regional hegemony would have between 1494 and 1945. The current global system leader, the United

States, has a vested interest then in suppressing efforts to control the Asian region.

But, at the same time, there is indigenous competition for regional leadership. China's position in East Asia is one thing. Its position in a broader version of Asia is something else. It has the regional leader in South Asia, India, with which to contend, although that competition is still forming. At some future point, Russia may return to the competition for regional lead state status, just as it was the main proximate competitor of Japan in the first half of the 20th century or China in earlier centuries.

We have tended to focus on competitions between the states ranked first and second in hierarchies. We also need to pay more attention to competitions between states ranked second and third, especially vis-à-vis regional dynamics. It may be that there is a process that is something like boxers fighting their way up through the ranks before they are allowed to take on the reigning champion. Regional contenders would then have to sort out their relative positions before taking on the global system leader. If so, we need to pay as much attention to China-India relations as to China-U.S. interactions.

Coalitions and Bipolarisation

In the European past, periods of intense conflict have been preceded by a period of change in alignments. Sensing a changing geopolitical landscape with some states improving and others losing position, decision-makers look for alliance partners to enhance their future security prospects and to cope with new or more acute threats. The partnerships that are sought may be novel because new partners are now available or because old conflicts no longer seem quite so compelling in the face of new threats.

In each case, old rivalries were ended (although not always permanently) and new alignments were solidified. The outcome was that all great powers were formally or informally on one side or the other. The alignment structure was highly bipolarised. Only one great power defected from this pre-war polarised array—and only after it seemed to make sense to join early the on going war on what appeared incorrectly to be the winning side. Bipolarised alignment structures do not force states to go to war. In World War I, Italy did nothing at first, Japan limited itself primarily to improving its Asian-Pacific position at Germany's expense and the United States stayed out of the war for nearly three years. But perceived options are limited and who threatens whom, is made only too clear. Once bipolarisation is achieved, it becomes more difficult to talk one's way out of short-term disputes and confrontations.

Increases in Economic Competitiveness

We tend to focus on territorial disputes and border clashes but the underlying tectonic shifts in the world system are more fundamental changes in economic productivity, with profound implications for international relations. First, the global hierarchy is based in some considerable part on technological innovation and prowess (Modelski and Thompson, 1996). Technological innovation tends to be monopolised for a period of time by the pioneer lead economy. In conjunction with the technological edge, the supporting development of global reach military capability, and a coalition win in global warfare, one state moves to the head of the global hierarchy. Its lead status is finite however. It either needs to be renewed (in terms of technological innovation) or it is likely to be replaced by another lead economy.

This process generates a fundamental tension in world politics at some point, as incumbent and main challenger begin to square off. While this is a global level process, it has important implications for regional stability if the main challenger is located in the region in question. Then the process tends to become fused with competition over regional leadership. The fusion of the regional and global leadership question, in turn, has tended to result in intensive combat in the European past, with both questions being fought out at the same time among broad coalitions of actors.

Yet prior to any showdown between the adversarial coalitions, there are heightened tensions among an ascending economic challenger, the system leader in relative decline, and other states affected by these shifts in productivity. Energy questions have become increasingly critical to ascent in the global hierarchy (Bunker and Ciccantel, 2005, 2007; Lepoire, 2009). Prior to the British industrial revolution, it sufficed to gain technological edges in harnessing wind and ocean currents for commercial advantages. The industrial revolution altered energy requirements for productivity improvements. Coal, electricity and petroleum in succession have served as a base for obtaining the additional energy needed to take the lead in economic innovation. One spin-off is increased competition for access to energy sources as ascending powers attempt to increase their share of the resource consumption pie. If increased competition occurs as the world economy is in the process of transiting from one energy regime to another, increased competition and increased resource scarcity can be expected to aggravate further already tense relations, both within the home region and in more distant regions.

Equally problematic are structural imperatives that dictate that economies competing at or near the technological frontier will depend to produce similar products. Economies of scale considerations strongly encourage external markets for these products. As a consequence, there is increased competition for markets and market share (Sen, 1984). The larger the number of states with ascending economies in a region, the greater the competition will be. Increased competition leads to various by products such as accusations of unfair practices (dumping, tariffs, and government subsidies tend to lead the list) which deepen the sense of commercial rivalry. More acute commercial threats, in turn, can contribute to perceptions of rivalry about security issues (Thompson, 2006).

Further framing these reasons for economic tensions are regular intervals of technological change along the lines of Schumpeter's creative destruction. Uncertainty requires investments with no guarantees. Some innovations will catch on while others will constitute dead ends. If we add to this technological uncertainty, the unease/euphoria of geographical shifts in economic activity—as in the earlier transition from north western Europe to North America to the ongoing one to Asia—the overall effect is destabilising. There will be winners and losers for sure but it is not clear which parts of the world will fall into one category or the other.

Multiple Rivalries "Ripening"

While we do not have a long history of taking rivalries seriously—that is as subjects of inquiry as opposed to innocuous labels that describe pairs of states that tend to keep bumping heads in international relations, it should not be surprising to suggest that protracted conflicts have something to do with expectations of future conflict and instability. In the past two decades of research, we have learned that conflict in world politics is traceable to a quite small set of dueling dyads. In a world of 200 states, there are nearly 20,000 dyads or pairs of states possible. Only a small handful, however, make trouble.[14] Yet the basic nature of rivalry means they are likely to make trouble repeatedly (Colaresi, Rasler and Thompson, 2007).

We are quite familiar with prominent rivalries in European history. That is exactly the point of emphasising multiple rivalries "ripening" for greater conflict propensities (Thompson, 2003). World War I was not about only one of these prominent rivalries. It was about them (and a number of others) all proceeding simultaneously in non-linear ways.

The point here is that while rivalries are few in number, they often possess various interdependencies that encourage them to function as a field or network of rivalries. Decision-makers focused on coping with their principal rival may not realise all of the implications for the whole field of rivalries when they act. What might seem reasonably straight forward in more narrow dyadic terms lead to rapid conflict escalation in multiple rivalries if they are sufficiently primed to act more belligerently.

In Asia, we have already observed the triangle linking the Soviet Union, China, and the United States in operation in the 1970s and 1980s. Future Asia will have more important players than three. The network will be more complex and, therefore, more difficult to predict. We already have rivalries between China and the United States but also other Chinese rivalries with Japan and India. But there are also other rivalries between India and Pakistan, Uzbekistan and Kazakhstan, Malaysia and Singapore, China and Taiwan, and the two Koreas. More may be on the way. Asia already has a sufficient number of operative rivalries for network complexity. The question, though, is not how many rivalries are in play but whether a number of them become more prepared to escalate hostilities at roughly the same time.

Weak Constraints on Conflict

One possibility is that a set of constraining forces have developed that make intense conflict less probable. There are actually two sets. One is the Kantian tripod of democratisation, economic interdependence, and international institutions (Russett and Oneal, 2001). The other is the high cost of industrialised warfare that includes our beliefs about nuclear deterrence. In both cases, the expectation is that the world has changed, thanks to the development of strong constraints that make warfare between industrialised states far less likely than once was the case.

Maybe the world has changed irrevocably. It is conceivable that decision-makers will balk at going to war with other industrialised states even if the stakes (regional hegemony and global pre-eminence) are extremely high. That could mean that traditional conflicts over regional and global pecking orders will play themselves out in some other way or, alternatively, that the pecking order conflicts will simply proceed without resolution until or unless one side wins by default.

In any event, two of the three Kantian tripod legs are apt to be less than fully operative. To the extent that democratisation works to constrain conflict between states, it is usually expected to work dyadically between two democratic states. There are reasons to think that we have exaggerated the effects of the democratic peace that cannot be pursued here.[15] Even so, the odds seem high that Asia will continue to be characterised by a mixed crew of democratic and authoritarian states for the indefinite future. The democratic peace does not apply to mixed dyads. On the contrary, its converse—that democratic and authoritarian states are more likely to become involved in conflict than are democratic or authoritarian dyads—would seem to be more applicable (Gleditsch and Hegre, 1997).

Global and regional institutions have not played salient roles in restraining interstate conflict in Asia. Why should we anticipate them doing so in the future (Rapkin and Thompson, 2009)? That leaves economic interdependence. Presumably, economic interdependence will expand in the future and could have some applicability in Asia. Basically, the expectation is that commerce builds mutual interests in maintaining a relatively pacific status quo as opposed to allowing turmoil and disruptions that interrupt the gains derived from economic exchange. People involved directly in these exchange gains are expected to lobby with their own governments for avoiding the costs of conflict. But even decision-makers who are distant from economic transactions should have to think twice about giving up trade and investments.

It is not implausible that economic interdependence could have a conflict constraining effect. However, there are at least two problems. One is that it has not always worked in the past. Economically interdependent states are, on occasion, willing to sacrifice their economic activity in order to pursue other goals that are given higher priority. A second problem is that economic interdependence can also aggravate the nature of competition. If two economies produce similar products within a shared region, as is predictable when developing economies make the push into transforming their economies into industrialised ones, they must compete for market share. Intensified competition is likely to follow, with the possibility that commercial rivalry will contribute to perceived threats and spill over into security rivalries. Increased economic interactions can then promote conflict as much, if not more, than constrain it. Economic interdependence, thus, is something less than a sure thing when it comes to inhibiting conflict (Rapkin and Thompson, 2006).

The costs of industrialised warfare are difficult to deny (Levy and Thompson, 2011). The increased cost of war technology means that very few states are competitive at the highest ranks. The increased lethality of weaponry should also make victory a more dubious concept. The consequences of nuclear exchanges may suffice to preclude their use. It is comforting that 65 years have passed without their being used again. Here again, though, it is not clear that we have finally moved beyond warfare between industrialised states "merely" because the costs now clearly exceed the conceivable gains. The states that are most likely to be involved in Asian conflict are also the ones that are or will become the states most equipped to engage in conflict. It is not Cambodian-Thai artillery exchanges that need be worried about in this respect. It is increased conflict among the current major powers (China, Russia, the

United States), possible candidates for major power status in the near future (India, Japan), and the small states (North Korea, Pakistan) with some nuclear capability and also less predictability that should be the main focus. In the past, it is conflict among the major powers that has led to big wars. If all of the major powers are involved in various types of Asian disputes over spatial control and relative position, Asia, as a region, has a greater risk of experiencing more intensive conflict than do other regions.

We think nuclear deterrence was critical in keeping the Cold War cold. It may also have had something to do with the non-use of nuclear weapons in Sino-Soviet and Indo-Pakistani militarised disputes. But we really cannot be sure. Nor can we be sure that something that may have worked in the past will work in a future context. One of the hallmarks of European conflict of the past 500 years is that the intensification of conflict was rarely fully premeditated. States have not initiated global warfare so much as they escalated more minor clashes into larger and wider affairs under the right circumstances. Thus the question is less whether nuclear deterrence or the high costs of war can be relied upon to stave off a World War III. Rather, the question is whether these constraints can intervene to halt a spiralling escalation that begins without overtly invoking the high costs of industrialized warfare initially.

Conclusions

Challengers, bipolarisation, economic destabilisation, multiple and interdependent rivalries, and weak conflict constraints are not necessarily a package of problems that either all come together or not at all. They represent a list of factors that can be extracted in generalised form from the specifics of European regional history (1494-1945). It is possible that none of these will be applicable to the future of Asian international relations. Yet that seems most unlikely since they are already evident in various stages of progress. But they do not need to all be present. None are necessary or sufficient factors for regional instability. To the extent that they represent emerging Asian realities, however, the prospects for something like older European international relations, has a greater probability of occurrence.

The older European dynamics are not exactly alien processes being forced on Asia. They are the type of processes that have some likelihood of emerging if the appropriate mix of factors comes together. For reasons argued above, Asian international relations are not likely to work exactly as European international relations played themselves out. Not all of the mix of factors are identical. But Asia appears to be moving toward multipolar contests over regional hegemony. In that respect, Asia is becoming more like older Europe.

To the extent that the full list of factors conducive to regional instability turn out to be manifested in Asia in full strength, the most likely outcome is one of the possible scenarios depicted in Friedberg (2010). He imagines an Asia split into two camps—one centred on the mainland, relatively authoritarian, and linked closely to China; the other forming a rimland arc around the contours of Asia, stretching from India through Indonesia to Japan, and linked closely to the United States. If this configuration were to come to pass, it would seem to fit in seamlessly with the older East Asian pattern of initially eastern and then increasingly western pressures on China. We should expect to see other fusions of eastern and western dynamics along similar lines in the next 25 years.

Notes

1. See Kacowitcz (1998), Gleditsch (2002), Lemke (2002), Buzan and Waever (2003), and Miller (2007) for strong efforts.
2. The late Andre Gunder Frank (1998) used his interpretation of the history of the world economy to predict that China was likely to return to the centre of world wealth after its absence for only a few centuries. The nature of the analysis is similar to relying on regional geopolitical structure but most definitely did not rely on a regional focus.
3. The only roughly equivalent regional experience with which I am familiar is the Warring States period in ancient China that led to the Han unification.
4. Analyses of the European balance of power may be found in Levy and Thompson (2005, 2010).
5. As the rivalry died away, Franco-German relations, of course, have remained central to European regional integration.
6. With hindsight, we know that these powerful states were not always seeking regional hegemony but contemporary decision-makers thought they were and behaved accordingly.
7. The Johnston (1995) thesis is that there have been two strategic cultures discernible in Chinese history—A Confucian, relatively pacific and accommodating one and another one that is much more belligerent. Of the two, Johnston thinks the latter has been the more predominant one and cautions against subscribing to the idea that there is a strong historical legacy of a defensive strategic culture. He also singled out the Ming dynasty as a period in which the more belligerent code was clearly operative.
8. The point here is that the Song dynasty is the one native Chinese dynasty that was almost always in retreat against its adversaries losing first its

northern territory and finally its southern territory. A more judicious research design might have focused on relatively strong and weak phases within several dynasties, as opposed to pitting behaviour within two dynasties with strong "personalities" against each other.

9. This basic point is take from Dehio (1962). See, as well, Thompson (1992) and Rasler and Thompson (1994).
10. This Asian "eastern question" is not to be confused with the Ottoman eastern question which reduced to whether and when the Ottoman Empire should be partitioned among European major powers.
11. At times, of course, China developed sea power without shedding its commitment to land power orientations. Perhaps the closest China ever came to developing a genuine sea power perspective was when it was practically forced to in the Southern Song era.
12. Views vary widely on this issue. See, among many others, Rapkin and Thompson (2003), Chan (2008), Goldstein (2008), Levy (2008), Zhu Feng (2008), and Kugler and Tannen (2009).
13. These factors are framed differently in Thompson (2009a, 2009b).
14. For an overview of Asian rivalries, see Ganguly and Thompson (2011).
15. See, for instance, Rasler and Thompson (2005).

References

Bunker, Stephen G. and Paul S. Ciccantel (2005) *Globalization and the Race for Resources*. Baltimore, Md.: Johns Hopkins University Press.

Bunker, Stephen G. and Paul S. Ciccantel (2007) *East Asia and the Global Economy: Japan's Ascent, with Implications for China's Future*. Baltimore, Md.: Johns Hopkins University Press.

Buzan, Barry and Ole Waever (2003) *Regions and Powers: The Structure of International Security*. Cambridge: Cambridge University Press.

Chan, Steve (2008) *China, the United States and the Power Transition Theory: A Critique*. London: Routledge.

Chase, Kenneth (2003) *Firearms: A Global History to 1700*. Cambridge: Cambridge University Press.

Cohen, Warren I. (2000) *East Asia at the Center: Four Thousand Years of Engagement with the World*. New York: Columbia University Press.

Colaresi, Michael, Karen Rasler, and William R. Thompson (2007) *Strategic Rivalry: Space, Position, and Conflict Escalation in World Politics*. Cambridge: Cambridge University Press.

Frank, Andre Gunder (1998) *Global Economy in the Asian Age*. Berkeley: University of California Press.

Friedberg, Aaron L. (1993/94) "Ripe for Rivalry: Prospects for Peace in a Mulitpolar Asia." *International Security* 18: 5-33.

Friedberg, Aaron L. (2010) "The Geopolitics of Strategic Asia," in Ashley J. Tellis, Andrew Marble, and Travis Tanner, eds., *Strategic Asia 2010-*

11: Asia's Rising Power and America's Continued Purpose. Seattle, Wa.: National Bureau of Asian Research.

Ganguly, Sumit and William R. Thompson (2011) "Conflict Propensities in Asian Rivalries," with Sumit Ganguly in Sumit Ganguly and William R. Thompson, eds., *Asian Rivalries: Conflict, Escalation and Limitations on Two-Level Games.* Stanford, Ca.: Stanford University Press.

Gleditsch, Kristian S. (2002) *All International Politics is Local: The Diffusion of Conflict, Integration and Democratization*. Ann Arbor: University of Michigan Press.

Gleditsch, Nils Petter and Havard Hegre (1997) "Peace and Democracy: Three Levels of Analysis." *Journal of Peace Research* 41, 2: 283-310.

Goldstein, Avery (2008) "Parsing China's Rise: International Circumstances and National Attributes," in Robert S. Ross and Zhu Feng, eds., *China's Ascent: Power, Security and the Future of International Politics.* Ithaca, NY: Cornell University Press.

Johnston, Alastair Iain (1995) *Cultural Realism: Strategic Culture and Grand Strategy in Chinese History.* Princeton, NJ: Princeton University Press.

Kacowicz, Arie M. (1998) *Zones of Peace in the Third World: South America and West Africa in Comparative Perspective*. Albany: State University Press of New York.

Kang, David C. (2010) *East Asia Before the West: Five Centuries of Trade and Tribute*. New York: Columbia University Press.

Kugler, Jacek and Ronald L. Tammen (2009) "Implications of Asia's Rise to Global Status," in William R. Thompson, ed., *Systemic Transitions: Past, Present, and Future*. New York: Palgrave-Macmillan.

Ledyard, Gari (1983) "Yin and Yang in the China-Manchuria-Korea Triangle," in Morris Rossabi, ed., *China Among Equals: The Middle Kingdom and its Neighbors, 10th-14th Centuries*. Berkeley: University of California Press.

Lemke, Douglas (2002) *Regions of War and Peace*. Cambridge: Cambridge University Press.

LePoire, David J. (2009) "Exploration of Connections Between Energy Use and Leadership Transitions," in William R. Thompson, ed., *Systemic Transitions: Past, Present, and Future*. New York: Palgrave-Macmillan.

Levy, Jack S. (2008) "Power Transition Theory and the Rise of China,"in Robert S. Ross and Zhu Feng, eds., *China's Ascent: Power, Security, and the Future of International Politics*. Ithaca, NY: Cornell University Press.

Levy, Jack S. and William R. Thompson (2005) "Hegemonic Threats and Great Power Balancing in Europe, 1495-1999." *Security Studies* 16, 1: 1-33.

Levy, Jack S. and William R. Thompson (2010) "Balancing at Sea: Do States Ally Against the Leading Global Power?," *International Security* 35,1 (Summer, 2010): 7-43.

Levy, Jack S. and William R. Thompson (2011) *The Arc of War: Origins, Escalation and Transformations* (Chicago, Il.: University of Chicago Press).

Miller, Benjamin (2007) *States, Nations and the Great Powers: The Sources of Regional War and Peace.* Cambridge: Cambridge University Press.

Modelski, George and William R. Thompson (1996) *Leading Sectors and World Powers: The Coevolution of Global Economics and Politics.* Columbia: University of South Carolina University Press.

Rapkin, David P. and William R. Thompson (2003) "Power Transition, Challenge, and the (Re) Emergence of China." *International Interactions* 29: 312-342.

Rapkin, David P. and William R. Thompson (2006) "Economic Interdependence and the Emergence of China and India in the 21st Century," in Ashley Tellis and Michael Wills, eds., *Strategic Asia 2006-07: Trade, Interdependence, and Security*. Seattle, Wa.: National Bureau of Asian Research.

Rapkin, David P. and William R. Thompson (2009) "Kantian Dynamics and Systemic Transitions: Can International Organizations Influence U.S.-China Conflict?," in William R. Thompson, ed., *Systemic Transitions: Past, Present, and Future*. New York: Palgrave-Macmillan.

Rasler, Karen and William R. Thompson (1994) *The Great Powers and Global Struggle, 1490-1990*. Lexington: University Press of Kentucky.

Russett, Bruce and John Oneal (2001) *Triangulating Peace: Democracy, Interdependence and International Organizations.* New York: W.W. Norton.

Rasler, Karen and William R. Thompson (2005) *Puzzles of the Democratic Peace: Theory, Geopolitics and the Transformation of World Politics.* New York: Palgrave-Macmillan.

Sen, Gautam (1984) The Military Origins of Industrialization and International Trade Rivalry. London: Pinter.

Thompson, William R. (1992) "Dehio, Long Cycles and the Geohistorical Context of Structural Transitions." *World Politics* 45 (October, 1992): 127-152.

Thompson, William R. (2003) "A Streetcar Named Sarajevo: Catalysts, Multiple Causation Chains, and Rivalry Structures." *International Studies Quarterly* 47: 453-74.

Thompson, William R. (2006) "Commercial Rivalry and Global War." Paper presented at the annual meeting of the American Political Science Association, Philadelphia, Pa., August.

Thompson, William R. (2009a) "How Might We Know a Systemic Transition is Underway? Clues for the 21st Century," in William R. Thompson, ed., *Systemic Transitions: Past, Present, and Future.* New York: Palgrave-Macmillan.

Thompson, William R. (2009b) "Structural Preludes to Systemic Transition

Since 1494," in William R. Thompson, ed., *Systemic Transitions: Past, Present, and Future*. New York: Palgrave-Macmillan.

Wang, Yuan-kang (2010) *Harmony and War: Confucian Culture and Chinese Power Politics*. New York: Columbia University Press.

Zhu Feng (2008) "China's Rise Will Be Peaceful: How Unipolarity Matters," in Robert S. Ross and Zhu Feng, eds., *China's Ascent: Power, Security and the Future of International Politics*. Ithaca, NY: Cornell University Press.

2

A Necessary Ghost Hunt? West European Integration in East Asian Regionalism Studies

Zhang Zhenjiang

Introduction

East Asian regionalism (EAR hereafter) has been the most noticeable phenomenon of the post-Cold War world and is currently one of the most widely discussed and debated topics in the field of international relations. The policy and theoretical implications of West European integration (WEI hereafter) as the most successful example of regionalism and integration in the world are important and unavoidable for any policymaker and scholarly theorist of regionalism. Both East Asia and West Europe are independent geographic regions comprising several independent nation-states. Both have a long history of the international exchange of peoples and commodities and changes of frontiers resulting from conflicts and wars. The transition of nation-states in West Europe from conflict to cooperation, from enmity to friendship, from anarchy to union, from balance of power to regional integration sets an admirable example for any geographic region consisting of warring states and for regional international relations. In addition, West Europe's impressive success and achievements have set not only an admirable example but also a standard for any regionalising practice in the world. Thus, it is natural for scholars of EAR to bear WEI in mind consciously or unconsciously.

George Bernard Shaw once said that Britain and America are "two countries divided be a common language".[1] In the same way, EAR studies and WEI studies have long been academic fields divided by a common subject: the study of regionalism.[2] This is typically the case in China. Many universities and think tanks, including different social sciences academies, have separate centres and institutes of European and Asian studies. Interaction between them is rare although they are all conducting studies on regionalism. Two reasons may account for this. First, EAR is a new discipline, that was initiated after the Cold War and was formalised in the late 1990s. EAR scholars, being newcomers to regionalism, are generally busy describing the new, unstable and fast changing EAR and have devoted less time to comparative studies. Second, WEI scholars regard WEI as the most successful model of regionalism, that is not easy to replicate; they also look down upon EAR and EAR studies. Murray has pointed out to the mutual benefit to be derived from comparative studies of the two regions, because "Europe and Asia can learn from each other", and that WEI especially could learn from Asia's open regionalism, but that there is a "small body of literature on how Europe can learn from Asia".[3]

The landscape has, however, been changing over the past two to three years. There is a growing body of literature comparing EAR and WEI in both the Chinese and English academic communities.[4] As mentioned earlier, both EAR and WEI are case studies on regionalism and comparison and could shed light on each other. However, the comparison could be too much and improper. A British scholar comments: "How useful it is to have such a dominant presence [treatment of the EU as the archetypal model] when it comes to the construction of a truly comparative approach to studying regional integration is open to question."[5]

This paper focuses on the differences rather than the similarities between EAR and WEI. The first part is a brief overview of the regionalising process in East Asia and its achievements. The second part focuses on the unbridgeable gaps between the EAR and WEI in relation to the international context. The third part tries to tentatively theorise on the East Asian regionalising process over the last decades in terms of crisis-driven regionalism, open regionalism, small states-led regionalism, and competitive regionalism. Finally, this paper calls for a theoretical innovation in EAR studies.

Rising EAR and its Achievements

After World War II, following the economic integration of Western Europe, a worldwide trend towards regionalism movement commenced in the 1950s. However, partly due to the US-China confrontation and the United States' bilateral military alliance system, East Asia did not follow suit.[6]

After the Cold War, responding to the developing regionalisms in Europe, North America and Asia-Pacific, the Malaysian prime minister Mahathir bin Mohammad proposed the formation of an East Asian Economic Grouping (EAEG) in 1990, but the great powers, including America, Japan and China, did not show much interest in the idea. Through ASEAN efforts, the EAEG was modified into EAEC (East Asian Economic Caucus), put on to the APEC's agenda and also discussed annually in ASEAN. The South East Asian financial crisis of 1997 marked a turning point for the EAR. At the end of that year, ASEAN member countries and China, Japan and South Korea held their first Leaders' Summit and this led to the annual 10+3 summit, which became a symbol of EAR. In 2005, the first East Asian Summit (EAS, 10+6) was held in Kuala Lumpur, following which the EAR entered a new stage.[7] The ASEAN summit at the end of 2010 decided to welcome America and Russia into the EAR; thus 10+6 will become 10+8 in 2011.

Regional cooperation among the East Asian nations has progressed impressively since the birth of EAR. Briefly, this in the area of financial cooperation, East Asian Free Trade Area (EAFTA), and the rapid proliferation of multi-layer mechanisms for regional cooperation.

Regional Financial Cooperation

EAR began with the financial crisis, and its symbolic success was the Chiang Mai Initiative (CMI), a currency swap initiative established by 10+3 as a network of bilateral agreements to swap and repurchase central bank reserves among the ASEAN, China, Japan and Korea.[8] Although many analysts were sceptical about it, CMI has been a success. It started from a set of bilateral swap agreements, but beginning March 24, 2010 has eventually become the Chiang Mai Initiative Multilateralisation (CMIM)—a multilateral currency pool. The Asian Foreign Reserve Pool totals $120 billion: China, Japan, and South Korea contributed 80 per cent and ASEAN contributed 20 per cent. Japan and China contributed $38.4 billion each; Korea contributed $19.2 billion; Indonesia, Malaysia, Thailand and Singapore each contributed $4.76 billion; and the Philippines $3.68 billion (see table). CMIM has been recognised as representing "a material qualitative change that may lead to the creation of true regional institutions",[9] "a major step towards the inevitable fulfilment of the 1997 proposal to create an Asian monetary fund (AFM)" and "an important milestone signalling a maturing of ASEAN+3 cooperation."[10]

East Asian Free Trade Area (EAFTA)

The East Asian Free Trade Area was proposed by the East Asian Vision

CMIM: Signatories' Contributions and Access to Pool Resources

Countries		Financial contribution			Multiplier
		Billions of U.S. dollars	%		
China	38.4	less Hong Kong-34.2	32.0	28.5	0.5
		Hong Kong - 4.2		3.5	2.5
Japan		38.4	32.0	0.5	
Korea		19.2	16.0	1.0	
Three countries		96.0	80.0	–	
Indonesia		4.77	3.97	2.5	
Thailand		4.77	3.97	2.5	
Malaysia		4.77	3.97	2.5	
Singapore		4.77	3.97	2.5	
The Philippines		3.68	3.07	2.5	
Vietnam		1.0	0.83	5	
Cambodia		0.12	0.1	5	
Myanmar		0.06	0.05	5	
Brunei		0.03	0.02	5	
Laos		0.03	0.02	5	
ASEAN		**24.0**	**20.0**	-	
Total		120.0	100.0		

Source: www.mas.gov.sg/news_room/press_releases/2010/Joint_Press_Release_ CMIM_Comes_Into_Effect.html

Group (EAVG) and officially adopted and stated as a goal by the 10+3 summit.[11] Although FTA for the whole region of East Asia is a goal in pursuit, agreement on it will not be easy. In the matter of fact, there already has been established several solid FTA foundations for EAFTA. The ASEAN FTA (AFTA) has been virtually established. Through the Common Effective Preferential Tariff (CEPT) scheme for AFTA, ASEAN member countries have made significant progress in lowering intra-regional tariffs. More than 99 per cent of the products in the CEPT Inclusion List (IL) of the ASEAN six original members have been brought down to within the 0-5 per cent tariff range. The deadline for the four new members is 2015.[12] Other FTAs which might be useful stepping stones towards EAFTA are the complex FTA networks led by ASEAN plus China. Starting from 2002 and fully realised from 2010, China-ASEAN FTA opened up a combined market of 1.9 billion people accounting for a combined GDP of close to $6 trillion and a total trade value of $4.5 trillion, which is the largest FTA between developing countries in the world. The average tariff on goods from the ASEAN countries has been cut down to 0.1 per cent from 9.8 per cent. The six original ASEAN members—Brunei, Indonesia, Malaysia, the Philippines, Singapore and Thailand—will slash the average tariff on Chinese goods from 12.8 per cent to 0.6 per cent. By 2015, the policy of zero-tariff rate for 90 per cent of traded goods is expected to be extended among China and four new ASEAN members—Cambodia, Laos, Myanmar and Vietnam.[13]

At the beginning of the 1990s, there was essentially no Preferential Trade Agreement (PTA) or Free Trade Agreement (FTA) in East Asia, but over the last two decades, East Asian countries have signed more bilateral, multilateral or regional PTAs and FTAs than any other region of the world. Today there are a dozen of them. The current 10+3 FTA networks could facilitate an EAFTA: a virtual Asian community is already emerging from the complex web of FTAs.[14] (Fig. 2.1)

Research conducted by the Asian Development Bank (ADB) shows that in many spheres, including direct investment, policy cooperation, people-to-people contacts and macroeconomics, the intensity of cooperation in East Asia after the financial crisis in 1997 has been increasing.[15] (Fig. 2.2)

Rapid Proliferation of Multi-layer Mechanisms for Regional Cooperation

The multi-layer mechanisms for regional cooperation include the ten ASEAN member states, cooperation between China, Japan and South Korea in North East Asia, ASEAN-China, ASEAN-Japan, ASEAN-South Korea, and 10+3. All these mechanisms work independently and have their own independent agendas, leaders' summits, ministerial meetings, senior officials' meetings and different working and functional groups. The 10+3 is a good example. Since the beginning of the process in 1997, 10+3 cooperation has rapidly broadened and deepened, and is now being expanded to include food and energy security, financial cooperation, trade facilitation, disaster management, people-to-people contacts, narrowing the development gap, rural development and poverty alleviation, human trafficking, labour movement, communicable diseases, environment and sustainable development, and transnational crime and counterterrorism. By January 2010 there were 64 mechanisms coordinating 10+3 cooperation: One summit, 16 ministerial, 23 senior officials, 1 directors-general, 17 technical level meetings and 6 other tracks meetings.[16]

There is, in addition, the East Asian Summit (EAS). Since its inception in 2005, the EAS or the so-called ASEAN plus Six (10+6) has served as a forum for dialogue on broad strategic issues of relevance to East Asia as well as other regional and global issues, with focus on areas such as international terrorism, energy, infectious diseases, sustainable development and poverty reduction. The fifth EAS in 2010 accepted the ASEAN proposal to invite Russia and US to join it in 2011, to expand 10+6 to 10+8.[17]

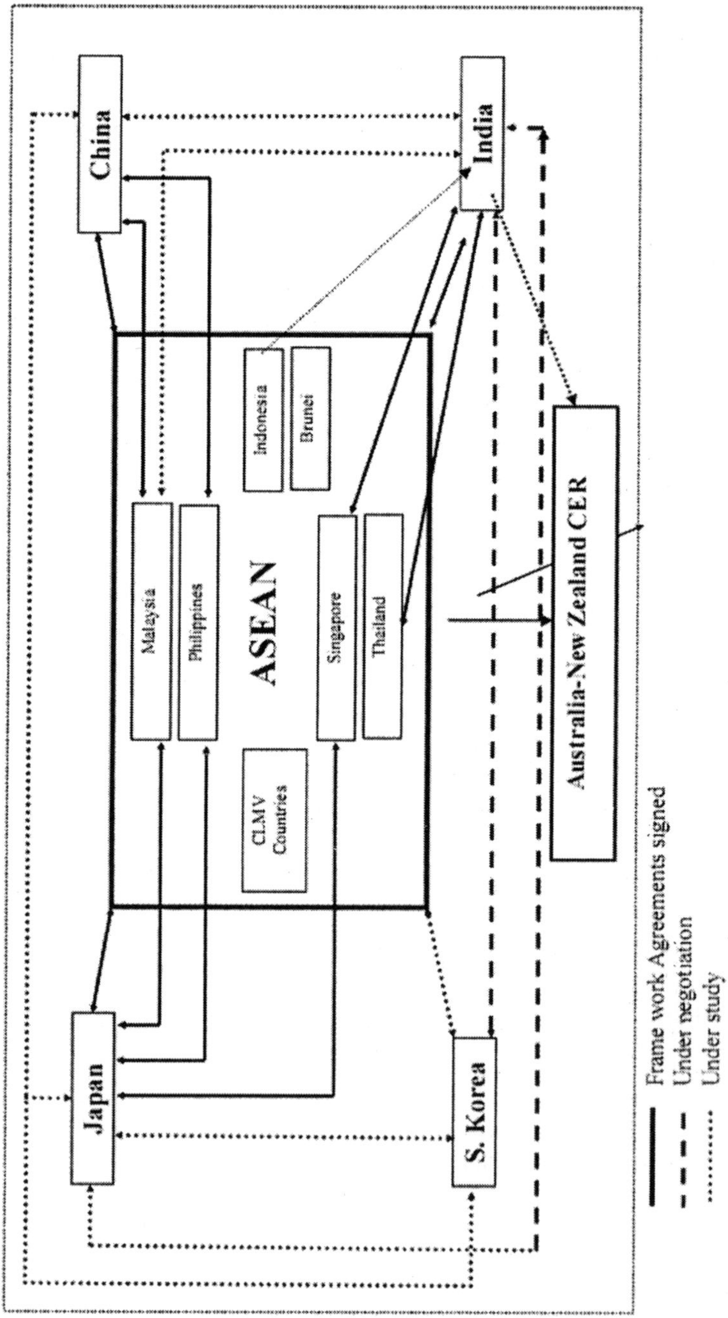
China
India
ASEAN
Malaysia
Philippines
Singapore
Thailand
Indonesia
Brunei
CLMV Countries
Australia-New Zealand CER
Japan
S. Korea
Frame work Agreements signed
Under negotiation
Under study

Fig. 2.1

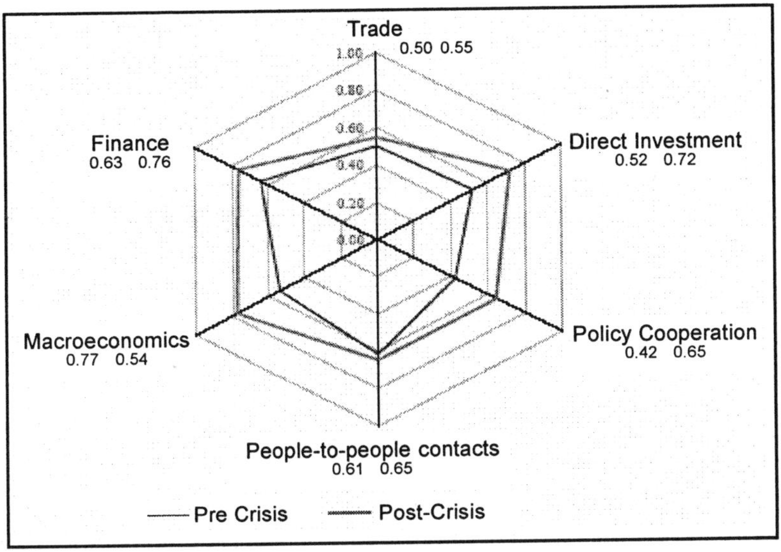

Fig. 2.2

The Issues of no Comparison between EAR and WEI

It has been pointed out that:

> It is not surprising that the experiences of the European integration project loom large in analyses of (and predictions for) regional integration elsewhere in the world. After all, the European Union (EU) is the best example we have of what happens when states agree to first move towards formal economic cooperation and coordination, and then some wider form of political integration.[18]

However, before making any comparison, it is important to realise that there are many differences between EAR and WEI. The most prominent difference is in terms of formal and informal institutionalisation. A Japanese scholar argues:

> Regional cooperation in Europe is extensive in scope and intensive in formal institutions and legal norms. Formal treaties or negotiations precede increased interaction in Europe—making regionalism in Europe politics-led or policy-induced. Asian regionalism, in contrast, has been driven by informal interaction and the growth of economic transactions without policy coordination or state-based negotiation.[19]

Another difference is in the forces that drive regionalism. EU has been driven by policy as a result of its specific history and geography. In contrast, Asian regional cooperation has been market driven.[20] The tradition of strong-government and weak-business in East Asia and a relatively balanced government-business relations in West Europe also makes a big difference. As a scholar argues: "European experiences indicate that the existence of business associations with the ability to set up a far-reaching agenda, and to exert pressure on the national governments for its realization, is one of the key factors to successful integration", unlike Asia at present.[21] Acharya suggests that the success of EU's regionalism came from its overcoming of nationalism, but in East Asia, the two processes of regionalism and nationalism have long "enjoyed a symbiotic relationship". The role of the nation-state in Asia differs considerably from Europe, as one of EAR's fundamental norms is the principle of non-interference in member states.[22]

These differences are significant for both the theory and practice of EAR. Apparently, there are greater economic, social and political disparities in East Asia than in Western Europe apart from the differences of religion, culture, economic development levels and political structures and norms. Katzenstein suggests that Asia and Europe have "regionally specific, systematically different patterns of politics and policies".[23]

Obviously, the differences between EAR and WEI are huge and endless. However, as to the comparison between EAR and WEI and what EAR could learn from WEI, this paper insists that their *different historical contexts* cannot be compared, learned or copied. This historical context includes the different international systems, the role of external powers, especially the role of America in regionalism, and the factor of economic globalisation.

The WEI started in the 1950s, when the Cold War was at its peak. Western Europe was at the forefront of the struggle between the US and the USSR. Under the umbrella of US protection, and also to counter the perceived threat from the Soviet Union, France and Germany tried to resolve their historic enmity.[24] In a way, both the US and the USSR encouraged the Western Europe regional initiatives. US encouragement, including the Marshal Plan, also fostered regionalism and integration in Western Europe. John Mearsheimer argues that it was not economic interdependence but the Cold War which was "the main reason that intra-EC relations have flourished". The Cold War caused the WEI "in two different but mutually reinforcing ways". West European countries' perceived common threat from the Soviet Union:

> Forced the Western democracies to band together to meet the common threat. Britain, Germany, and France no longer worried about each other, because all faced a greater menace from the Soviets. The Soviet threat muted concerns about relative gains from economic cooperation among the EC states by giving each Western democracy a vested interest in seeing its alliance partners grow powerful, since each additional increment of power helped deter the Soviets. The Soviet threat also muted relative-gains fears among West European states by giving them all a powerful incentive to avoid conflict with each other while the Soviet Union loomed to the east, ready to harvest the gains of Western quarrels... America's hegemonic position in NATO, the military counterpart of the EC, mitigated the effects of anarchy on the Western democracies and facilitated cooperation among them... America not only provided protection against the Soviet threat, but also guaranteed that no EC state would aggress against another... With the US serving as night watchman, relative-gains concerns among the Western European states were mitigated and, moreover, those states were willing to allow their economies to become tightly interdependent.[25]

In contrast, the EAR started in the post-Cold War era when America became the only superpower.[26] America's attitude and policy towards regionalism changed significantly, even with regard to European integration after the Cold War. Some scholars have commented that America is the ghost at the feast in both regions. Its role in the two regions differs considerably, with regard to at least: "first, the nature of the US response to regionalising tendencies in each region, and, second, in terms of its hard power status in each region over time."[27] Generally speaking, for EAR in the post-Cold War era there is no encouragment from the great powers. Some scholars have argued that EAR has risen while America was busy focusing on anti-terrorism.[28] Also, "there have already been a series of skirmishes between the United States and Asian countries over the budding initiatives toward East Asian regionalism".[29] The first skirmish was America's negative response to Mahathir's EAEG proposal in 1990; the second was its sharp opposition to Japan's initial proposal for an Asian Monetary Fund in 1997 as a potentially fatal disruption of the IMF. Even about the Chiang Mai Initiatives America continues to insist that any such regional arrangement must be consistent with the global institutional order and "wants CMI lending to be linked to IMF programmes and their conditionality which is what some Asians want to escape." In the field of trade, the US has grown increasingly uneasy over the prospect of an East Asian bloc, which could discriminate significantly against US exports. One US response has been to counter the intra-Asian network of FTAs with bilateral FTAs of its own with Singapore and Korea.[30]

From opposing the EAR from the first EAS in 2005, the US government has now begun to seriously deal with it and trying to develop Asia-Pacific initiatives to co-opt or replace the East Asian efforts. Strategically, George W. Bush formally called for establishing a Free Trade Area of Asia Pacific (FTAAP) at the APEC Summit in 2006. Because of America's position and efforts, the FTAAP was eventually included in the APEC's agenda and became a long-term goal at the following APEC summit. On the policy side, the US government started negotiations to join the Trans-Pacific Partnership (TPP) in 2009, which is a more realistic way of joining Asia-Pacific and possibly to break the current East Asian grouping which excludes the US.[31]

The EAR is also becoming a reality in the era of economic globalisation. Economic globalisation started in the 1970s, but was at its peak post the Cold War. The *New York Times* columnist Thomas Friedman says that "the world is ten years old",[32] referring to the integration of national markets with foreign, regional, international and global markets which includes the exchange of goods, services, capital and labour. A prerequisite of economic globalisation is the practice of market economy by most, if not all, members of the international community. In short, economic globalisation requires nation-states to have a global perspective. Logically, global perspective conflicts with national and regional perspectives.

The impact of economic globalisation on EAR has been paramount. On the one hand, East Asian states are trying to meet the needs of regional economic interdependence. Regionalism is also a way of dealing with the challenges of economic globalisation. Some scholars have pointed out that the new regionalism including EAR itself is a response to globalisation.[33] In contrast to the European integration being seen as "internally" oriented, EAR is driven by "external" forces. On the other hand, while proceeding with regionalism, economic globalisation also requires the East Asian states keep an eye on the external sphere and remain open to global markets.[34]

Both West Europe and East Asia are entering into inter-regional economic arrangements to deal with the challenges of globalisation; but their effectiveness varies since West Europe has basically accomplished its regional integration while East Asia is struggling on both intra-regional and interregional fronts at the same time.[35] Economic globalisation not only defines the functions of EAR but also determines its form. Combined with the role of the United States, the EAR, a creation of the new international context in the post-Cold War era, has take the new form of "open" regionalism.

How to Theorize EAR?

EAR being the product of very different international context, how do we theorise its process? What is EAR and how do we theorise it academically? Keeping in view the differences between the EAR with WEI, this paper tentatively argues that crisis-driven regionalism, open regionalism, small states-led regionalism, and competitive regionalism could describe and theorise the current EAR.[36]

Crisis-driven Regionalism

The idea of EAEG proposed by Mahathir in 1990 was a product of a sense of crisis and a response to the rising regionalism around the world. From the formation of a single market in the late 1980s, regionalism in West Europe entered a new phase, which some scholars regarded it as the beginning of "new regionalism". The rising trading blocs at that time not only included West Europe but also the hotly debated idea of the North America Free Trade Agreement (NAFTA) and the Asia-Pacific Economic Cooperation (APEC) proposed by Australia. The rise of other regionalising initiatives posed a challenge to East Asia. The end of the Cold War not only ended the pressure stemming from the bipolar conflict and released the states' and regional validity, but also created a sense of crisis: if others were grouping together, East Asia risked being isolated. Though the regional great powers were opposed to or were not interested in this proposal, ASEAN kept on discussing and pushing and this preparation led to its official birth in 1997.

The Asian financial crisis of 1997/98 taught the important lesson that East Asia needed to strengthen its regional economic cooperation in order to sustain dynamic economic growth and stability. National efforts alone were not enough to strengthen individual economic fundamentals; Global or regional efforts were essential for dealing with the crisis. The global initiative to strengthen the international economic system was unsatisfactory. Especially, the stagnant GATT Uruguay Round discouraged any multilateral arrangements. At the same time, the indecisiveness and ineffectiveness of both APEC and ASEAN in proposing appropriate bailout measures severely hurt the East Asian leaders' confidence in them. The general sentiment in Asia was that the region must establish its own self-help mechanism for economic management. Thus, the Asian financial crisis nurtured the sense of a "region" with a common set of challenges.[37] As a scholar has put it, the crisis produced a wedding of 10+3 and two funerals of ASEAN and APEC.[38]

This crisis not only produced the EAR, but also drives it along. According to one scholar, the spread of Severe Acute Respiratory Syndrome (SARS) in 2003, a typical regional and later an international crisis, gave a new impetus to EAR. Regional frameworks like 10+3 and ASEAN, with strong encouragement from each member state, began to be actively involved in the campaign against SARS. This scholar finds that "economic factors are no longer sufficient to bring a closer regional integration in this region" and concludes that EAR is changing from economics-driven regionalisation to crisis-driven regionalism.[39]

Unfortunately for the world economy but fortunately for EAR, the international financial crisis in 2008 again raised concerns about regional financial stability. Largely because of the crisis itself and the crisis-driven needs of regional cooperation, the finance ministers' at the February 2009 ASEAN+3 meet finally reached an agreement to transform the existing Bilateral Swap Arrangements (BSAs) of Chiang Mai Initiative into a regional foreign reserve pool of $120 billion to "address short-term liquidity difficulties in the region and to supplement existing international financial arrangements."[40] CMIM was formally put into operation in March 2010.

Open Regionalism

"Open regionalism" became a popular term in international political economy after C. Fred Bergsten's analysis. He said that the concept itself: "seeks to assure that regional agreements will in practice be building blocks for further global liberalisation rather than stumbling blocks that deter such progress".[41] In this sense, EAR is a typical example of open regionalism. Any regionalising effort by the East Asian countries does not inhibit but helps them to expand their connections with external and global players.

Using APEC as the main case, Bergsten called for turning "open regionalism" into "an operational and highly significant policy programme".[42] The term "open regionalism" is used in this paper to describe the EAR process, which is not only an opening to external countries but also offers an opening for its own members to pursue their own bilateral, multilateral, regional and even global projects inside or outside of the region.

Theoretically, any arrangement that covers the whole region is ideal. However, most of the current EAR arrangements in the region remain partial. The FTAs in East Asia are a good example. An EAFTA is difficult; even reaching a consensus on a long-term goal may take much effort and many years. However, bilateral, multilateral and sub-regional FTAs have been proliferating in the region in the past ten years. In 2000, there were only three FTAs in effect (including AFTA), one was under negotiation, with another three at the proposal stage. But in just nine years, these figures increased more than tenfold; in January 2009, East Asia was at the forefront of the world's FTA activity, with 37 FTA in effect and another 72 at various stages of preparation. All of the countries in the region are players. Japan has implemented bilateral Economic Partnership Agreements (EPAs) with eight countries and a general FTA with ASEAN. In addition, Japan is negotiating agreements with Korea, Vietnam, India and Australia. China and South Korea have implemented FTAs with ASEAN. ASEAN recently concluded FTA negotiations with India, Australia and New Zealand, and is considering negotiating with the EU. Some ASEAN members, such as Singapore and Thailand, are actively pursuing a host of bilateral FTAs.[43]

East Asian FTA networks are not limited to the East Asia region only, but are expanding globally. As a precondition for signing FTA with ASEAN, Australia, New Zealand and India joined an EAR with the first EAS (10+6) in 2005. Surprisingly, America and Russia were granted entry tickets in 2010 and the 10+8 EAS will be held at the end of 2011. In keeping with Bergsten's concept, EARs FTAs are stepping stones to broader regional, even global FTA arrangement. In this sense, the EAR is very "open".

Small States-led Regionalism

A distinguishing feature of EAR is that it has been led throughout by ASEAN, a group of small states.[44] Although there are many debates regarding its effectiveness and sustainability, but from a descriptive rather than a normative perspective there is agreement on ASEAN's leading role in the process. Different terms characterise it: "leader", "driver", "hub and spoke", "core", "centre", etc.[45]

ASEAN's leading role in EAR is reflected in several ways. First, its initiator, Malaysia, a core member of ASEAN, proposed the idea of EAEG in 1990; ASEAN persisted with the idea in APEC and ASEAN meetings till 1997. Secondly, although 10+3 and 10+6 are general regional arrangements, they all consist of working bilateral agreements, with ASEAN as the hub and the great powers—China, Japan, Korea, Australia, New Zealand, India, America and Russia—as spokes. Thus, all 10+3 or EAS summits have to be held in South East Asian countries; the exception are the ASEAN+X meetings, which are held alternatively in South East Asia and the next year in X.

Thirdly, ASEAN not only chairs the summit and various related conferences, but also sets the conference agenda.[46] The 10+3 Foreign Ministers' Conference in Vietnam, which granted America and Russia entry into EAS is a good example of ASEAN's power of agenda setting. Last but most importantly, ASEAN has laid down three criteria for the membership of the EAS: the state must be a dialogue partner of ASEAN; it must accede to the ASEAN Treaty of Amity and Cooperation in South East Asia (TAC), and it must have "substantive relations" with ASEAN.[47]

Because of ASEAN's leading role and its significant impact on EAR, different regional mechanisms including 10+1, 10+3, EAS, ASEAN Regional Forum (ARF), and even APEC are adopting the "ASEAN way". The ASEAN way is gradually becoming the "Asian Way" and the "Asia-Pacific Way".[48]

Competitive Regionalism

"Competitive regionalism" refers at least to two things. One is that the EAR was born out of the competitive regionalisms worldwide. EAR was a late comer in regionalism because of the Cold War. Starting from the European single market in the late 1980s, new regionalism has been quickly proliferating with the ending of the Cold War. The EAR belongs to this new regionalism. One feature of the new regionalism is its proliferation in many regions simultaneously. All the regionalism(s) stimulate and influence each other and form the new wave of regionalisms in the whole world. The EAR was stimulated by "fears of exclusion from, or loss of bargaining leverage to, the European Union and NAFTA, but also shaped by learning from their examples."[49] Many researchers notice that proliferating and complicated FTAs arouse the rising competitive regionalism in East Asia.[50]

Competitive regionalism also means that there is not one regionalism, but several regionalisms in East Asia. The 10+3 was formed in 1997. After that, other than a joint declaration on East Asian cooperation issued in 1999 and slow development of bilateral swap agreements under the Chiang Mai Initiatives, the EAR did not make much progress. In 2002, China signed the TAC and China-ASEAN and declared that they were planning to establish China-ASEAN FTA in 2011. These bold initiatives created a chain reaction in East Asia. Japan and South Korea quickly follow suit, although in a different form, to strengthen their economic linkages with ASEAN. Traditionally, Japan and Korea have politically significant agriculture sectors, that have been their Achilles heel in their pursuit of free trade.[51] With the swift proliferation of bilateral, multilateral and regional FTAs, the EAR gathered new impetus and entered a new stage.

Great power competition for ASEAN did not stop. After uneasy efforts, India, Australia and New Zealand have successfully joined EAR by joining the first EAS in 2005—which means that they have to be dialogue partners of ASEAN—to accede to the ASEAN TAC, and have "substantive relations" with ASEAN[52] and America and Russia have followed suit making East Asia the playground for all the great powers in the world. The great powers are all trying to compete for close relations with ASEAN making ASEAN not only the hub of EAR but also for the whole world where great powers compete not only for FTAs and economic gains but also for security and strategy as in ARF.

In conclusion, the snowball effect initiated by 10+1 of China-ASEAN triggered an East Asian "competitive regionalism" consisting of varied layers of regionalisms, including 3, 10, 10+1, 10+3, 10+6 and the future 10+8.

Concluding Remarks

First, it is necessary to have better studies on EAR, especially by scholars living in Asia because WEI has already provided many experiences, lessons and theories. A famous Chinese historian more than a century ago is supposed to have remarked:

> As for Chinese studies and Western studies, one prospers then both prosper, one declines then both decline. It is already time for mutual reinforcements and benefits. For today's academic community, it is impossible to have prosperous Chinese studies with declining Western studies, and it is also impossible to keep declining Chinese studies with rising Western studies.[53]

In short, if we want to have good EAR studies, we must have good WEI studies. This is a precondition for a "fruitful dialogue" between EAR and WEI.[54]

Secondly, it is a challenging task to make good comparisons and draw proper lessons from WEI. Evelyn Huber, then President of the Organised Section in Comparative Politics, American Political Science Association, highlighted the challenges and great benefits of studies comparing different regions by calling for "refinement of concepts and theories" and pointed out that "cross-regional comparison deserves a central place among our research designs".[55] Obviously, her call has been responded to, though slowly, in EAR studies. The West European experience is being increasingly discussed in EAR studies and more and more scholars are arguing that EAR can draw on WEI. This development should be greatly encouraged and welcomed.

Third, before we start on this great enterprise, it is most important to know what cannot be compared and copied from WEI; wrong comparisons currently prevail.[56] Relying too much on WEI standards, EAR scholars are becoming pessimistic about EAR's past processes and its future; are becoming very critical of EAR and over looking East Asian context and its relative achievements.[57] EAR and WEI are different developmental processes and have "occurred for different reasons, in different ways and with different outcomes".[58] The external context also matters and cannot be replicated copied from WEI to EAR. We are living in a new historic era of globalisation, but also of regionalisation. Globalisation in sync with regionalism determines the nature of EAR as a product; WEI, a product of the Cold War era but sustainable in the new era, is very different. "How best to use the EU experience as one example of integration rather than the only experience and/or archetypal model identified"[59] is an open question.

Originally theoretical innovation is required for describing and explaining the process, achievements, motivations and problems in EAR. There is a theory poverty when extrapolating regionalism from the European experience.[60] Concepts such as crisis-driven regionalism, open regionalism, small states-led regionalism and competitive regionalism used in this paper are a bold attempt in this direction. It is critically inspiring in this regard to study WEI theories and the endless efforts of scholars to produce theories including functionalism, neo-functionalism, federalism, inter-governmentalism, institutionalism, etc. Bearing WEI and other regional studies in mind, borrowing from Peter J. Katzenstein, the future of EAR studies depends on "analytical eclecticism".[61] In this sense, this paper has only raised the question, not answered it.

Notes

1. Anthony Jay, ed., *The Oxford Dictionary of Political Quotations*, 1996, p. 337.
2. This paper refers mainly to regionalism studies in Chinese. Even in English, comparative regionalism studies are new enterprises. The publication years in note 4 show that the work has been growing just in the past two to three years.
3. Philomena Murray, "East Asian Regionalism and EU Studies", *European Integration*, 32(6) (November 2010): 610-11.
4. Chinese literature includes: Lei Jianfen, "East Asian Cooperation does

not need to copy European Integration", *Global Times*, 4 September 2009; Zhao Huaipu, "Some Implications of European Integration to East Asian Cooperation", *Journal of College of Diplomacy*, Vol. 2, 2005; Xiong Guangkai, "East Asian Cooperation and European Experiences", Keynote Speech at the conference on "Regional Cooperation: European Experiences and East Asia", 10-11 October 2006; Zhang Xiaowen, "East Asian Expereices of Integration", *Newspaper of Chinese Social Sciences*, 17 March 2010; Wang Keyan & Li Rong, "Factors for Promoting European Integration and its implication for East Asian Regional Cooperation", *Modern Commercial and Trade Industry*, Vol. 2, 2009; ChenYugang, "European Experiences and East Asian Cooperation", *World Economy and Politics,* Vol. 5, 2006; Zhang Zhenjiang, "David Mitrany's International Cooperation Thoughts and its Implication for East Asian Cooperation", *Diplomatic Review*, Vol. 2, 2008.

English literature includes: Philomena Murray, "East Asian Regionalism and EU Studies", n. 3, pp. 597-616, p. 598; S. Levine, "Asian Values and the Asia Pacific Community: Shared Interests and Common Concerns", *Politics and Policy,* 35(1) (2007): 102-35; G. Capannelli, "Asian Regionalism: How does it Compare to Europe's?", *East Asian Forum*, 21 April 2009; G. Capellini, and C. Filippini, "East Asian and European Economic Integration: A Comparative Analysis", Asian Development Bank, Working Paper Series on Regional Economic Integration No. 29, May 2009; Y. Hidetaka, "Regional Integration and Business Interests: A Comparative Study of Europe and Southeast Asia", *European Journal of East Asian Studies* 6(2) (2007): 217-43; H. Giessmann, "Regionalism and Crisis Prevention in (Western) Europe and (Eastern) Asia: A Systematic Comparison", *Asia-Pacific Review* 14(2) (2007): 62-81; P. Murray, "Should Asia Emulate Europe?", in W. Moon and B. Andreosso-O'Callaghan (eds.), *Regional integration: Europe and Asia compared*, Ashgate: Aldershot, 2005, pp. 197-215; P. Murray, "The EU's Norm Diffusion in the Promotion of Regional Integration in East Asia", Paper presented at the International Conference on the 'Transformative Power of Europe', Otto-Suhr-Institute for Political Science, Freie Universität Berlin, 10-11 December 2009; P. Murray, "Comparative Regional Integration in the EU and East Asia: Moving beyond Integration Snobbery", *International Politics* 47(3-4) (May-July 2010): 308-29; T. Okagaki, "Europe as a Model of Regionalism in the Asia-Pacific", *East Asian Forum*, 6 May 2009, http://www.eastasiaforum.org/2009/05/06/relativising-europe-as-a-model-of-regionalism-for-theasia-pacific; Y.C. Park and C. Wyplosz, "Monetary and financial integration in East Asia: the relevance of European experience", European Commission. European Economy, Economic Papers 329, September 2008, http://ec.europa.eu/economy_finance/publications.

5. Shaun Breslin, "Comparative theory, China, and the future of East Asian regionalism(s)", *Review of International Studies*, 36, 2010, pp. 709-29.
6. Regionalism in South East Asia is led by ASEAN. However, it is a by-product of the Cold War since ASEAN's direct goal was trying to keep South East Asia's independence from the rivalry of the great powers.
7. M. Beeson, "ASEAN Plus Three and the Rise of Reactionary Regionalism", *Contemporary Southeast Asia: A Journal of International and Strategic Affairs*, 25(2) (August 2003): 251-68; E.J. Lincoln, *East Asian Economic Regionalism*, New York, 2004; D.K. Nanto, "East Asian Regional Architecture: New Economic and Security Arrangements and U.S. Policy", CRS Report for Congress, Order Code RL33653, 4 January 2004; C. Roberts, *ASEAN Regionalism: Cooperation, Values and Institutionalization*, Routledge, 2010; R. Stubbs, "ASEAN Plus three: Emerging East Asian Regionalism", *Asian Survey*, 42(3) (May/June 2002): 440-55; D. Webber, "Two funerals and a wedding? The ups and downs of regionalism in East Asia and Asia-Pacific after the Asian crisis", *Pacific Review*, 14(3) (August 2001): 339-72.
8. Some scholars refer to EAR and financial regionalism. See Joel Rathus, "The Chiang Mai Initiative: China, Japan and financial regionalism", http://www.eastasiaforum.org/2009/05/11/the-chiang-mai-initiative-china-japan-and-financial-regionalism/; Shintaro Hamanaka, "Re-considering Asian Financial Regionalism in the 1990s", Working Papers on Regional Economic Integration 26, Asian Development Bank, http://aric.adb.org/pdf/workingpaper/WP26_Reconsidering_Asian_Financial_Regionalism.pdf.
9. Douglas W. Warner, Paul Lelot and Wei Wang, "Assessing East Asian Financial Cooperation and Integration", *Singapore Year Book of International Law and Contributors* 12, 2008, pp. 1-42, p. 36.
10. William W. Grimes, "The Asian Monetary Fund Reborn?: Implications of Chiang Mai Initiative Multilateralisation", *Asian Policy,* No. 11, January 2011, pp. 79-104, http://www.nbr.org/publications/asia_policy/AP11/AP11_ E_CMI.pdf
11. *Final Report of the East Asia Study Group*, http://www.aseansec.org/viewpdf.asp?file=/pdf/easg.pdf
12. The AFTA agreement was signed on 28 January 1992 in Singapore by Brunei, Indonesia, Malaysia, Philippines, Singapore and Thailand and was joined by Vietnam in 1995, Laos and Myanmar in 1997 and Cambodia in 1999. AFTA primarily seeks to increase ASEAN's competitive edge as a production base in the world market through the elimination, within ASEAN, of tariffs and non-tariff barriers; and attract more foreign direct investment to ASEAN. For details, see http://www.aseansec.org/12021.htm.
13. Vincent Wei-cheng Wang, "The Logic of China-ASEAN Free Trade Agreement: Economic Statecraft of 'Peaceful Rise'", paper for "China

in the World, the World in China International Conference 'Implications of a Transforming China: Domestic, Regional and Global Impacts'", 5-6 August 2007, Institute of China Studies, University of Malaya. http://web.rollins.edu/~tlairson/mba/chinaaseanfta2.pdf

14. Nagesh Kumar, "ASEAN and Broader Regional Connectivity in Asia-Pacific: The case of Economic and Financial Integration", http://www20100324.chula.ac.th/chulaglobal/images/stories/Asean%20Connectivity/Dr.%20Narkesh%20Kumar.pdf
15. Srinivasa Madhur, "Perspective on Asian Regional Cooperation and Integration", ADB-ITD Seminar on Asian Trade and Investment Integration and Implications for Thailand, 9 June 2008, http://www.adb.org/Documents/Presentations/TRM/Perspective-Asian-RCI.pdf
16. "ASEAN Plus Three Cooperation", http://www.aseansec.org/16580.htm
17. Kor Kian Beng, "Proposal for Asean+8", *Straits Times*, 10 April 2010, http://www.straitstimes.com/BreakingNews/SEAsia/Story/STIStory_512798.html; "East Asian Summit: the International Politics Implications of Moving from 10+6 to 10+8", *China Daily*, 29 October 2010, http://www.chinadaily.com.cn/hqpl/2010-10/29/content_11474431.htm.
18. Breslin, "Comparative Theory, China, and the Future of East Asian Regionalism(s)", n. 5.
19. Okagaki, "Europe as a Model of Regionalism in the Asia-Pacific", n. 4.
20. Capannelli, "Asian regionalism: how does it compare to Europe's?", n. 4; Okagaki, *ibid.*; D. Nair, "Regionalism in the Asia Pacific/East Asia: a frustrated regionalism?", *Contemporary Southeast Asia,* 31(1) (2008): 110-42.
21. Y. Hidetaka, "Regional Integration and Business Interests: A Comparative Study of Europe and Southeast Asia", *European Journal of East Asian Studies,* 6(2) (2007): 217-43, 243.
22. A. Acharya, "Made in America? Agency and Power in Asian Regionalism, Roundtable: Peter J. Katzenstein's Contributions to the study of East Asian Regionalism", *Journal of East Asian Studies* 3 (2007): 370-8, 373.
23. P. Katzenstein, "Regionalism Reconsidered", *Journal of East Asian Studies* 3 (2007): 395-412, 396.
24. Thus, European integration is seen as "internally" oriented. G. Capannelli, "*Asian Regionalism: How does it Compare to Europe*'s?", n. 4.
25. John J. Mearsheimer, "Back to the Future: Instability in Europe after the Cold War", *International Security*, 15(1) (Summer 1990): 5-56.
26. Charles Krauthammer, "The Unipolar Moment", *Foreign Affairs*, 70(1), America and the World 1990/91, 1990/1991, pp. 23-33; Charles Krauthammer, "The Unipolar Moment Revisited", *National Interest*,

Vol. 70, Winter 2002, pp. 5-17; Michael Mastanduno, "Preserving the Unipolar Moment: Realist Theories and U.S. Grand Strategy after the Cold War", *International Security*, 21(4), Spring 1997, pp. 49-88.

27. Philomena Murray and Nicholas Rees, "European and Asian regionalism: Form and function", *International Politics* 47 (2010): 269-275, 270.
28. Zhang Zhenjiang, "Free Trade Area of Asia Pacific: America's Strategy and China's Policy", *World Economy and Politics*, Vol. 4, 2009.
29. C.F. Bergsten, "China and Economic Integration in East Asia: Implications for the United States", 2007, http://www.iie.com/publications/pb/pb07-3.pdf
30. *Ibid.*
31. Zhang Zhenjiang, "Free Trade Area of Asia Pacific ...", n. 29.
32. Thomas L. Friedman, "Opening Scene: The World Is Ten Years Old", in *The Lexus and the Olive Tree: Understanding Globalization*, New York: Anchor Books, 2000.
33. Zhang Zhenjiang, "Regionalism: New and Old", *Jinan Journal (Social Sciences)*, Vol. 4, 2008.
34. Capannelli, "Asian Regionalism: How does it Compare to Europe's?", n. 4.
35. Vinod K. Aggarwal, "Explaining Trends in EU Interregionalism" in Vinod Aggarwal and Edward Fogarty (eds.), *European Union Trade Strategies: Between Globalism and Regionalism*, London: Palgrave, 2004; Mathew Doidge, "From Developmental Regionalism to Developmental Inter-regionalism? The European Union Approach", NCRE Working Paper No. 07/01, July 2007. http://www.europe.canterbury.ac.nz/publications/pdf/Doidge_workingpaper0701_developmentalregionalism.pdf
36. There is much research trying to answer these questions in different places and from different perspectives. I also did some analysis in some papers in Chinese. See Zhang Zhenjiang, "ASEAN's Role in East Asian Cooperation", *Southeast Asian Studies*, vol. 3, 2004.
37. Masahiro Kawai, "Emerging Asian Regionalism: Ten Years after the Crisis", Paper prepared for the International Conference organised by ADB and the Thai Ministry of Finance, in collaboration with the Thai Development Research Institute, in Bangkok, 18 July 2007, http://aric.adb.org/updates/ear%20updates/Bangkok%20conference/session1/Kawai%20EA%20 Regionalism%20Ten%20Years%20After%20ADB-ThMOF%20Bangkok% 20rev%2007-17-2007.doc
38. Webber, "Two funerals and a wedding? ...", n. 7.
39. Victor C.M. Chan, "A New Pathway for Regional Governance in East Asia: From Economic-driven Regionalization to Crisis-driven Regionalism", *Ritsumeikan Journal of Asia Pacific Studies,* Vol. 21, pp. 149-61, 2006.

40. Joint Media Statement from the Finance Ministers Conference in Phuket, Thailand, "Joint Media Statement—Action Plan to Restore the Economic and Financial Stability in the Asian Region", 22 February 2009, http://www.mof.go.jp/english/if/as3_090222.pdf
41. C.F. Bergsten, 1997, "Open regionalism", Institute International Economics, Working Paper 97-3, http://www.iie.com/publications/wp/wp.cfm?ResearchID=152#TABLE1.
42. *Ibid.*
43. Masahiro Kawai and Ganeshan Wignaraja, "The Asian 'Noodle Bowl': Is It Serious for Business?", *ADBI Working Paper 136*, April 2009, http://www.eastasiaforum.org/wp-content/uploads/2009/04/adbi-wp136asian noodlebowlseriousbusiness.pdf
44. Several states in South East Asia are not small. However, "small-states" here refers to their relative comparison with China and Japan. The term is roughly but widely used in International Relations.
45. C.M. Dent, *East Asian Regionalism*, London: Routledge, 2008; N. Morada, "ASEAN at 40: prospects for community building in Southeast Asia", *Asia-Pacific Review,* 15(1) (2008): 36-55; A. Jetschke, "Institutionalizing ASEAN: Celebrating Europe through network governance", *Cambridge Review of International Affairs,* 22(3) (2009): 407-26.
46. George Modelski identifies great powers' power of agenda-setting in international relations. George Modelski, "From leadership to organization: the evolution of global politics", in V. Bornschier and C. Chase-Dunn (eds.), *The Future of Global Conflict*, London, 1999, pp. 11-39.
47. *The Straits Times*, 29 December 2005.
48. Amitav Acharya, "'Ideas, Identity, and Institution-Building: From the 'ASEAN Way' to the 'Asia-Pacific Way'", *Pacific Review,* 10(3) (1997): 319-46.
49. Gregory W. Noble, "Explaining Regionalism: A Brief Overview", 2005. http://project.iss.u-tokyo.ac.jp/crep/pdf/dp/dp1.pdf
50. Shujiro Urata, "An Economic Analysis of Competitive Regionalism in East Asia", paper presented at the symposium on "Competitive Regionalism: Strategic Dynamic of FTA negotiations in East Asia and Beyond" at Waseda University, Tokyo, Japan, 30-31 May 2008, http://www.waseda-giari.jp/sysimg/imgs/wp2008-E-15_Urata.pdf; Mireya Solís, Barbara Stallings and Saori N. Katada (eds.), *Competitive Regionalism: FTA Diffusion in the Pacific Rim*, Palgrave Macmillan, 2009.
51. Yoichiro Sato, "Free-Trade Agreements in the Asia-Pacific: Competitive Aspects Of Sub-Regional Trade Institution Building", http://www.apcss.org/Publications/Edited%20Volumes/RegionalFinal%20chapters/Chapter14Sato.pdf

52. *The Straits Times*, 29 December 2005.
53. Wang Guowei's remarks, cited in Yu Yingshi, *Humanities, Democracy, Intelligence*, Hai Tun Press, 2010, p. 21.
54. Murray, "East Asian Regionalism and EU Studies", n. 3, pp. 598-9, 603.
55. Evelyn Huber, "Letter from the President: The Role of Cross-regional Comparison", *APSA-CP Newsletter* 14(2) (2003): 1-6, 1-3.
56. For example, Moravcsik argues that the West European experiences are no different from EA; that the regionalizing experiences in the two regions are same. Andrew Moravcsik in the Chinese translations version of his *The Choice for Europe: Social Purpose and State Power from Messina to Maastricht*, Ithaca, NY: Cornell University Press, 1998, in Chinese Edition 2008. A Chinese scholar says the EAR could learn how to handle America's role in WEI. Li Dongqi, "East Asian Regionalization How to Handle its Relations with America—Comparing Analysis with European Integration", *Diplomatic Review*, Vol. 2, 2010, pp. 82-92.
57. Nair, "Regionalism in the Asia Pacific/East Asia: A Frustrated Regionalism?", n. 21; H. Katsumata, "Establishment of the ASEAN Regional Forum, Constructing Talking Shop or A 'Norm Brewery'", *The Pacific Review,* 19(2) (2006): 181-98; Baogang He, "East Asian Ideas of Regionalism: A Normative Critique", *Australian Journal of International Affairs*, 58(1) (2004): 105-25.
58. P. Murray, "Comparative Regional Integration in the EU and East Asia: Moving beyond Integration Snobbery", *International Politics,* 47(3-4) (May-July 2010): 308-29.
59. Breslin, "Comparative Theory, China, and the Future of East Asian Regionalism(s)", b.5, p. 711.
60. Dario Battistella, *Théories des relations internationales*, Chinese Translation edition 2010, p. 258 footnote 2.
61. J.J. Suh, Peter J. Katzenstein and Allen Carlson (eds.), *Rethinking Security in East Asia: Identity, Power and Efficiency,* Stanford University Press, 2004.

3

Globalisation and the Rise of Asia: Regional Cooperation to Sustain Economic Stability in Asia

Niklas Swanström

Asia experienced severe economic turmoil barely over a decade ago, but has fared better than most during the latest financial crisis: to a large degree the Asian states seem to be currently more resilient than many others. This is partly due to the lessons learned from the past and the relatively sound economic policies in some Asian states. However, regional cooperation does not in any substantial way explain Asia's resilience-cooperation and coordination having notably been absent during the crisis.

The reality is that despite increased intra-Asian trade there is little regional engagement to sustain economic stability. Many of the great number of attempts at cooperation in the region remain empty shells that have little impact on economic stability. In fact, the growing number of economic initiatives has created a virtual noodle bowl where each structure is unconnected with the rest and the complexity is more a hindrance to stability than an asset. This is especially true of initiatives between the three economic giants—China, India and Japan. In many ways the sub-regions have multiplied their cooperative structures, but the cooperation is shallow.

On the other hand, there are some reasons optimism, such as: the improved relations between China and Japan, the cross-straits relationship between the mainland and Taiwan as well as Sino-Indian relations, which have greatly improved the cooperative climate in the region.[1] Both China and Japan have confirmed their support for multilateral processes in North East Asia and an extended East Asia. Former Prime Minister Hatoyama of Japan has spoken positively about the opportunities stemming from multilateralism. In China, in 2002 at the 16th Party Congress the Chinese Communist Party extolled multilateral cooperation, a position that was further strengthened at the 17th Party Congress. Even within the South Asian Association for Regional Cooperation (SAARC), which has the lowest intra-regional integration, there have been significant improvements between 2003 and 2007.[2] There is no doubt that Asia attaches great importance to multilateral processes, even if their success is debatable.[3] Despite ongoing trans-regional cooperation within the ASEAN Regional Forum (ARF), SAARC, Doha Round, Association for South East Asian Nations (ASEAN) and ASEAN+3 there are very few similar positive developments in the broader regional context for cooperation and integration.

Intra-regional trade costs are much higher in Asia than in many other regions—they are particularly high within SAARC and lowest in ASEAN, which has had notable success in decreasing trade costs.[4] China has been most interested recently in opening up multilateral mechanisms for reducing economic instability, but many of the other states are reluctant—fearing the Chinese economic boom (and its political influence) and its domination of a number of economies.

This paper will raise questions regarding the types of organisations that exist today; the reasons for the failure to coordinate more effectively, and who in Asia is currently performing below potential. It will further raise the question of how Asia *should* and *could* cooperate in the future: the two questions do not necessarily overlap. The focus will be on the political problems that have prevented economic coordination and the effective measures necessary to promote economic stability.

Existing Structures and Problems

Asia has witnessed unprecedented growth in the last decade, particularly North East Asia and India. China and India have in many ways led this development and today many fear or look forward to the emergence of China as the world's economic hegemon. However, the speed and impact of Asia's growth should not be exaggerated. The GDP per capita both in China and India is still low, at desperate levels in some cases. It will take them decades and possibly centuries to overtake the US and European economies on a per capita level. Investments in military and social infrastructure will take decades to mature before they are even close to competing with what has been developed in "the West". Secondly, the growth of the Asian economy is directly intertwined with the West (and vice versa), so it is no longer possible to speak in terms of a singular economy, and the growth of the Asian markets will undoubtedly further European and American interests and growth. This international integration has been tremendously important both for Asian and European markets, but the limitations of intra-regional trade as a driving engine should be recognised. The strength of the regional structures is primarily that they are closer to the problems they face and the members share common problems—not least that transaction costs in trade should potentially be much lower. However, it is undeniable that all regional attempts are dependent on other regional structures and international regimes, and ASEAN and the EU are no exception.

Economic growth notwithstanding, there is a lack of regional integration and stability instruments in Asia; intra-regional trade has increased but cooperative efforts are lacking in much of the region, especially in areas where there is a growing need for supranational institutions. The most important trade relationship in Asia, China-Japan, can illustrate this: despite very high levels of economic exchange there are very few real institutionalised cooperative mechanisms. In fact, the old saying that relations are economically hot but politically cold applies. This is also very much the case in South Asia, even though the situation is slightly better in South East Asia. From 1991 to 2008, intra-regional trade grew faster in Asia than in the North America Free Trade Agreement (NAFTA) zone, but this is not only because the original level was lower but also very much in spite of the lack of institutionalisation and high trade costs in the region.[5] Much of intra-regional trade is done through the myriad of multilateral institutions in the region that have been established both formally and informally; but more often the trade is growing despite institutional and political hindrances.

The economic sphere has the most multilateral institutions in Asia, but even this field there is a lack of primarily multilateral cooperative organisations. Most trade agreements are typically bilateral and have grown and developed to such an extent that Richard Baldwin has compared the situation to a noodle bowl.[6] This network of bilateral agreements forms a regional structure of overlapping trade agreements, but according to Baldwin the attempt fails to achieve true regionalism and has in fact increased costs. This is largely due to the political tensions that are so evident in Asia as well as in all the sub-regions. It is almost an axiom that there is no shortage of issues to complicate bilateral and multilateral relations, and no shortage of good reasons for cooperation. Despite all the historical tensions and power struggles that exist in the region, it is striking that tension has decreased among governments (excluding North Korea and Pakistan-India relations). Statistics show that trade has increased significantly in all areas, but there is also a countertrend that populations distrust and fear each other more. Distrust and negative perceptions of other states and peoples are major issues in the way of relation building in all regions, even if South East Asia has fared much better than South and North East Asia where we can see positive developments.

Trans-regional or sub-regional multilateral structures for managing economic stability are far more important than the existing multilateral regional structures, mainly because of the dependency on the international economy, but arguably even more on account of the way in which all regional structures tend to be politicised. The sub-regional cooperation of some growth triangles (GT) has managed to avoid much of the politicisation but these structures have a very limited impact on the economy at large and on economic stability in particular.[7] The potential for economic cooperation and economic stability is far greater than the political will to coordinate economic policies and create mechanisms for management of economic instability. The numerous existing regional mechanisms are weak and shallow. Created to manage specific issues, they are in many ways competing with each other rather than reinforcing economic stability. Bilateral economic agreements and interstate cooperative organisations are still most effective.

There is no longer an intellectual vacuum or open military conflict in Asia to prevent further integration and cooperation. On the contrary, there is a vibrant exchange of ideas and information that assists in consolidating recent positive developments in Asia at large.[8] The problem may be that there is too much development and too many competing ideas about how to structure regional integration. The noodle bowl of economic structures is there, but there is more emphasis on the amount of noodles rather than the taste (effectiveness) of the dish. The governments in the region and its sub-regions will have to focus on decreasing the number of agreements and structures and develop those that can make some serious impact at the regional level.

Asia's failure to manage its regional structures effectively and decrease costs in trade and improve economic stability is becoming much more problematic, since with the economic contraction in Europe and the US, Asia is increasingly left to rely on itself or even assist the West in strengthening its economies. This is particularly evident in the much-needed Chinese intervention in Europe and its direct engagement in Spain, Greece and Portugal. As the world economy becomes increasingly integrated, the risks are also increasing and we will need more effective (sound) mechanisms to handle future economic instability.[9]

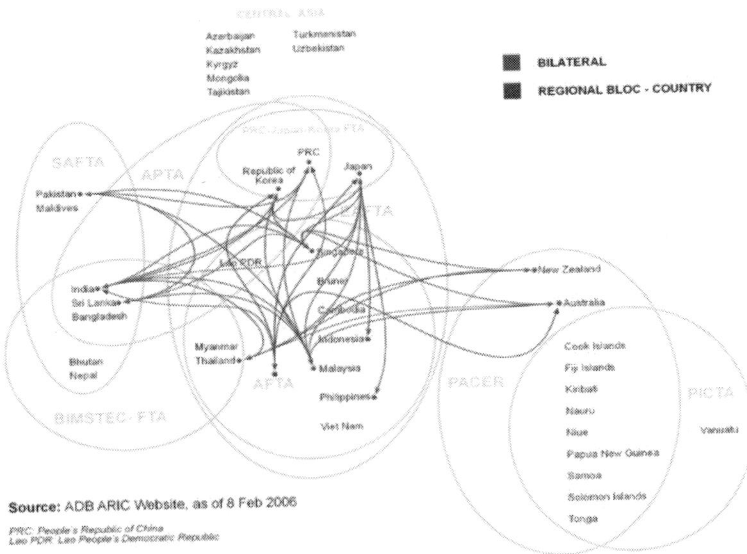

Current Institutions

Currently, the Asia Pacific Economic Cooperation (APEC) is the most prominent organisation, but the Comprehensive Economic Partnership in East Asia (CEPEA), Asia Free Trade Area (AFTA), East Asia Free Trade Area (EAFTA), South Asia Free Trade Area (SAFTA), East Asian Summit (EAS) and ASEAN Regional Forum (ARF), ASEAN+1 and ASEAN+3 are among the multitude of organisations that are all important regional and trans-regional structures competing with sub-regional and bilateral structures in the region (some of them are exemplified in the diagrams). Their impact has been modest, even if they have created a new alphabet soup of abbreviations in Asia. More comprehensive structures that are able to have an impact and bring about change need to be established. This is not an argument for delinking the Asian economy from the outside world: the Asian economy is already integrated into the world economy and is a crucial component for its development. On the contrary, this is an argument for more effective regional structures that could deal with regional instability and more effectively connect to other regional structures, such as NAFTA and the EU.

Asian regional structures would need to cooperate better with other international structures, as the former have not developed the technical capacity or political trust for effective surveillance and monitoring or credible implementation of support and rescue programmes. Drysdale has argued that the EU cooperates closely with the IMF in developing rescue packages for the troubled European economies due largely to its capacity to cooperate and analyse the situation.[10] The Asian economies could teach the European and American economies a thing or two about fiscal responsibility; even so, it is apparent that the existing Asian regional structures are flawed and there is a need to further strengthen them in order to realise the economic potential of Asia and on which the world economy will be increasingly dependent.

Political sensitivity and lack of trust are two major hindrances to developing more effective regional structures. There is a serious trust deficit in both South Asia and North East Asia and between Japan, India and China. Notably in South East Asia, where trust is more developed, successful cooperation has been possible. The growing inequality that has followed the rise of China and India is very much the reason for this trend. This will continue to be a problem, as the rise of China and India will increase the distrust between them and the smaller states that will end up in a dependency relationship with a perceived relative loss of influence and power.

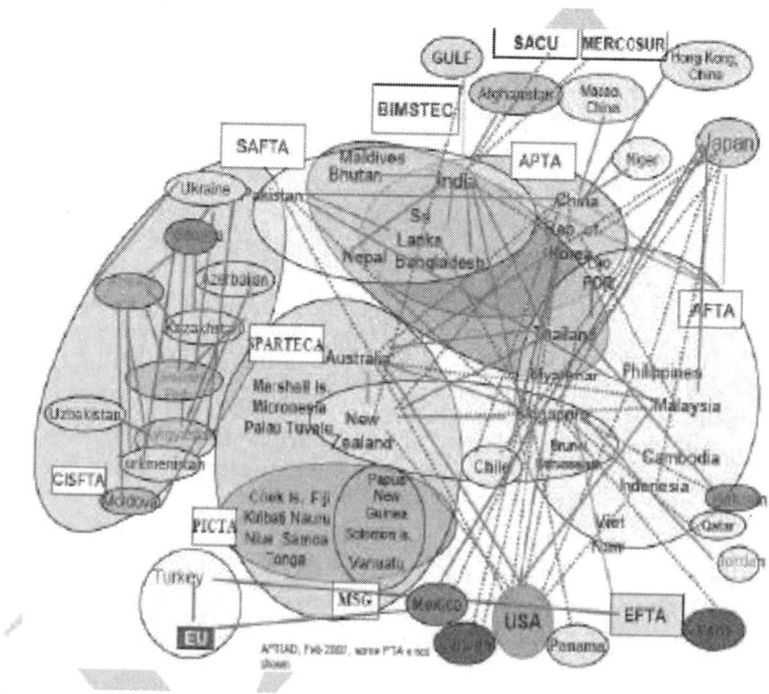

Source: http://en.reingex.com

Despite the large number of multilateral structures that have been established and the verbal support for further regional integration and strengthening of an Asian identity, realist (*Realpolitik*) perspectives continue to lurk behind all these strategies and national agendas remain predominant. Little or no national sovereignty has been surrendered to a regional organisation that could potentially deal more effectively with economic and security challenges. While not always a bad policy, in many areas the region will underperform and in some cases fail to increase regional, and by extension national, security. Despite the need for regional security mechanisms and trade-oriented regional mechanisms, most of the current important structures are trans-regional in character. There are several reasons for this—among them being leadership, differing visions of cooperation and finally old-fashioned power politics.

Leaders and Structural Challenges

A major challenge the region and its sub-regions face is the question of leadership. Will China, India, Japan or even an external actor such as the US lead the regional structure or could collective leadership work? It is evident that there is a conflict of interest between the major powers over the leadership role. A certain state may not be willing to take the lead but would be unwilling to accept the leadership of another. The increased confidence and willingness of China and India to lead has led to a similar decline in confidence among competing states such as Japan and Pakistan. China is the regional economic hub but it is also the state with which most actors in the region have the most difficulty, not to mention a similar distrust in the West, particularly in the US;[11] which would effectively disqualify China as the regional leader. Similarly, India is greatly distrusted in both Pakistan and China. Historical issues—and more importantly—current issues such as political and economic influence make it difficult for any of these states to be accepted as a regional leader. The single successful actor in the regional context is ASEAN, and its model has been partly exported to North East Asia in the form of ASEAN+1 and +3. This has not only earned ASEAN a great deal of credit but also shown it to be a functional partner, which means that reliance on regional organisations such as ASEAN could potentially function even at a broader regional level.[12]

Apart from the leadership issues there are structural problems with the possible construction of a regional security order. The smaller economies are looking for solutions that decrease the influence of the major actors, especially China and India. China in turn thinks that deepening of current structures should be fundamental and is not interested in expanding the membership of existing East Asian structures to include South Asia, and especially not the US. Japan and India on the other hand are more open to allowing in more members and have not excluded possible inclusion of the US. India is potentially interested in opening up membership to Russia as well, something that both China and Japan would see as problematic. In essence, China together with Malaysia, is more interested, in focusing on ASEAN+3; Japan and India desire an expansion of the structure for a broader East Asia community.[13] In East Asia this is perceived as the growing importance of China and the relative decline of Japan in regional affairs. It would be very much in the interests of Japan and India to include more members to balance Chinese growth and the US is key in this. China on its part would like to bide its time—as time is in its favour as it continues to grow, politically, economically as well as militarily—but with increased cooperation to reduce instability in economic affairs. It is telling that China's contacts with India and the US (but also EU) in these issues are primarily bilateral, a situation in which it would have more leverage. This is not to say that China is not being constructive: quite the opposite. China is the economy in the region with the lowest trade costs and has been the quickest to open up with favourable terms for its regional partners—but it does have problems in managing crisis situations.

On the positive side, seen from a multilateral perspective, the end of the Cold War, globalisation (and regionalisation) as well as the increased importance of Asia have broken down the longstanding barriers to regional identity.[14] Leading the development has been economic integration, which has forged closer cooperation between the people of the region, supplemented by modest but important educational and cultural exchanges. The interaction has increased significantly and there have been the more important changes in their perceptions of each other over time.[15] However, it is still too early to speak about an Asian identity or for that matter about sub-regional identities.

Smaller is Beautiful

The main problem today in Asia is not the lack of vision in the region—as various impressive structures exist on paper—but implementation and management. While there have been problems in finding regional solutions that are acceptable to all actors, seemingly more difficult is the acceptance that these structures demand an independent role or at least the ability to function outside of total state control, which could be accomplished through confidence-building measures (CBMs). The sovereignty of each state is deemed crucial, despite the fact that a minimal surrender of sovereignty could translate into a tremendously positive development for security in the region at large. The informal cooperation in many fields is, however, impressive and in the fields of combating organised crime, rescue operations and human trafficking there healthy doses of formal cooperation are evident. However, it is striking that most structures in Asia have a strong informal component and there are virtually no formal organisations that could override bilateral and international engagements.[16] Bilateral agreements tend to take precedence over multilateral agreements resulting in the neglect of multilateral structures that would benefit the region over the long term. Even among the few successful attempts to increase financial stability, bilateral agreements remain predominant.[17]

Paradoxically, there are too many structures and organisations competing for influence and competence on the regional scene. In terms of Free Trade Agreements (FTA) an estimated 400-plus are already in existence. In many ways the current structures are prevented from taking independent action by other regional structures with overlapping and competing competencies. A problem area is that there is still a "left behind" syndrome in Asia—a tendency to assume that Asia has been left behind in trade liberalisation and integration; and the solution has been to focus on the number of organisations rather than the depth of cooperation.[18] This is not to say that the attempt should be to create a regional community similar to the European Union but only that there is a need for more effective cooperation. It could on the other hand be a strategy to "overfeed" the region with competing structures just to keep the actual influence down but at the same time publicly support the idea of regional integration. This would eventually increase transaction costs for trade rather than decrease them, which was the initial purpose of the FTA regimes.

All the regional actors' attempts to create regional visions and structures that could resolve specific problems do not take into account that too many competencies and too wide an involvement will only create competing organisations and bureaucracies and fail to accomplish the final objective of cooperation. There is a therefore a strong need for a common vision on how a regional structure should look and what it should encompass. It would be better for it to take on a more modest role rather than the all-encompassing function of the European Union's legal and treaty-based system. It is apparent that regional institution-building needs increased and improved cooperation that could facilitate discussion and implementation of regional ideas and efforts.[19] In particular, there is a need to develop an Asian Financial Stability Dialogue and a discussion concerning exchange rates (and eventually mechanisms that could regulate both of them). It is the reluctance to formalise anything that has resulted in the many number of structures that have some, but not sufficient, power to act regionally.

This has not, in any way, dimmed all hopes for more effective mechanisms and more coordinated structures to deal with economic instability. Gao Quanli, vice president of the Chinese Academy of Social Sciences, for example, notes:

> The current global financial crisis may provide a historic opportunity for North East Asian states to raise the level of economic cooperation, because developing complementary resources, technologies and products among concerned states is crucial for regional development.[20]

The situation continues, however, to be bleak even in the field of economic cooperation because of the unilateral and bilateral agreements that have prevented further Asian economic integration and the much-needed coordination on a number of economic issues.[21] That the many Asian states missed out on this opportunity during the financial crisis is worrisome and there is pessimism about the prospects of deeper regional integration. It is telling that in the worst of times, when economic instability threatens the very fabric of regional and international trade and domestic stability, political considerations and perceptions of *political* opponents prevail.

Rising Engagement and Decreasing Distrust?

The most important obstacle that Asia needs to overcome to further economic cooperation and coordination is the great-power rivalry and tensions not only between a number of regional states, but also with external actors. China-Japan, China-India and India-Pakistan would arguably be problematic; but more difficult still are the tensions on the Korean Peninsula and in Myanmar where external actors are a greater problem than regional actors as well as the situation across the Taiwan Straits and the tension in Kashmir (both domestic issues with strong international implications). For creating more integrative and all encompassing structures it is essential that all actors that is of economic interest (Taiwan and Myanmar) are included but also that areas of instability that could threaten further economic growth in the larger region (such as divided Kashmir but also nations such as North Korea) are included in some form, depending on their political status. This should not be seen as a reward for bad behaviour or a breach of sovereignty. The reality is that regional structures need to be inclusive and not shy away from the very real problems that do exist in the region.

Intra-regional trade is currently underutilised. Increased intra-regional trade would have a great benefit income generation and poverty alleviation, which are major challenges in many states in Asia.[22] Trade and economic integration could very well spur that extra percentage that is so important in many states. Moreover, there is political gain to be derived from further trade and integration. Increased trade will inevitably create new linkages between adversaries, e.g. the economic integration between Taiwan and mainland China has made the military conflict almost imperceptible, even if political tension still keeps the economic relationship below its potential.

A basic trust must be established so that it can percolate to the regional level—it is obvious that bilateral relationships prevent further integration and cooperation. The rise of China and India is a concern for Japan, Pakistan and the US, but also for many of the smaller states that fear that a rising China and India will increasingly have an impact on their internal affairs and business. This is not necessarily in military terms but more often in economic terms as the economic development and cheap labour of the two emerging powers are real economic threats. Their rise has also overshadowed the general growth of Asia. The reality is that many Asian states have experienced impressive growth and, instead of seeing China and India as threats, they should look to the possibilities that these two states offer a growing Asia. China—and to a lesser degree India—are seen as the key to intra-regional trade and economic growth regionally and increasingly internationally.[23] Asia as a region is closely integrated—even if not closely cooperating—as it is with the international economy at large. De-coupling from either the international economy or from the Chinese or Indian economies is not possible.[24]

The rise of multilateralism could offer a great advantage to Asia both in economic and security terms. We have seen that the rise of multilateralism in the world at large and in South East Asia has brought great advantages in terms of trade and security. North East Asia—and even more so South Asia—have been reluctant to adopt this trend; this could have some serious implications in the future. A crucial component here is how ASEAN has chosen to develop its cooperation and focus on intra-regional trust building before any greater tasks are taken on. Sovereignty is equally important for all sub-regions in Asia but South East Asia is the only region where enough trust has developed to take a number of smaller steps towards true multilateralism and cooperation at the expense of total sovereignty.

The failure to establish an Asian structure dealing for fostering economic stability and continued growth, much less a formal regional organisation, will prolong the role of the US and EU in the region, even if ASEAN emerges as a regional hub for cooperation. An important reason for this failure is the lack of leadership. Single states or combinations of states in North East and South Asia are not an option at the moment, mostly because of the lack of trust between the major actors. ASEAN as a driver of regionalism is more acceptable to many Asian states than a single leading state. A rotating chair could be an option but that is too sensitive in the short term, but could become necessary over time. There is a continued need for China, India and Japan to take a regional responsibility, but leadership from these states is not feasible in the short and intermediate term. ASEAN and its model seems a perfect panacea for the problems that Asia is now facing.

Another question is: what will drive a regional organisation or even cooperation forward? Economic crises have had a tendency to integrate states and create a regional sense of emergency, something that was noted in South East Asia during the financial crisis of 1997/98 and which led to much deeper and more successful economic cooperation. We have not seen the same tendency in Asia during the current economic meltdown in spite of some efforts to create regional mechanisms for economic crisis management in the form of currency swaps or even discussions on currencies or economic stability at a regional level. What is needed is a greater effort to prevent crises from erupting in the first place, something that requires a much greater political will and more financial integration than any of the states at present have the desire to commit to. It will have to be accepted that a minimum loss of sovereignty will result in a huge tremendous increase in economic stability, even if it will by no means be fully protected against bad economies, as has been very much apparent in Europe.

It should be noted that despite the modest progress made in the economic field there has been even less success in the security field, an area in which some Asian nations will face some of the more devastating threats. Issues of terrorism, separatism and piracy along with organised crime, are the rising storms on the horizon. If Asian states are not able to coordinate their efforts in the economic field they will face even greater difficulties in managing traditional and non-traditional security threats.

Notes

1. Niklas Swanström and Ryosei Kokubun (eds.), *Sino-Japanese Relations: The Need for Conflict Prevention and Management*, Newcastle Upon Tyne: Cambridge Scholars Publishing, 2008; Niklas Swanström and Ryosei Kokubun (eds.), *Sino-Japanese Relations: Rival or Partner for*

Regional Cooperation, Singapore: World Scientific Publishing, forthcoming 2011.
2. Staff Working Paper, "Intraregional Trade Costs in Asia: A Primer", United Nations Economic and Social Commission for Asia and the Pacific, 28 June 2010.
3. Mingjiang Li, "China's Participation in Asian multilateralism: pragmatism prevails", in Ron Huisken (ed.), *Rising China: Power and Reassurance*, The Australian National University, E Press, 2009; Richard E. Baldwin, "Managing the Noodle Bowl: The Fragility of East Asian Regionalism", *The Singapore Economic Review* (*SER*), 53(3) (2008): 449-78; Joseph Stiglitz, "Contagion, Liberalization, and the Optimal Structure of Globalization", *Journal of Globalization and Development*, 1(2) (2010).
4. Staff Working Paper, "Intraregional Trade Costs in Asia: A Primer", n. 2, p. 10. Of the three largest economies (China, India and Japan) China has the most open economy with the lowest trade cost and India the highest by far. This is in spite of India having taken the lead in reducing trade costs in all regions except ASEAN and EU, where China has taken the lead.
5. Durgesh Rai, "Asian Economic Integration and Cooperation—Challenges and Ways Forward for Pan-Asian Regionalism", *German Institute of Global and Area Studies Working Paper, No. 152,* November 2010.
6. Baldwin, "*Managing the Noodle Bowl ...*", n. 3.
7. Niklas Swanström, *Regional Cooperation and Conflict Management: Lessons from the Pacific Rim*, Department of Peace and Conflict Research, Uppsala University, 2002.
8. Swanström and Kokubun (eds.), *Sino-Japanese Relations*: *The Need for Conflict Prevention and Management,* n. 1; Niklas Swanström (ed.), *Asia 2018-2028: Development Scenarios*, Stockholm: Institute for Security and Development Policy, 2008.
9. Stiglitz, "Contagion, Liberalization, and the Optimal Structure of Globalization", n. 3.
10. Peter Drysdale, "Asia's Global Responsibilities: Delivering through global and regional arrangements", *East Asia Forum*, 10 October 2010.
11. Staff Working Paper, "Intraregional Trade Costs in Asia: A Primer", n. 2; Paul Gruenwald and Masahiro Hori, "Intra-Regional Trade Key to Asia's Export Boom", *International Monetary Fund, Survey Magazine: Countries and Regions*, 6 February 2008.
12. Baldwin, "Managing the Noodle Bowl ...", n. 3, p. 2.; Niklas Swanström, "Regional Structures in Northeast Asia: Whither and What", in Swanström and Kokubun (eds.), *Sino-Japanese Relations: Rival or Partner for Regional Cooperation*, n. 1.

13. Aurelia George-Mulgan, "Hatoyama's East Asia Community and Regional Leadership Rivalries", *East Asia Forum*, 13 October 2009.
14. John Ravenhill, "A three bloc world? The new East Asian regionalism", *International Relations of the Asia-Pacific*, vol. 2 (2002): 167-95.
15. Takashi Terada, "Constructing an 'East Asian' concept and growing regional identity: from EAEC to ASEAN+3", *The Pacific Review*, 16(2) (2003): 251-77; Mark Beeson, "East Asian Regionalism and the End of Asia-Pacific: After American Hegemony", *Asia-Pacific Journal: Japan Focus*, 12 January 2009; Thomas Berger, "Overcoming a Difficult Past: The history problem and institution building in Northeast Asia", in Martina Timmermann and Jitsuo Tsuchiyama (eds.), *Institutionalizing Northeast Asia: Regional Steps towards Global Governance*, Baltimore: The Brookings Institution, 2008; Takashi Oshimura, "The Function and Dysfunction of Identity in Institutionalizing Process: The case of Northeast Asia", in Timmermann and Tsuchiyama (eds.), *ibid.*
16. Huaigao Qi, "China's New Neighboring Diplomacy", *Contemporary International Relations*, 19(5) (2009): 92; Yasumasa Komori, "Regional Governance in East Asia and the Asia-Pacific", *East Asia*, 29(4) (2009): 321-41; Swanström, *Regional Cooperation and Conflict Management* ..., n. 7.
17. Hiren Doshi, "Multilateral vs. Bilateral trade agreements," UPI Asia.com, 29 January 2008; Ran Chen, "Rosy Prospects for Northeast Asian Cooperation", *Beijing Review*, 16 June 2009.
18. Baldwin, "Managing the Noodle Bowl ...", n. 3; Doshi, *ibid.*
19. For an excellent argument for institutionalisation of the region see Timmermann and Tsuchiyama (eds.), *Institutionalizing Northeast East* ..., n. 15.
20. Ran Chen, "Rosy Prospects for Northeast Asian Cooperation", n. 17.
21. Baldwin, "Managing the Noodle Bowl ...", n. 3.
22. Nin Merchant-Vega and Katherine Loh, "Can Greater Regional Economic Cooperation Unite 'Two South Asias'?", In *Asia*, The Asia Foundation, 5 January 2011; Swanström, "Regional Structures in Northeast Asia ...", n. 12.
23. Gruenwald and Hori, "Intra-Regional Trade Key to Asia's Export Boom", n. 11.
24. Prema-chandra Athukorala and Archanun Kohpaiboon, "Inter-regional Trade in East Asia: The Decoupling Fallacy, Crisis, and Policy Challenges", Asian Development Bank Institute, No. 177, December 2009; Taro Ichikawa, "ASEAN Urged to Facilitate Intraregional Trade", *IDN-InDepth News Analysis,* June 7,2010.

Section Two

RISING CHINA

4

The Rise of China, the Rise of India, and the Changing Geopolitics of Asia: Contending Perspectives on India-China Relations

Vincent Wei-cheng Wang

Introduction

An important global development in the first decade of the 21st century is the rise of large developing countries. Grouped as BRICs,[1] these fast-growing nations have catapulted over the "emerging economies" status and are poised to play greater roles on the world stage. Table 1 establishes the increasing importance of BRICs vis-à-vis established great powers. It also shows long-term trends that portend the changing balance of power in world political and economic affairs. Of particular note is the rise of China and India—the two most populous nations on earth that together make up nearly two-fifths of humankind. The sheer magnitude of their ascendance caused one prominent former Singaporean ambassador to the United Nations to proclaim the "irresistible" shift of global power to Asia.[2]

The rise of China and the rise of India are unmistakable. In 2009, China's and India's economies had already become the second—and fourth-largest, respectively. In PPP (purchasing power parity) terms, China's economy is already larger than Japan's and is next only to that of the United States; India's economy is also rapidly closing the gap with Japan's economy (although in official exchange rate terms both China and India are still relatively poor). The two Asian developing giants are riding on phenomenal growth rates (13.34 per cent per annum in 1990-2007 for China and 7.4 per cent for India) and their economies performed much better than established industrial democracies in the wake of the global financial crisis that began in 2008. Persistent differentiated rates of growth impact the global balance of power. The current financial crisis accentuates the *relative* decline of the United States (or the West in general) and the rise of Asia. China, and to some extent India, also become important traders, foreign exchange holders, military powers, and greenhouse gas emitters. Their stature and influence on the world stage will almost certainly rise.

However, whether an "Asian century" will finally arrive after five centuries of Western dominance of world affairs depends importantly not only on whether India and China can continue their respective rises but also on how they deal with their own and the other nation's ascent. These two proud nations are keenly aware of the other's rise and naturally compare themselves with it (more on the Indian side).[3] Despite their common aspiration to play larger global roles, they demonstrate sharp contrasts in their political systems, economic models and social structures. They have also maintained a very complex relationship that is weighed down by history but also offers promising opportunities in an era of globalisation (Table 1).

While the implications for the rise of China have been debated in various contexts, thin scholarly attention has been devoted to the rise of India, and how these two Asian great powers perceive the ascendancy of each other.

This paper contributes in this regard by examining the key factors influencing India-China relations and analysing elite perspectives in either nation on this relationship. It is divided into six sections. Section One begins with an overview of China's assessment of its security environment and its evolving grand strategy. Chinese security writers have developed the concept of "comprehensive national power" as a convenient way to frame the debates on China's security assessment and external strategy. Section Two introduces this concept and elucidates Chinese perspectives on the rise of their country and potential peer competitors such as India. Section Three analyses the key elements that comprise the complex Indo-Chinese relationship, including history, geography, territorial disputes, mutual threat perception and alignment patterns, economic partnership and competition; it focuses on the Indian elites' perspectives on the rise of China. Section Four summarises Chinese security analysts' perspectives on a rising India in the light of the changing bilateral relations. Section Five provides a conceptual framework by categorising Indian elites' perspectives on the rise of China and Indo-Chinese relations into three

Table 1: Emerging vs. Established Great Powers: Select Indicators

	China	*India*	*Brazil*	*Russia*	*USA*	*Japan*	*EU*
GDP (ppp, $ trillion); world rank in 2009	8.78; 3rd	3.57; 5th	2.013; 10th	2.11; 8th	14.14; 2nd	4.15; 4th	14.43; 1st
GDP (official exchange rate $ trillion)	4.909	1.236	1.574	1.255	14.26	5.068	16.24
GDP per capita (ppp, $); world rank in 2009	6600; 128th	3100; 163rd	10,100; 107th	15,100; 72nd	46,000; 11th	32,700; 40th	32,500; 42nd
GDP real growth rate (%); 2009 world rank	9.1; 4th	7.4; 10th	–0.2; 114th	–7.9; 206th	–2.6; 157th	–5.3; 189th	–4.1; 179th
GDP average annual growth rate % (1990–2007)	13.34	7.63	6.12	4.92	5.27	2.21	n/a
Population (number); world rank	1.34 billion; 1st	1.16 billion; 2nd	198.74 million; 5th	140.04 million; 9th	307.21 million; 3rd	127.08 million; 10th	492 million
Trade volume ($ billion)	2158.3	432.7	280.7	495.2	2609	1042	3642
Exports ($); world rank in 2009	1.204 trillion; 2nd	164.3 billion; 22nd	153 billion; 26th	303.4 billion; 13th	1.046 trillion; 4th	542.3 billion; 5th	1.952 trillion; 1st
Imports ($); world rank in 2009	954.3 billion; 4th	268.4 billion; 14th	127.7 billion; 26th	191.8 billion; 19th	1.563 trillion; 2nd	499.7 billion; 6th	1.69 trillion; 1st
Foreign Exchange Reserves ($ billion); rank	2648.3 (Sep 2010); 1st	287.74 (Sep 2010); 6th	282 (Oct 2010); 8th	476 (Oct 2010); 4th	129 (Jul 2010); 15th	1019 (Jul 2009); 2nd	726 (Eurozone, Aug 2010); 3rd
Military spending ($ billion); world rank in 2008	98.8; 2nd	36.6; 10th	27.1; 12th	61.0; 5th	663; 1st	46.9; 7th	n/a
Defence expenditure (% GDP)	4.3 (2006)	2.5 (2006)	1.7 (2009)	3.9 (2005)	4.06 (2006)	0.8 (2006)	n/a
Internet users (million); world rank in 2008	298; 1st	81; 4th	64.948; 5th	45.25; 8th	231; 2nd	90.91; 3rd	247
% global emission of greenhouse gases, 2005; rank	16.36; 1st	4.25; 7th	6.47; 4th	4.58; 6th	15.74; 2nd	3.17; 8th	12.08; 3rd

Sources: CIA, *World Factbook*.

paradigms: geopolitical, geo-economic, and geo-civilisational. Section Six concludes with three possible scenarios for the future of China-India relations.

China's External Strategy Since the End of the Cold War

In the two decades since the end of the Cold War, Chinese analysts have been constantly assessing their country's external security environment and debating appropriate responses and necessary adjustments.[4] Most analysts agree that China's security environment has markedly improved with the dissipation of Cold War confrontation, but some still see various external threats and internal challenges for China. Today's China is sanguine that large-scale military conflicts involving great powers are unlikely to occur and that China is likely to be increasingly secure from traditional security threats (military threats by a foreign power against China's territory or the physical security of China's population), but like other major countries, it is not immune from non-traditional security threats.

The disintegration of the Soviet Union and the end of the Cold War occurred during a critical juncture of China's post-Mao development. Chinese leaders concluded that their country needed a peaceful international environment for another two decades—a period of "strategic importance" to concentrate on further developing its economy. Economic development became the overriding lynchpin to increasing China's wealth, power, prestige and international standing.

For China, besides its own self-strengthening, managing relations with the US and navigating an international system that reflects Western (especially American) values and strengths would be crucial. The US, with its overwhelming military capabilities that could check China, but also the technologies and markets crucial to China's economic growth, could play a decisive role in China's aspirations.

China's former paramount leader Deng Xiaoping gave guidance to its foreign and security policy apparatus that, collectively, has come to be known as the "24-character" strategy: "observe calmly; secure our position; cope with affairs calmly; hide our capacities and bide our time; be good at maintaining a low profile; and never claim leadership" (*lengjing guancha, zhanwen jiaogen, chenzhuo yingfu, taoguang yanghui, shanyu shouzhuo, juebu dangtou*). Added later was, "make some contributions (*you suo zuo wei*)". This 24-character maxim has guided China's security and foreign policies since the early 1990s. The Pentagon concludes that taken as a whole, Deng's strategy remains instructive in that it suggests "both a short-term desire to downplay China's capabilities and avoid confrontation, and a long-term strategy to build up China's power to maximise options for the future."[5] Following Deng's fundamental strategy, his successors calibrated their tactics. While third-generation leaders (centred on Jiang Zemin) restored China to international respectability after the 1989 Tiananmen crackdown, certain actions they took also helped fuel the so-called "China Threat" debate.[6]

China's fourth-generation leaders (centred on Hu Jintao) initially tried to rectify some part of the legacy of their predecessors. Domestically, they promoted the concept of a "harmonious society" (*hexie shehui*) to address some of the side effects of rapid and single-minded growth, such as social unrest, income inequality and environmental degradation. Internationally, they pursued a policy of "peaceful rise" (*heping jueqi*)[7] or "peaceful development" (*heping fazhan*) that relies on reassurance (good-neighbour policy) and incentives (lucrative trade or investment deals), by leveraging China's expanding economy. China's more polished foreign policy exudes confidence and poise.[8] While Western scholars have generally accepted that China's rise is perhaps inevitable, as of now there is no consensus on the implications of China's rise for the rest of the world.[9]

China's "peaceful rise" policy contains several interlocking elements:

1. It is based on an embracing of globalisation as part of the solution to China's growth imperatives. It relies both on China's domestic economy and the international marketplace to sustain and fuel growth.
2. To achieve the goal of great-power status, China must secure a peaceful international environment, which is crucial to sustaining its economic development and augmenting its power. Ensuring stability in China's periphery and avoiding a premature showdown with the US are thus essential.[10]
3. The new diplomacy is marked by greater international engagement. Whereas China used to distrust "multilateralism" for fear that multilateral institutions could be used to constrain or punish it, now Chinese leaders recognise that deeply engaging these organisations helps promote China's trade and security interests and limit American power.[11] China can be pragmatic on contentious issues. [12] It has sought to reassure its nervous neighbours and has more actively engaged in regional affairs (e.g., hosting of the Six-Party Talks).
4. The main instrument used for advancing China's objectives is its economic power—buoyed by its phenomenal economic

growth, rapidly expanding domestic markets, and driven by its voracious appetite for raw materials needed for its economic development.

To sum up, China's "peaceful rise" is a comprehensive long-term strategy leveraging globalisation as a catalyst to accelerate its own economic development and elevate its power and stature. The language is peace and stability; the style is constructive diplomacy; and the substance is economics—at least for now.[13]

Comprehensive National Power: China Tracks its Rise

To help conceptualise the structure of the international system, track the major countries' changing fortunes, and evaluate the results of "peaceful rise", Chinese analysts have developed a "scientific" method—Comprehensive National Power (*zonghe guoli*)—to predict power relations among the major countries. Michael Pillsbury, a noted authority on the Chinese military, terms the CNP as a "unique aspect of China's strategic assessments of the future security environment".[14]

CNP consists of tangible and intangible indicators of national power. Some writers arrive at the CNP by compiling the absolute numbers for each major country. Others use a relative number (e.g., a country's CNP as a percentage of the CNP of the United States). The goals of the different methods are to (1) highlight the *pecking order* of the major powers, and (2) reveal the *gaps* between them (most importantly, the gap between the US and the next few major powers that follow it).

Chinese writers also show the evolution of CNPs over time, including future projections. CNP serves as a handy measure for tracking China's relative standing at a particular time and its rise and fall vis-à-vis other major countries over time. Table 2 as an example shows the relative importance of each component making up CNP. Table 3 compares two different calculations and projections of CNP scores.[15]

A detailed discussion of CNP is beyond the scope of this paper. However, for an overview of China's security assessment and external strategy, several findings are instructive:

1. For the foreseeable future, the United States will remain the most powerful country. CNP trends confirm that the international structure since the end of the Cold War has been characterised by "*yi chao, duo qiang*" (one superpower, many great powers), although the gaps between "number one" and "numbers two and three" have narrowed.

Table 2: Weighted Coefficients of Major Component Factors

National Power Factor	*Weighted Coefficient*
Total CNP	1.00
Natural resources	0.08
Economic activities capability	0.28
Foreign economic activities capability	0.13
Scientific and technological capability	0.15
Social development level	0.10
Military capability	0.10
Government regulation and control capability	0.08
Foreign affairs capability	0.08

Source: Wang Songfen (ed.), *Shifie zhuyao guojia zonghe guoli bijiao.yanjiu [Comparative Studies of the Comprehensive National Power of the World's Major Nations]*(Changsha: Hunan chubanshe, 1996), p. 169.

Table 3: CNP Scores and Ranks over Time
(as Percentage of US CNP by Year, US = 100) (rank)

Country	*1970*	*1980*	*1989/90*		*2000*		*2010*		*2020*	
	CASS	*CASS*	*Huang*	*CASS*	*Huang*	*CASS*	*Huang*	*CASS*	*Huang*	*CASS*
US	100 (1)	100 (1)	100 (1)	100 (1)	100 (1)	100 (1)	100 (1)	100 (1)	**100 (1)**	100 (2)
China	25 (9)	33 (8)	37 (6)	34 (8)	53 (5)	42 (7)	72 (3)	52 (7)	**97 (2)**	61 (5)
Germany	42 (3)	52 (3)	64 (3)	58 (3)	68 (3)	67 (3)	72 (2)	77 (3)	**77 (3)**	85 (3)
Japan	34 (4)	50(4)	62 (4)	58 (3)	66 (4)	76 (2)	69 (4)	97 (2)	**73 (4)**	119 (1)
India	15 (10)	19 (10)	24 (9)	18 (10)	34 (8)	22 (10)	44 (6)	26 (10)	**57 (5)**	30 (10)
France	33 (6)	46 (5)	47 (5)	46 (5)	47 (6)	59 (4)	48 (5)	70 (4)	**48 (6)**	82 (4)
Brazil	–	–	26 (8)	–	33 (9)	–	39 (7)	–	**47 (7)**	–
England	34 (4)	42 (6)	36 (7)	42 (6)	34 (7)	48 (6)	33 (8)	54 (6)	**32 (8)**	60 (6)
Canada	33 (6)	35 (7)	23 (10)	36 (7)	22 (10)	38 (8)	21 (9)	40 (8)	**20 (9)**	42 (8)
Australia	26 (8)	29 (9)	19 (11)	28 (9)	18 (11)	29 (9)	17 (10)	31 (9)	**17 (10)**	32 (9)
USSR	64 (2)	77 (2)	65 (2)	66 (2)	79(2)	–	–	–	–	–
Russia				(50)	–	54 (5)	–	57 (5)	–	56 (7)

Source: The scores for 1989 and 2000 are from Huang Shuofeng, *Zonghe guoli lun* (On Comprehensive National Power) (Beijing: Zhongguo shehui kexue chubanshe, 1992), pp. 220-1. Scores for 2010 and 2020 were generated by Pillsbury. Table combines Tables 9 and 10 in Pillsbury, *China Debates the Future Security Environment*, n. 4, pp. 248-9. Huang = projections to 2020 of AMS (Academy of Military Science) GNP statistics. CASS = Chinese Academy of Social Sciences. Ranks for the 1970 and 1980 score contain ties because Pillsbury only provided percentages (of US CNP) for those years; actual CNP scores were provided for later years.

2. Despite the variations in the CASS and AMS studies—the former gave Japan's economic power the kind of weight ("Japan as Number One") that was once common; the AMS study gave more credit to China's rise and better conforms to the popular Western image of China catching up—both studies show that China's CNP has improved (absolutely and relatively) from 1970 to 2000, and is projected to improve further. By 2010, China's CNP will become the third-or fifth-highest, and by 2020 the second-or fifth-highest in the world.
3. Until its disintegration, the Soviet Union was the second most powerful nation. This confirms that the international system during the Cold War was essentially *bipolar.*
4. Germany's and Japan's CNPs were consistently around the third- or fourth-highest. Yet as "one-dimensional powers", their limited military profile (as a result of their aggression and defeat during World War II) limited their CNP.
5. Over time, developing countries like China, India and Brazil have played increasing roles in international affairs, and their weights are expected to eclipse Britain, Canada and Australia. This adds to the "multipolarity" (*duojihua*) envisioned for the 21st century.

One interesting question is the geopolitical significance of the 2008 global financial crisis that originated in the US. A 2009 CNP study puts the US and China as Number One and Two respectively.[16] The Chinese have begun to vigorously debate questions[17] like how should China adjust its behaviour as its capabilities continue to grow. Should China continue to "hide its capacities and bide its time" (*taoguang yanghui*); or should it begin to "make contributions" (*yousuo zuowei*)? Should China take a more active approach in its external strategy? Would its interests be best served by focusing on playing the role of being "Number Two" (*lao er*)? This implies accepting a Western-directed world order and benefiting from it—the "responsible stakeholder" scenario.[18] An alternative scenario is for China to play both hands as a stakeholder and challenger—working within the existing system (cooperate and soft-balance if necessary) while also challenging US pre-eminence.[19]

This article examines one important aspect of China's security assessment and external strategy—Sino-Indian relations—that has not received as much attention of Chinese analysts as it should.

Indian Perspectives on Sino-Indian Relations: Key Factors

One of the most significant developments of the 21st century is the rise of China and India. The economic takeoff of the world's two most populous nations is occurring simultaneously, portending far-reaching implications. As one of the growing numbers of books on this subject[20] put it, "rarely has the economic ascent of two still relatively poor nations been watched with such a mixture of awe, opportunism, and trepidation."[21]

While the implications of the rise of China have been debated in various contexts (global or systemic, regional, bilateral), little scholarly discussion has been devoted to either the rise of India[22] or how they perceive the ascendancy of the other. Yet as constructivists[23] would argue, how these two very different Asian giants perceive and deal with each other will be important for scholarly interest and policymaking. This section analyses this complex relationship and examines how Indian elites—in political, security, and economic arenas—perceive the rise of China.

History

Although China and India were two adjoining civilisations, there was remarkably little historical evidence of direct political interaction between them.[24] However, there was mutual intellectual fascination. Many Chinese scholars visited India in the first millennium to study Buddhism and other subjects, and spent a decade or more in India. Chinese monks such as Faxian in the fifth century and Xuanzang in the seventh played important roles in introducing Buddhism to China and bridging the two cultures. Many Indian scholars also went to China and worked there between the first century and the eleventh.[25]

Religion was not the only relationship between the two. Trade was also important. Indian intermediaries facilitated trade between China and Western Asia for centuries.[26] A branch of the famous Silk Road extended into the plains of northern India. Yet for the most part there was little interaction—mostly indirect—between China and India before the arrival of Western imperial powers.

Colonialism afflicted both India and China and pitted the two civilisations against each other. During the Opium Wars (1839-1842), Britain tried to forcibly sell the opium from its East India Company in China. The two nations' shared colonial experience made for empathy—an Asian anti-colonial pride—between them. Nehru and Gandhi were friendly with the Nationalist Chinese leader Chiang Kai-shek. India gained independence from Britain in 1947. When Mao Zedong established the communist regime in China in 1949, India was among the first to recognise the People's Republic of China on April 1, 1950. Nehru, typical of Indian leaders, personally invested in maintaining friendly ties with China and cultivated personal relationships with Chinese leaders, especially Chinese Premier Zhou Enlai. Nehru, who promoted the slogan "*Hindi-Chini bhai-bhai*" (India and China are brothers), reportedly said, "China was my most admired nation."[27] An Indian security analyst said, "From the 1950s on, we have looked at China from an Asian solidarity standpoint—whether it was nuclear weapons (China's 1964 explosion) or the United Nations (PRC's entry in 1971)."[28]

However, the goodwill was short-lived. The colonial legacy also sowed the seeds of discord. The McMahon Line—a demarcation line drawn on the map referred to in the 1914 Shimla Accord, signed between Britain and Tibet—was to form the boundary between British India and Tibet, over which China claimed suzerainty. While Britain and Tibet considered the agreement binding, China disputed the McMahon Line. India considered the line an international boundary. It was the root of the thorny and persistent border dispute between India and China (to be discussed later). Figures 1 and 2 show the disputed borders on the eastern sector (today's Indian state of Arunachal Pradesh, formerly North East Frontier Agency) and on the western sector (today's Chinese region of Aksai Chin).

In 1950 China's People's Liberation Army (PLA) entered Tibet and took control of the vast region that had historically served as a buffer (in strategic and cultural terms) between India and China. As former Indian army chief of staff General Ved P. Malik put it, "The first time we (Indians) came into direct contact with the Han Chinese was after 1950, when the PRC occupied Tibet. We suddenly became neighbours."[29] In 1959, after the failed uprising against the PRC, the fourteenth Dalai Lama, Tibet's highest religious and political leader, fled to India. Nehru in 1960 offered Dharamshala as a location for the government of Tibet in exile. The Tibetan refuge in India became another irritant in the bilateral relationship.

In 1962 the small skirmishes not uncommon along the disputed border escalated into open military confrontation. War erupted on October 20, 1962 when Chinese troops forcibly evicted Indian troops from the Dhola post in the eastern sector. Over the next month the Chinese troops easily overwhelmed ill-prepared Indian troops in all sectors along the McMahon Line. Then on November 21, the Chinese government announced a unilateral withdrawal to points where it considered the territorial boundaries to be. Although the war did not change the status quo of the border, India essentially had lost the war, suffering territorial loss and national humiliation. Ever since then, the 1962 war has cast a long shadow over the Indo-Chinese relationship, and India's defeat has coloured the Indians' perceptions of China.

The worsening Indo-Chinese relations became entangled in the regional alignment during the Cold War, with the Soviet Union and India on one side and China and Pakistan (and later the US) on the other. China's successful nuclear test in 1964 deepened Indian apprehensions. If the 1962 war taught India the importance of indigenous conventional deterrence, India's nuclear test in 1974 sought to respond to China's nuclear capabilities. From 1962 to 1976 China and India were mired in a tense cold war. It was not until 1976 that the two countries again exchanged ambassadors. History clearly casts a long shadow on Indo-Chinese relations (Fig. 1, 2 and 3).

Geography

Historically, China and India each had their own geographic orientation: China toward East Asia and India toward South Asia. But modern Tibet after China's entry in 1950 connected these two spheres. The development of missile technologies, made possible by the two countries' economic growth, had the effect of "shrinking the strategic chessboard".[30]

In recent years, China has expanded its influence in the Central and South West Asian areas by organising and promoting the Shanghai Cooperation Organisation (SCO);[31] India has pursued a Look East policy of strengthening its relationships with countries in East and South East Asia. Both countries seek to play a greater role in areas adjacent to their own, and even farther. China and India thus manoeuvre on overlapping "strategic spaces".

Ranjit Gupta, a former Indian ambassador to five countries, thinks that China has always treated India with hostility, adopting a "systematic plan" to hem in India with the support of Pakistan, and its influence in Myanmar, Nepal and Bangladesh, and military activities in Tibet. He argues that historically China has behaved like an imperial power, expanding when the empire was strong.[32]

In the past decade, to ensure its energy security and shore up its oil supply route,[33] China has constructed facilities and secured access to ports around India (e.g., Gwadar in Pakistan, Hambantota in Sri Lanka, Chittagong in Bangladesh and Sittwe in Myanmar), prompting some foreign analysts to label this a "String of Pearls" strategy (see Figure 3). Some sensitive Indians warn that China is turning the Indian Ocean into a "Chinese Lake".[34] In 2009 China dispatched destroyers to the Gulf of Aden to protect Chinese merchant ships from Somali pirates operating in that area. The flotilla's passage through the Indian Ocean caused some concern in India.

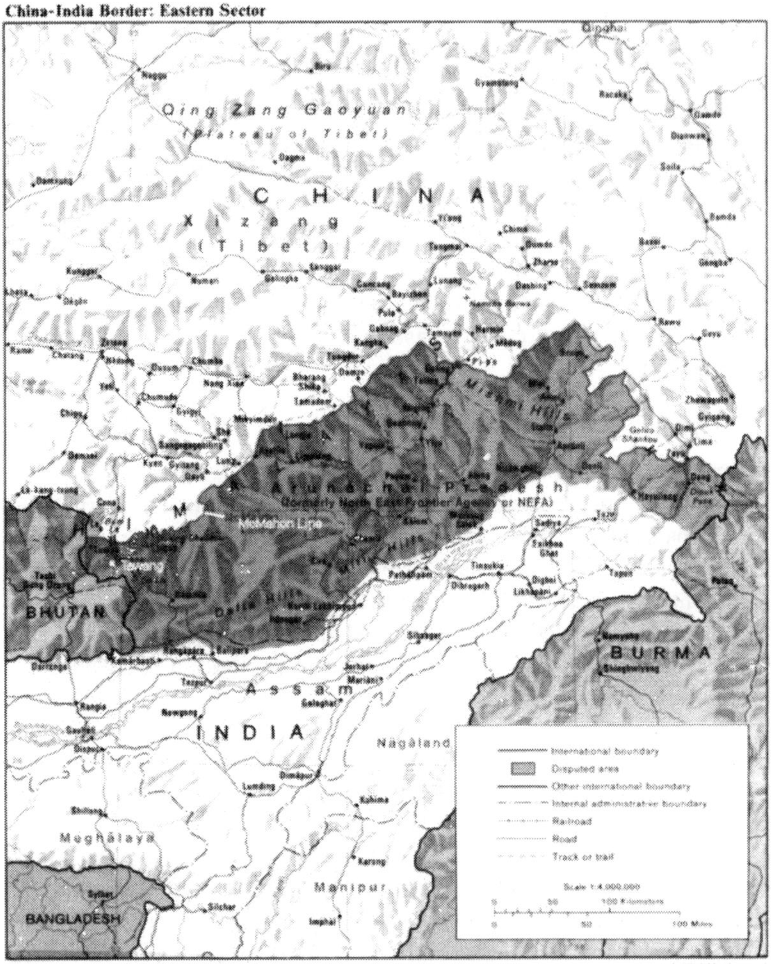

Figure 1: China-India Border: Eastern Sector

Source: http://en.wikipedia.org/wiki/File:China_India_eastern_border_88.jpg

Territorial Disputes

Among all the issues separating China and India, the territorial disputes arising from the undemarcated border significantly inform the Indian perspectives of China. Almost every Indian informant during my field research in 2008 raised the border issue as a major obstacle to better Indo-Chinese relationship. They feared that the potential of a flare-up still exists.[35]

China-India Border: Western Sector

Figure 2: China-India Border: Western Sector

Source: http://en.wikipedia.org/wiki/File:China_India_western_border_88.jpg

As mentioned earlier, the border disputes can be traced back to the McMahon Line. After the 1962 war, the two sides largely observed the Line of Actual Control (LAC) in the eastern sector and the Line of Control (LOC) in the western sector. The results are that China claims the Indian-controlled Arunachal Pradesh and India claims the Chinese-controlled Aksai Chin. The Chinese claim is partially based on Tawang, the birthplaceof the sixth Dalai Lama. The Chinese argue that Tawang is

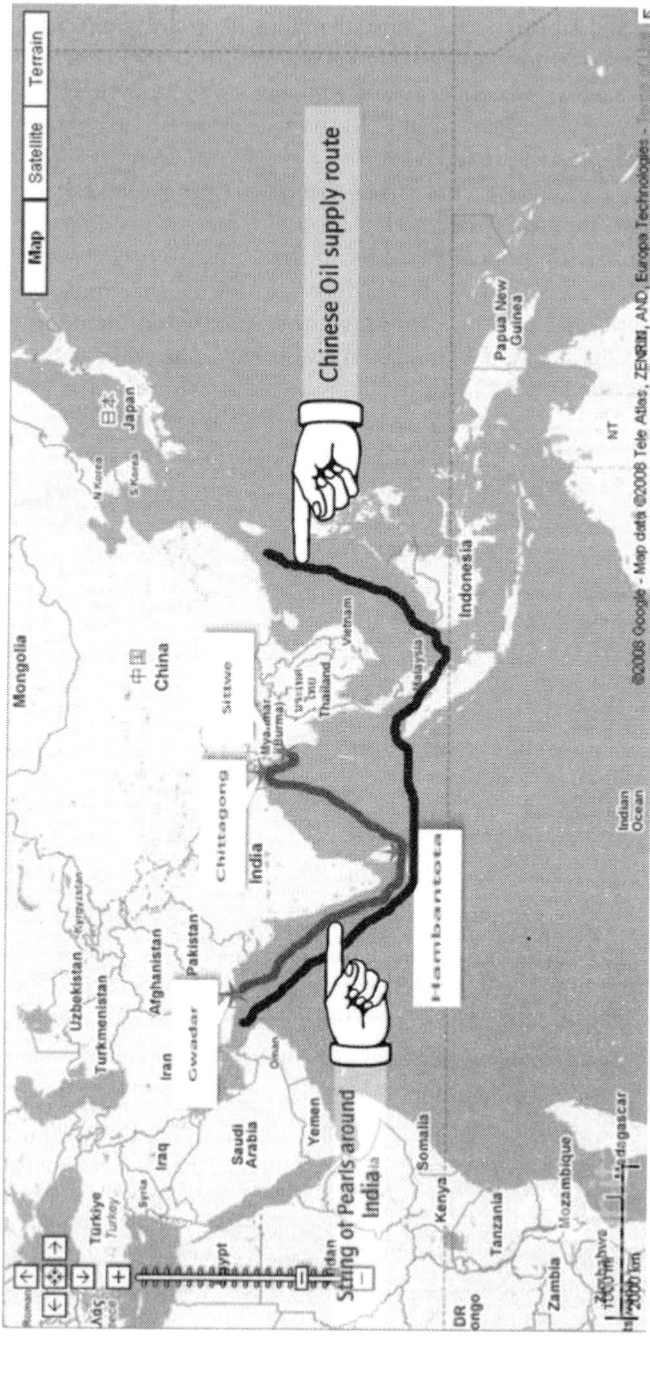

Source: http://www.marinebuzz.com/2008/01/23/china-garlands-india-with-string-of-pearls/

Figure 3: "String of Pearls"

Tibetan territory and Tibet is part of China; therefore, the entire Arunachal Pradesh is Chinese territory. India claims Aksai Chin, which connects Tibet and China's north-western province Xinjiang, as the eastern-most part of its Jammu and Kashmir state.[36] Kashmir itself was partitioned three-way by India, Pakistan and China.

Occasionally the Chinese emphasis on their legal titles provokes the Indians. Just days before President Hu Jintao's state visit to India in November 2006, the Chinese ambassador to India Sun Yuxi declared, "In our position the whole of what you call the state of Arunachal Pradesh is Chinese territory and Tawang is only one place in it and we are claiming all of that. That's our position."[37] In 2007 the Chinese Embassy in Delhi decided to strengthen its position by declining a visa to an Indian official from Arunachal Pradesh on the grounds that he did not need one as he was a "Chinese citizen".[38] The respective statuses of Tibet and Sikkim, the latter of which India incorporated in 1975 as its twenty-second state, also add to the complexity.[39]

In reality, this issue is mainly a placeholder and its impact will be "bounded". In recent years, the two sides have set up working groups to deal with the issue and try to resolve it peacefully, but have little to show for it. They have also sought to "compartmentalise" this issue from the overall improvement of the bilateral relationship.[40] As an American diplomat aptly put it, "The border issue is unlikely to be a serious problem in the relationship, because both sides benefit from this 'festering' that allows them to justify more military spending and certain postures."[41]

Mutual Threat Perception and Triangular Strategic Relationships

As Table 4 shows, both countries have substantial military capabilities. Over time, each has deployed certain weapons against the other. As mentioned before, India's 1974 nuclear test was spurred by China's successful test in 1964. India, under a government led by the Bharatiya Janata Party (BJP), in 1998 again conducted nuclear tests. India's minister for defence, George Fernandes, specifically rationalised this action as a counter to the threat India perceived from a rising China and closer Sino-Pakistani alliance.[42]

China has always loomed large in India's defence and foreign policies. The 1962 war, the border dispute, the complex *ménages à trois* (China-India-US and China-India-Pakistan) and each nation's ambitions all play a role, causing each to suspect the true intentions of the other. Some Indians viewed the SCO and the String of Pearls with concern. China's military assistance to Pakistan allows the latter to act as a proxy to "weigh down" India.

A hardnosed Indian analyst asserts, "China and India are natural rivals in Asia for geo-strategic, economic, and ideological (democracy vs. autocracy) reasons. In every aspect, we are contrasts. Our interests clash. We also compete for the same resources in Africa. Such rivalry is not easily reconcilable."[43] Many Indians feel that a rising China may make it harder for India to ascend.[44]

Economic Partnership and Rivalry

In many aspects, China's economic data is more impressive than India's (see Table 4): China has achieved higher growth rates, higher income levels, larger economy, greater trade volumes, and has attracted more foreign investment. But their economies are also complementary. China's success stems from being the manufacturing base of foreign multinationals with global sales networks; India's is domestically oriented, focusing on engineering and service.[45] China's hardware proficiency can complement India's software prowess. Some Indians and (fewer) Chinese envision the two nations merging into a giant "Chindia"—a formidable economic partnership with the world's largest populations and complementary economic strengths.[46]

Yet their two economies also compete, especially over the energy resources each needs for economic development. While many in the Indian business community see an economically rising China as an *opportunity* (for Indian products or services, for business alliance possibilities), more see it as a *threat.*

During my field trip to Mumbai and Delhi in May-June 2008, I sought to study the impact of Chinese products on Indian companies and consumers by direct observation and elite interviews. Indian companies that exclusively serviced the domestic market often complained about the inexpensive Chinese goods flooding the Indian market. Consumers were more ambivalent: While they generally liked the low-cost Chinese goods, they were also concerned about food and product safety as well as the quality of the goods. Indian companies that sell to international markets invariably faced strong competition from their Chinese counterparts. Some executives wondered that the incredibly low prices of the Chinese products, which undermined the Indian companies, were because of the Chinese government's help.[47] In this regard, India's experience is not too different from that of other countries with a backlash against cheap and unsafe Chinese products.

Table 4: China vs. India: Rise of Two Asian Giants

Indicator (Unit), information as of 2010	*China*	*India*
GDP (with PPP) ($ bn)	9872	4046
World Rank	2	4
GDP Real Growth Rate (%)	10.3	8.3
GDP (official exchange rate) ($ bn)	5745	1430
GDP Per Capita (with PPP) ($)	7400	3400
Exports ($ bn)	1506	201
Imports ($ bn)	1307	327
Main export partners (%)	US (20.0), Hong Kong (12.0), Japan (8.3), S Korea (4.6)	UAE (12.9), US (12.6), China (5.6)
Main import partners (%)	Japan (12.3), Hong Kong (10.1), S Korea (9.0), U.S. (7.9)	China (10.9), US (7.2), Saudi Arabia (5.4), UAE (5.2)
Stock of foreign direct investment at home ($ bn)	574.3	191.1
Foreign exchange reserves ($ bn)	2622	284.1
Population (mn)	1321.9	1129.9
Armed Forces (number)	2,105,000	1,288,000
Main Battle Tanks (number)	7660	4059
Artillery (number)	17,600[d]	5680
Surface Combatant Vessels (number)	75	48
Submarines (number)	62	16
Aircraft Carriers (number)	0	1[d]
Combat Aircraft (number)	2554	599
Attack Helicopters (number)	39	20
Nuclear weapon status	Confirmed	Confirmed
Chemical weapon status	Probable	Confirmed
Biological weapon status	Suspected	Unknown
Short-range ballistic missile status	Confirmed	Confirmed
Medium-range ballistic missile status	Confirmed	Confirmed
Intermediate-range ballistic missile status	None	Confirmed
Submarine-launched ballistic missile status	Confirmed	Probable
Intercontinental ballistic missile status	Confirmed	None
Strategic bomber status (text)	None	None

(*Contd.*)

Indicator (Unit), information as of 2010	China	India
Strategic submarine status (text)	Confirmed	None
WMD Commitments (text)	BTWC, CWC, NPT[b]*	BTWC, CWC

Notes: All economic data are as of end of 2010 and all military data are as of 2007.

Sources: Central Intelligence Agency, *World Factbook*, https://www.cia.gov/library/publications/the-world-factbook/ and National Bureau of Asian Research *Strategic Asia* Database Query, http://strategicasia.nbr.org/data/DataSheet/Criteria.aspx; accessed 20 April 2011.

Cooperation and Competition in Global Governance

The two aspiring great powers maintain an extremely complex relationship regarding global governance (the respective role of each country and the role of the Global South in general). On several global economic issues they cooperate. The two led the "uprising" of the Global South at the 2003 Cancún meeting which led to the halting of the Doha Round of the World Trade Organisation (WTO). They often criticise the "democratic deficit" of Bretton Woods institutions, such as the WTO, International Monetary Fund and the World Bank, and advocate expanded roles and higher profiles of the Global South (especially, large emerging countries such as China and India). They also accuse the Global North of being hypocritical and self-serving for wanting to pry open the service, investment and IT sectors of the Global South while protecting the politically influential agricultural sectors at home.

In the December 2009 at the Copenhagen conference on global climate change China and India also adopted similar positions, including insisting on the developing countries' right to economic development, declining to accept mandatory cuts in pollution levels, and instead proposing to decrease the energy intensity of industries. These positions belie the fact that China and India are now the world's largest and third-largest emitters of carbon dioxide respectively.[48] Both members of the so-called BRICS, China and India, have tried to coordinate their policies on severa issues related to global governance.

However, the two also clash on other key issues. China is a permanent member of the United Nations Security Council. India lobbies for "UN reform" including expansion of the Security Council in the name of greater democracy and fairer representation. India has argued that it deserves a permanent seat because it has the world's second-largest population (which will become the largest in a few decades) and nuclear weapons, and there is only one Asian country and one developing country on the SC. China has been decisively cold to the idea of UNSC expansion. It jealously guards its own seat. It also rebuffs several other developing-country potentials, such as India, Brazil, Indonesia and South Africa, because admitting any of them would also lead to the admission of Japan and Germany, which provide important financial support to the UN. China is most opposed to Japan's seat. Consequently, it must also reject India's bid. China can thus ensure it will have more global influence than India.

Being among the world's fastest growing economies, China and India both have huge energy needs and have elevated the importance of "energy security" in their external strategies, including global sourcing and transport of oil. While China is more aggressive in this regard, the two share many similarities in their attempt to achieve energy security. This could be another area of potential conflict.

Chinese Perspectives on a Rising India

Until now Chinese elite discourse on India has been predominantly informed by *realism*. This observation based on my field research consists of several aspects: (1) war and territorial disputes, (2) spheres of influence, (3) alignment, (4) power considerations,[49] and (5) mistrust.

The 1998 Indian nuclear tests ironically became a turning point in Chinese perspectives about, and policies toward, India. As the CNP comparisons in Table 3 show, throughout the Cold War and until the late 1990s, China was, as Avery Goldstein pointed out, not convinced that relations with India would be as strategically significant for it as the relations it was cultivating with other major powers.[50] The generally sceptical or dismissive view of India held by Chinese elites stemmed resulted from several reasons: (1) China's confidence stemming from its military victory over India in 1962, (2) China's more impressive economic performance compared to India's, (3) India's ethnic, religious and linguistic diversity (or cleavage), which Chinese analysts generally view as a handicap, and (4) India's domestic politics (federal system, extremely fragmented party politics, a chaotic democracy lacking efficiency), which they also generally view as a serious impediment to India's future prospects—the Indian form of democracy is anathema to China.

After India's 1998 nuclear tests, however, official relations have warmed. In June 2003 Prime Minister Atal Bihari Vajpayee made a historic visit to China, the first in over a decade. The two countries have elevated their relationship to one of "strategic and cooperative partnership" for peace and stability. During President Hu Jintao's visit to India in November 2006, the two sides adopted a ten-point strategy to further strengthen the bilateral relationship. Jingdong Yuan, a China expert, quoted a Chinese diplomat: "Beijing now views its relationship with India as one of global and strategic importance that is long-term, all around, and stable."[51]

Diplomatic pleasantries notwithstanding, Chinese perspectives on India in the early decade of the twenty-first century embody the following elements:

- While China must accomplish its goal of "peaceful rise", it must also realise that India is also rising and has ambitions to play a greater role in regional and global politics and economics.
- China must "manage" India's rise by reducing the threats it will pose to China and by selectively cooperating on issues of mutual interest.
- China should reduce or eliminate the chance that India may harm China's interests by compartmentalising the border disputes, containing the Tibet issue, and keeping alive the "Pakistan card".
- China must carefully monitor the implications of India's military modernisation and India's growing security and overall relations with the US lest they harm China's interests or aspirations.
- To enhance cooperation, China should increase trade with India. It should also attempt to cajole India into taking its same side on various international issues.

While Indian-Chinese relations have changed—for the better in many respects—the above complex motivations and calculations show that bilateral relations will remain mainly *instrumental* and *pragmatic*, rather than effective, and there remain limits to cooperation and potential for conflicts. To sort out the alternative scenarios, we need to not only inquire into the distribution of the various *types* of elites (realists, liberals, constructivists; so far the realists dominate both sides), but also can benefit from three contrasting paradigms.[52]

Contending Paradigms

Geopolitics

As soon as India and China came into direct contact through the Tibet nexus, geography has conditioned their relations. China and India are neighbours. The Chinese have a saying, "A distant relative is less useful than a proximate neighbour." Friends can change, neighbours cannot. "You can't change geography", says an Indian think-tank analyst.[53] So the logic goes, India must get along with China. Indeed, various Indian leaders have made this a priority, although many Indians feel that India's goodwill is not reciprocated. A third neighbour—Pakistan—further complicates this relationship.

The strategic spaces of these two Asian giants overlap and they both have ambitions to become major regional, if not world, powers. The Chinese have a saying, "The same mountain cannot accommodate two tigers." From the Indian perspectives, India cannot accept Chinese hegemony. A rising China makes India's ascent more difficult, if not impossible. This can also partly explain why the Indians felt compelled to sign a landmark nuclear agreement with the US.

In the geopolitical paradigm, the logic of balance of power prevails. Competition, mutual suspicion, alliance and military build-up—standard tenets of realism—have heavily conditioned Indo-Chinese relations. Power is important in this paradigm. Tan Chung depicts power politics as horizontal expansion, which leads to border disputes. As stated, historically China and India did not have border disputes. China did not occupy Tibet until 1950. Modern concepts of sovereignty and territorial integrity have ensnarled both China and India. Many of my informants seemed to accept certain basic realist premises and their arguments confirmed the geopolitics paradigm. Viewing Indo-Chinese relations through the geopolitics paradigm will, however, have a negative impact on the relationship.

Geo-economics

Yet at the same time, China and India are both rising economically. And there exists a complementarity between their economies. In the geo-economics paradigm, the logic is interconnectivity and mutual dependence. This creates space (complementarity) and turns the zero-sum competition in the first paradigm into a win-win situation. An increasing number of books champion this prospect:[54] China's hardware combining with India's software; China's *yang* blending with India's *yin*. Judging from the relatively still moderate trade between the two[55] and the fact that neither is a key trading partner of the other, there exists immense potential for a closer economic partnership to gradually emerge, which would help ameliorate the overall bilateral relationship. However, the emergence of a "Chindia" requires a leap of faith that is not supported by evidence.[56] While several of my informants thought Chindia was a good idea, almost nobody predicted it would happen.

Geo-civilizations

The third paradigm is not the mainstay of Western international relations theories. It is reflectivist rather than rationalist. Its logic is affinity rather than material interests.

Economic historian Angus Maddison opined that in the past one thousand years China's population had constituted one-third to one-sixth of the world's population and India's population had sometimes been larger than China's. Tan Chung opined that this meant that these two countries were most hospitable. He described the relationship between the two as "made in heaven". With population congregating, wealth was created. With their shared origins in the Himalayas, the Ganges and the Indus gave rise to the Indian civilisation and the Yellow and the Yangtze to the Chinese civilisation. As Sen pointed out, before the modern era there was a lot of mutual admiration between China and India. In the twentieth century the two also shared Asian pride and anti-colonial solidarity. Their mutual suspicion and antipathy is a more recent phenomenon.

Mao Zedong in his lifetime only visited two "countries": the Soviet Union and the Indian embassy. Nehru, whose affection for China was legendary, was welcomed by 500,000 people when he visited China. Every Chinese believes that when they die, they "return to the west" (India). Buddhism originated in India but flourished in China. One Indian scholar hailing from Ladakh summarised his visits to China this way, "People conjure up India as 'the land of the Buddha' or land of poverty."[57] Although some Indians rightly feel that the Chinese may have behaved in a condescending or overbearing way toward the Indians, China's current advantage is not preordained nor can it be expected to last forever.

This paradigm will call for a total re-conceptualisation of the Indo-Chinese relationship. It may be far-fetched to think of an Indo-Chinese partnership that is as cordial or close as the US-UK bond. But appreciating each other's civilisational attractiveness can create a deeper and more enduring bond that is currently missing in the Indo-Chinese relationship.

Based on my fieldwork in 2006 and 2008, an overwhelming majority of Chinese informants (85 per cent) adopt the geopolitical paradigm, a minority geo-economic, and virtually nobody geo-civilisational. In India, roughly 60 per cent of informants adhere to the geopolitical paradigm, 30 per cent geo-economic, and 10 per cent geo-civilisational. This rough estimate of the distribution of elite beliefs helps inform the future of Indo-Chinese relations.

Prospects: Whither?

What would the future hold for Indo-Chinese relations? Three scenarios emerge. The first is continued, even heightened, *rivalry*—guided by the logic of the geopolitics paradigm. Indications of this are not difficult to find. China figures prominently in Indian defence planning. China's growing military and economic power may deeply unsettle India. With newly accumulated wealth from almost two decades of fast growth, India may devote greater resources to the military. It will become more aligned with the US—in a reversal of its stance during the Cold War. The US-Indian nuclear agreement epitomised this trend. China may enhance its support of Pakistan and increase its influence in the South Asian continent, the Indian Ocean and South West Asia.

The second possibility is "*Chindia*"—driven by the logic of the geo-economics and geo-civilisations paradigms—to jointly promote a multipolar world and a more equitable global order (e.g., reforming the United Nations). However, an Indo-Chinese entente aimed at the US is unlikely, as each derives many benefits by maintaining a good relationship with the US.

The third possibility is a *pragmatic* management of their relationship, seeking solutions to their unresolved disputes while exploring areas of cooperation. Compared to the hot war of 1962 and the cold war that ensued, Indo-Chinese relationship has shown promise of normalisation. However, irritants still exist. The two sides should not be satisfied with prolonged but indecisive talks on settling the border issue. The Chinese had border disputes with just about every one of their land neighbours. For long periods of time, China typically remained stuck in principled positions without any real progress, but it has shown in a number of cases that it can make concessions and conclude an agreement.[58] Both China and India need to show greater political will for settling the border dispute (beginning with mutual recognition of each other's actual control). Other confidence-building measures, such as greater Chinese sensitivity to Indian concerns about China's support to Pakistan and more transparency and better communication to prevent accidents or misperceptions, would help. For a truly solid relationship, the two can benefit from the insights of the geo-civilisation paradigm.

China and India are two great nations, both making remarkable a transformation. Their choices, including the interpreting the other's intentions, will importantly shape our future world. Like Alexander Wendt cogently said, "Anarchy is what states make of it."[59] The future of Indo-Chinese relationship is not condemned to rivalry and hostility; nor will a Chindia naturally result just because it "makes sense". To return to the constructivist's axiom, it depends on the evolving structure of elite identities and preferences, informed by the three paradigms and socialised through interactions.

Notes

1. BRIC (or BRICs) stands for Brazil, Russia, India and China. The term was coined in 2001 and subsequently made popular by Goldman Sachs. See Dominic Wilson and Roopa Purushothaman, "Dreaming with BRICs: The Path to 2050", Goldman Sachs *Global Economics Paper*, No. 99 (1 October 2003), http://www2.goldmansachs.com/ideas/brics/book/99-dreaming.pdf. In 2011 the group added South Africa and is now called BRICS.
2. Kishore Mahbubani, *The New Asian Hemisphere: The Irresistible Shift of Global Power to the East* (Public Affairs, 2009).
3. See Jonathan Holslag, *China + India: Prospect for Peace* (New York: Columbia University Press, 2010); Mohan Guruswamy and Zorawar Daulet Singh, *Chasing the Dragon: Will India Catch up with China?* (Delhi: Longman, 2010); and John Garver, *Protracted Contest: Sino-Indian Rivalry in the Twentieth Century* (Seattle: University of Washington Press, 2002).
4. For discussion of Chinese analysts' perception of their country's external security environment, especially along its periphery, see Michael Pillsbury, *China Debates the Future Security Environment* (Washington, DC: National Defense University Press, 2000); Yang Chengxu (ed.), *Zhongguo zhoubian anquan huanjing toushi* [*Perspectives of China's Security along its Boundaries*](Beijing: Zhongguo qingnian chubanshe, 2003); and Zhu Tingchang (ed.), *Zhongguo zhoubian anquan huanjing yu anquan zhanlue* [*China's Security Environment along the Periphery and Security Strategy*](Beijing: Shishi chubanshe, 2002).
5. U.S. Department of Defense, *Annual Report to Congress: Military Power of the People's Republic of China, 2007*, p. 7, http://www.defenselink.mil/pubs/pdfs/070523-China-Military-Power-final.pdf
6. See Michael Brown *et al.* (eds.), *The Rise of China* (Cambridge, MA: MIT Press, 2000). The two books that helped set the stage for the "great debate" over China were Richard Bernstein and Ross H. Munro, *The Coming Conflict with China* (New York: Knopf, 1997) and Andrew J. Nathan and Robert S. Ross, *The Great Wall and the Empty Fortress* (New York: Norton, 1997). Herbert Yee and Ian Storey (eds.), *The China Threat: Perceptions, Myths, and Reality* (London: Routledge Curzon, 2002) assembles a range of views on this issue.
7. For an exposition of the concept of "peaceful rise" aimed at Western audiences, see Zheng Bijian, "China's 'Peaceful Rise' to Great-Power

Status", *Foreign Affairs,* 84(5) (September/October 2005): 18-24. Zheng is one of the most important advisers to Hu.

8. For an account on China's more confident diplomacy, see Evan S. Medeiros and M. Taylor Fravel, "China's New Diplomacy", *Foreign Affairs*, 82(6) (November/December 2003): 22-35; and Bates Gill, *Rising Star: China's New Security Diplomacy* (Washington, DC: The Brookings Institution, 2007).
9. Bonnie Glaser and Lyle Morris, "Chinese Perceptions of U.S. Decline and Power", *China Brief,* IX(14) (9 July 2009): 4.
10. China's "charm offensive" toward South East Asia is exemplary. See Vincent Wei-cheng Wang, "The Logic of China-ASEAN FTA: Economic Statecraft of 'Peaceful Ascendancy' ", in Ho Khai Leong and Samuel C.Y. Ku (eds.), *China and Southeast Asia: Global Changes and Regional Challenges* (Singapore: Institute of Southeast Asian Studies, 2005), pp. 17-41; Jane Perlez, "China is Romping with the Neighbours (U.S. is Distracted)", *New York Times* (3 December 2003): A1; and Joshua Kurlantzick, *Charm Offensive: How China's Soft Power is Transforming the World* (New Haven, CO.: Yale University Press, 2007).
11. For China, the word "multilateralism" now sounds like a coded opposition to America's "unilateralism"; China prefers a "multipolar" world (in which China acts as a great power) to a "unipolar" world founded on US hegemony.
12. For example, on the South China Sea issue, China acceded to ASEAN's Treaty of Amity and Cooperation (TAC) and promoted peaceful dialogue over territorial disputes. On Taiwan, China has replaced its military bluster with economic enticement.
13. Wang, "The Logic of China-ASEAN FTA ...," n. 10, pp. 32-4.
14. Pillsbury, *China Debates the Future Security Environment*, n. 4, p. 203.
15. For more details and the various calculations, see Pillsbury, *China Debates the Future Security Environment*, n. 4, chapter 5.
16. A 2009 CNP study by the Korean *Chosun Daily* puts the US and China as Number One and Two with CNP of 69.15 and 54.73, respectively. Hangzhou.com, 18 August 2009, "Hanguo gongbu G20 guoli pinggu, zhongguo zonghe guoli paimin shijie dier" (Korea Unveils G20 National Power Assessment; China's GNP is Ranked Second in the World", http:/ /money.hangzhou.com.cn/20090728/ca1767694.htm; accessed 18 April 2011.
17. These questions were summarized from the author's interviews with two dozen Chinese think-tank analysts and international relations scholars in Beijing and Shanghai in June 2006.
18. Robert Zoellick, former US Trade Representative and current World Bank President, coined the phrase "responsible stakeholder" to signify America's policy on and expectations from a rising China. "USTR Says

China's Power Must Be Integrated into World Community" (Zoellick's 25 February remarks to Asia Society), e-mail update from "USINFO East Asia" <iipgeap@ STATE.GOV>. Political scientist John Ikenberry thinks this strategy is both possible and preferable for the West. "The Rise of China and the Future of the West", *Foreign Affairs*, 87(1) (January/February 2008): 23-37.

19. Liselotte Odgaard, "The Art of Walking on Two Legs: Stakeholder and Challenger Elements in China's National Security Strategy after the Cold War", lecture at Wilson International Center for Scholars, 20 May 2009.
20. Bill Emmott, *Rivals: How the Power Struggle Between China, India, and Japan Will Shape Our Next Decade* (Harcourt Books, 2008); Pete Engardio (ed.), *Chindia: How China and India are Revolutionizing Global Business* (New York: McGraw-Hill, 2007); Garver, *Protracted Contest*, n. 3; Tarun Khanna, *Billions of Entrepreneurs: How China and India Are Reshaping Their Future and Yours* (Boston: Harvard Business School Press, 2007); Robyn Meredith, *The Elephant and the Dragon: The Rise of India and China and What it Means for All of us* (New York: W.W. Norton, 2007); and Waheguru Pal Singh Sidhu and Jing-dong Yuan, *China and India: Cooperation or Conflict?* (Boulder, CO: Lynne Rienner, 2003).
21. Engardio, *Chindia ..., ibid.*, p. 16.
22. See Arvind Panagariya, *India: The Emerging Giant* (Oxford University Press, 2008) and Stephen G. Cohen, *India: Emerging Power* (Washington, DC: Brookings Institution Press, 2001).
23. See Alexander Wendt, "Anarchy is What States Make of it: The Social Construction of Power Politics", *International Organization,* 46(2) (Spring 1992): 391-425.
24. Sidhu and Yuan, *China and India: Cooperation or Conflict?,* n. 20, p. 9.
25. Amartya Sen, *The Argumentative Indian: Writings on Indian History, Culture, and Identity* (New York: Picador, 2005), 161.
26. *Ibid.,* p. 166.
27. Speech by Prof. Tan Chung at National Taiwan University, Taipei, 20 May 2008.
28. Interview with Prof. Phunchok Stobdan, Senior Fellow, Institute for Defence Studies and Analyses, New Delhi, 2 June 2008.
29. Interview with General Ved P. Malik, President, Institute of Security Studies, Observer Research Foundation, New Delhi, 2 June 2008.
30. Paul Bracken, *Fire in the East: The Rise of Asian Military Power and the Second Nuclear Age* (New York: HarperCollins, 1999).
31. Shanghai Five was founded in 1996 by the leaders of China, Kazakhstan, Kyrgyzstan, Russia and Tajikistan. The group was renamed the SCO in

2001 with Uzbekistan joining it. The SCO currently has four observers: India, Iran, Mongolia and Pakistan. See http://www.sectsco.org/

32. Interview with Amb. Ranjit Gupta (retired), New Delhi, 26 May 2008. Ambassador Gupta admitted that his viewpoints on China reflect that of the security community and are uncommon among the Indian foreign service.
33. Kenneth Lieberthal and Mikkal Herberg, "China's Search for Energy Security: Implications for U.S. Policy", *NBR Analysis* 17(1) (April 2006): 3-42.
34. "India's ocean is Chinese lake: 'String of pearls' threatens India", http://www.zimbio.com/President+Mahinda+Rajapakse/articles/185/India+ocean+Chinese+lake+String+pearls+threaten, accessed 12 February 2009.
35. Interview with Brigadier Gurmeet Kanwal (retired), Director, Centre for Land Warfare Studies, New Delhi, 3 June 2008; interview with Rajeswari Pillai Rajagopalan, Senior Fellow, Observer Research Foundation, New Delhi, 28 May 2008.
36. Historically, Aksai Chin was part of the Himalayan Kingdom of Ladakh. Ladakh was annexed from the rule of the local Namgyal dynasty by the Dogras and the princely state of Kashmir in the nineteenth century. It was subsequently absorbed into British India.
37. Randeep Ramesh, "International: Himalayan Dispute: Beijing Insists on Claim to Territory in Runup to President Hu's Visit to Delhi: Last Vestige of Old Tibetan Culture Clings on in Remote Indian State", *The Guardian* (20 November 2006), p. 23. Obtained through *Lexis-Nexis*.
38. Ravi Velloor, "China Reaffirms its Claim to Disputed State; No Visa for Indian Official from Arunachal as he is a 'Chinese Citizen' ", *The Straits Times* (28 May 2007). Obtained through *Lexis-Nexis*.
39. In 2000 the seventeenth Karmapa Urgyen Trinley Dorje, who was endorsed by both the Dalai Lama and the Chinese government, made a dramatic escape from Tibet to the Rumtek Monastery in Sikkim. China eventually recognized Sikkim as an Indian state in 2003, on the condition that India accepted Tibet as a part of China. This mutual recognition led to a thaw in Sino-Indian relations.
40. Interviews with Sibi George, Political Counsellor, and S.D. Sharma, Head of Chancery, Embassy of India, Washington, DC, 9 May 2008.
41. Interview with Joel Ehrendreich, First Secretary, Political Affairs, US Embassy, New Delhi, 4 June 2008.
42. "Defence Minister Calls China 'Potential Threat Number One' ", *PTI News Agency*, New Delhi, 3 May 1998 (*BBC Summary of World Broadcasts*, 5 May 1998). Obtained through *Lexis-Nexis*.
43. Interview with Bharat Karnad, Research Professor, National Security Studies, Centre for Policy Research, New Delhi, 4 June 2008.

44. Interviews with Rajesh Rajagopalan, Professor in International Politics, Jawaharlal Nehru University, New Delhi, 27 May 2008, and Nandan Unnikrishnan, Director, Eurasian Studies, Observer Research Foundation, New Delhi, 28 May 2008.
45. Yasheng Huang and Tarun Khanna, "Will India Overtake China?" *Foreign Policy,* no. 137 (Jul-Aug 2003), pp. 74-81, and Engardio, *Chindia*, n. 20.
46. Engardio, *Chindia*. The term "Chindia" is a portmanteau word that refers to China and India together in general, and their economies in particular. The credit of coining the now popular term goes to Jairam Ramesh, who has been India's Minister of State for Environment and Forests since May 2009. Jairam Ramesh, *Making Sense of Chindia: Reflections on China and India* (Delhi: Research Press, 2005).
47. Interviews on 24 May 2008 with Mumbai industrialists: Anil Tiwari, Shree Krishna Engineering Works; Bhavin Shah, Alkon Plastics Pvt. Ltd.; Prayag Thakkar, Mutual Industries; Vinay Sawar, Teryai Equipment.
48. As of 2007, China, the United States and India were responsible for 22.3 per cent, 19.9 per cent, and 5.5 per cent of the world's emission of carbon dioxide, respectively. See http://en.wikipedia.org/wiki/List_of_countries_by_ carbon_dioxide_emissions
49. China has always been very conscious of its relative standing in the world vis-à-vis India's, and has usually regarded itself as enjoying a comfortable lead. It has taken measures to widen that gap (for example, by arming Pakistan to wage a kind of "proxy war") and tends to regard India's rising power warily.
50. Goldstein described his Chinese interlocutors (during 1998-2000) often expressing rather sceptical, even dismissive, views of India's prospects. Avery Goldstein, *Rising to the Challenge: China's Grand Strategy and International Security* (Stanford, Calif.: Stanford University Press, 2005), p. 168. During my own field research in Beijing and Shanghai in summer 2006, I got essentially the same impression (albeit less dismissive).
51. Jingdong Yuan, "The Asian Triangle: Chinese Perspectives on a Rising India", *World Politics Review* (30 November 2008), http://www.worldpoliticsre view.com/article.aspx?id=2983; accessed 2 December 2008.
52. The following section benefits from Prof. Tan Chung's lecture, 20 May 2008.
53. Interview with Nandan Unnikrishnan.
54. Engardio, *Chindia*, n. 20; Meredith, *The Elephant and the Dragon*, n. 20.
55. Trade between China and India crossed $10 billion at the end of 2004. In 2008 two-way trade was $40.2 billion. Although trade relations have

been fast growing, they have not been balanced. China was India's third-largest export destination, absorbing 9.3 per cent of Indian exports ($155 billion), largest import source (11.1 per cent of $232.2 billion) and overall largest trading partner. India suffered a trade deficit of $11.4 billon. Conversely, India was not as important to China in trade as China was to India, not appearing in the lists of top five export or import partners. Data calculated from CIA, *World Factbook*, https://www.cia.gov/library/publications/the-world-factbook/geos/in.html, 15 February 2010. Chinese Premier Wen Jiabao had called for bilateral trade to increase to $60 billion by 2010.

56. See Yasheng Huang, "The Myth of Economic Complementarity in Sino-Indian Relations", *Journal of International Affairs,* 64(2) (Spring/Summer 2011): 111-24.
57. Interview with Phunchok Stobdan.
58. "Russia and China Settle Eastern-Border Dispute", *International Herald Tribune,* 22 July 2008, p. 3.
59. Wendt, "Anarchy is What States Make of it ...", n. 23.

5

The United States, the Rise of China, and US-India Relations

Robert J. Art

Introduction

In this paper, I address two questions, focussing the bulk of the analysis on the first. The questions are: what will future US-China relations look like, and how will those relations affect the US-India relationship?[1]

In this analysis I do not mean to imply that America's future relations with China will wholly determine America's relationship with India. That would be absurd. But what I do argue is that the course of Sino-American relations will have a pronounced effect on US relations with India. The prism that I employ in analysing US-China relations and in drawing the implications of those relations for India is a realist one, and I rely heavily on balance of power analysis.

My argument consists of the following five points: (1) the US and China are not doomed to have a wholly conflictual relationship as China continues its rise, but neither are cooperative relations between them assured; (2) there is a large overlap between American interests in East Asia and those of China, but also some clear and important differences; (3) as a consequence, wise policy choices by both countries can enhance the chances that their relationship will be more cooperative than conflictual; (4) if the United States and China experience highly conflictual relations (but not war), then the United States, according to balance of power logic, will "woo" India, seeking to offset China's power by allying with India or having close military relations with it, and, conversely, if China and the United States experience generally cooperative relations, even if China's military power is increasing, then the United States will feel less pressure to court India; and (5) if the highly conflictual scenario turns out to be the case, this will likely yield India leverage in its relationship with the United States; if the cooperative scenario is the case, India will have less leverage over the United States. In either case, the state of Sino-American relations will have a significant bearing on US-India relations.

What, then, will US-China relations look like over the course of the next decade or two? Historically, the rise of one great power at the expense of the dominant one has nearly always led to intensely conflictual relations between the two, and, more often than not, eventually to a war between them that has dragged in other great powers.[2] Is this dismal history of rising-versus-dominant great power competitions, including great power war, the future for US-China relations?

Clearly, there will be political and economic conflicts and political frictions between the United States and China as China's economic and military power in East Asia, and its global economic and political reach, continue to expand. Clearly, there will also be some arms racing between China and the United States as each jockeys for advantage over the other, and as each is driven by their respective military necessities of intimidating and defending Taiwan, and as the United States responds to China's growing maritime power projection capabilities. Historically, dominant powers have not readily ceded their positions to rising challengers, and rising challengers have always demanded the fruits that they believe their growing power entitles them to. There is no reason to expect that things will be different in this regard between China and the United States. Thus, they will not be able to avoid a certain degree of conflict and political frictions over the next several decades.

Are political friction and conflictual relations, and even war, the chief things that these two powers have to look forward to, or are there also some significant shared interests and hence bases for cooperation in both the medium and the longer term; such that the peace-inducing aspects of the US-China relationship could come to overshadow the conflict-producing ones? No one can say for certain which will be the case. However, if we believe that there are distinct elements in the Sino-American relationship that differ from the dominant-rising power dyads of the past, then the dismal history of such dyads need not be the future of this one. If this is so, then the right policy choices by both countries can keep the two on a path that has more cooperative than conflictual elements to it, thereby avoiding the doom-and-gloom scenario portrayed by too many of today's analysts.

In order to answer these questions, I proceed as follows. First, I review three previous cases of rising versus dominant powers to assess whether the current US-China case differs in from them in any significant ways.

Second, I review the state of economic interdependence between the United States and China. Third, I analyse the degree of overlap between American and Chinese interests in East Asia. Finally, I draw a few brief implications of this analysis for US-India relations.

Historical Perspectives on Power Transitions and Security Dilemma Dynamics

The first way to assess the future course of U.S.-China relations, then, is to look for cases in the recent past where a rising power has challenged a dominant one, and then to see how the US-China relationship compares to them. I do this in Table 1, which compares the current Sino-American competition with the three most important rising-power-versus-dominant-power competitions of the last one hundred years—those that resulted in either a great power hegemonic war or a sustained and intense political-military competition for hegemonic dominance between two great powers.[3] The first two competitions—Britain versus Germany in the decade before World War I and Britain versus Germany from 1933-1939—resulted in war; the third—the United States versus the Soviet Union during the Cold War—resulted in an uneasy peace between the two powers, punctuated by intense arms races and numerous proxy wars.[4]

I focus on three variables in order to explain the outcome of these cases: (1) the level of security that both powers enjoyed, or believed they enjoyed, in general and vis-à-vis one another in particular, with security defined as protection of the state's homeland from physical attack, invasion, or conquest and its political sovereignty from severe infringement; (2) the extent of economic interdependence between them, with interdependence defined in terms of the level of economic interactions—especially trade—between the two states; and (3) the degree and intensity of ideological competition that they experienced. I do not present these three variables as a full-blown deductive theory about war and peace between rising and dominant great powers—such theories have been presented by others—although with somewhat contradictory results. Instead, I argue that these three factors—and especially the first, as made clear below—are the most important ones to look at in order to determine the level and intensity of hostility and conflict, and hence the likelihood of war, between two great powers that believe they are experiencing, or may soon experience, a fundamental power shift between them.

All other things being equal, the intensity of competition and the likelihood of war between a dominant power and a rising challenger should vary as follows. First, the lower the level of security each enjoys or believes it enjoys vis-à-vis the other or in general, the more likely are serious

Table 1: Dominant Power Versus Rising Power Competitions

Dominant-Rising Power Dyad	*Security Enjoyed by Both Powers Vis-Vis One Another*	*Level of Economi c Inter-depen-dence*	*Ideo-logical Com-petition*	*Outcome*
United Kingdom-Germany pre 1914	In 1914 security thought to be low for the future if corrective action (war) not taken	High	Low	War
United Kingdom-Germany pre 1939	Low in the 1930s due to presumed airpower threat	Low to medium	Medium to high	War
US-Soviet Union during the Cold War	Initially believed to be low; turned out later to be high	Low	High and Intense	Cold War; serious crises at first; then an uneasy peace
US-China Today	High for US, but problematical for China because Chinese nuclear forces at present are potentially vulnerable to a US first strike	High	As yet, low to non-existent	To be determine d

security dilemma dynamics, intense arms races, and war between the two. Conversely, if these states feel that they are relatively secure from attack and that their political sovereignty is not being compromised (or will not be compromised) by the actions of the other, then they can experience a greater level of hostility that may arise from conflicts over non-security issues without war being the result. There are many reasons why two states can wage war against one another; however, if each believes it is relatively safe from attack by the other, then one of the most powerful historic incentives for war—insecurity—is removed.

Second, under the right conditions, the higher the level of economic interdependence between two states, the less likely will security competitions and war between them take place. The three conditions under which high levels of economic interdependence can be peace-inducing are: (1) when two states believe that they can more profitably resort to economic rather than military means in order to prosper; (2) when they believe that the economic vulnerabilities that result from economic interdependence cannot be quickly and easily turned to their military disadvantage; and (3) when they believe that should the first two conditions change, they can readily protect themselves or find allies who will.[5]

Finally, the greater the ideological differences between the two states, the more intense will be their competition and the more likely they are to experience arms races, intense security dilemmas, and war. Security concerns have been powerful factors for war, but they are not the only factors. Ideological hostilities—conflicts over how social power should be organised within states—have also contributed to hostility and war between states. The Cold War, for example, was not simply about US-Soviet security, it was also about values—democratic market capitalism versus communism—and which would prevail globally.

Close inspection of Table 1 reveals that neither economic interdependence nor ideological competition is a good predictor of whether a dominant and rising power will go to war or remain at peace. In 1914, for example, Germany was England's second best customer for its exports, and England was Germany's best customer for its exports.[6] Before the war, they did not experience intense ideological competition, and in fact shared many political similarities. Yet, the two ended up at war. Throughout the 1930s, England and Germany did not have as high a level of economic interdependence with one another as they had before World War I, and they did experience some ideological competition.[7] At the outset of, and during the course of, the Cold War, the United States and the Soviet Union experienced a low level of economic interdependence with one another, and their ideological competition was high and intense.[8] Yet they remained in an uneasy peace, although one that was much more fragile during the first part of the Cold War than during the second. Thus, the relationship between levels of economic interdependence and intensity of ideological competition, on the one hand, and war or peace, on the other, is indeterminate. High and low levels of interdependence are correlated with war, and high and low levels of ideological competition are correlated with both war and peace.

Therefore, it is the degree of security enjoyed by these pairs of states vis-à-vis one another, and especially the severity of the security threat that the dominant state perceives emanating from the rising state, that constitutes the most important variable to predict whether a hegemonic struggle will result in war.[9] When security was low or believed to be low, intense crises and war were more likely; when security was high or believed to be high, better relations and peace tended to prevail. Economic interdependence and ideological competition are not irrelevant to producing peace and war, but they are only "helper variables." They can reinforce peaceful trends when the security enjoyed by both states vis-à-vis one another is high, but they cannot override the deleterious effects produced when the security enjoyed by both states vis-à-vis one another is low.

In 1914, Germany needed England to stand aside so that it could shore up its failing great power ally Austria-Hungary and deal with the growing power of Russia. Diplomatic isolation for Germany (the dissolution of Austria-Hungary) meant the political encirclement and eventual strangulation of Germany in the eyes of Germany's leaders. If Britain stood aside while Germany attained continental hegemony, however, then Britain's security would have been at risk if a hegemonic Germany deployed the resources of the European continent against it. So, in classic security dilemma terms, Britain's security would be at risk by the continental hegemony that Germany believed it needed for its security, and, similarly, Germany's leaders believed that Germany's security would be at risk if Britain denied it continental hegemony. The same logic applies to the British-German struggle during the 1930s. A Nazi Germany that had defeated Russia and vanquished the continent could aggregate its resources and turn them into an invasion force to crush England's defences. The threat that German hegemony posed to Britain brought on the two World Wars.

Security factors were also central in explaining why the Cold War stayed "cold." The United States and the Soviet Union were in an intense ideological competition for most of the Cold War, but that competition never turned into a direct war between the two primarily because of the restraining effects of nuclear deterrence. True, the stability-instability paradox did not work to prevent all serious crises between the two, and true, one of these crises—and the Cuban missile crisis of 1962—brought the United States and the Soviet Union closer to war than either would have liked.[10] However, the workings of the stability/instability paradox were ultimately beneficial: it worked to prevent those crises that did occur from escalating into war, and it worked, ultimately, to reduce dramatically the frequency and the severity of crises after the Cuban missile crisis. In fact, one could make a strong argument that with one exception, serious security crises between the United States and the Soviet Union disappeared after 1973.[11]

China does not present the type of security threat to the United States that Germany did to Britain, or Britain to Germany. America's nuclear forces make it secure from any Chinese attack on the homeland. Moreover, China clearly presents a potentially different type of threat to the United States than the Soviet Union did during the Cold War because the geopolitics of the two situations are different. The Soviet geopolitical (as opposed to the nuclear) threat was two fold: to conquer and dominate the economic-industrial resources of western Eurasia and to control the oil reserves of the Persian Gulf. Europe and the Persian Gulf constituted two of the five power centres of the world during the Cold War—Japan, the Soviet Union, and the United States being the other three. If the Soviets had succeeded in dominating Europe and the Persian Gulf through conquest or political-military intimidation, then it would have controlled three of the five power centres of the world. That would have been a significant power transition.

China's rise does not constitute the same type of geopolitical threat to the United States that the Soviet Union did. If China ends up dominating the Korean peninsula and a significant part of continental South East Asia, so what? As long as Japan remains outside the Chinese sphere of influence and allied with the United States, and as long as the United States retains some naval footholds in South East Asia, such as in Singapore, the Philippines, or Indonesia, China's domination of these two areas would not present the same type of geopolitical threat that the Soviet Union did. As long as Europe, the Persian Gulf, Japan, India, and Russia (once it reconstitutes itself as a serious great power) remain either as independent power centres or under US influence, Chinese hegemony on land in East and South East Asia will not tip the world balance of power. The vast size and central position of the Soviet Union in Eurasia constituted a geopolitical threat to American influence that China cannot hope to emulate.

If judged by the standards of the last three dominant power-rising power competitions of the last one hundred years, then, the US-China competition appears well placed to be much safer. Certainly, war between the two is not impossible because either or both governments could make a serious misstep over the Taiwan issue or with their navies at sea. War by miscalculation is always possible, but the possession of nuclear weapons by both sides has to have a restraining effect on both by dramatically raising the costs of miscalculation, thereby increasing the incentives not to miscalculate. Nuclear deterrence should work to lower dramatically the possibility of war by either miscalculation or deliberate decision (or if somehow such a war broke out, then nuclear deterrence should work against its escalation into a large and fearsome one.) Apart from the Taiwan issue or some serious incident at sea, for example, over access to the sea lanes in the South China Sea, it is hard to figure out how to start a war between the United States and China. There are no other territorial disputes of any significance between the two, and there are no foreseeable economic contingencies that could bring on a war between them. Finally, the high economic interdependence and the lack of intense ideological competition between them help to reinforce the pacific effects induced by the condition of mutual assured destruction.

The workings of these three factors should make us cautiously optimistic about keeping Sino-American relations on the peaceful rather than the warlike track. The peaceful track does not, by any means, imply the absence of political and economic conflicts in Sino-American relations, nor does it foreclose coercive diplomatic gambits by each against the other. What it does mean is that the conditions are in place for war to be a low probability event, if policymakers are smart in both states, and that an all-out war is nearly impossible to imagine. By the historical standards of recent dominant-rising state dyads, this is no mean feat.

In sum, there will be some security dilemma dynamics at work in the US-China relationship, both over Taiwan and over maritime supremacy in East Asia, should China decide eventually to contest America's maritime hegemony, and there will certainly be economic and political conflicts, but nuclear weapons should work to mute their severity because the security of each state's homeland will never be in doubt as long as each maintains a second strike capability vis-à-vis the other. If two states cannot conquer one another, then the character of their relationship and their competition changes dramatically than would be the case if they could.

Economic Interdependence in Sino-American Relations

The second way to assess the future state of Sino-American relations is to consider more carefully the nature of the economic relations between the two states. Alone, economic interdependence, as argued above, cannot produce peace, but it is by no means irrelevant as a peace-inducing factor. When security conditions between a rising and a dominate state are stable—that is, when both feel their homeland is secure from invasion and conquest—then economic interdependence reinforces that stability. Economic interdependence is a "helper" variable in this sense. China and the United States today are in a condition of high economic interdependence. Consider these facts.

In 2008 (the most recent figures available), China (which includes mainland China, Hong Kong, and Macao) exported $364 billion to the United States, or 20 per cent of its total exports of $1.8 trillion for the year.[12] In 2008 China's GDP (in current dollars) was $4.3 trillion.[13] This means that in 2008, 8.5 per cent of China's GDP was exported to the United States, an astoundingly high figure and one that looks as if it creates a huge dependency of China on access to the US market.[14] This dependency is large, but not quite as large as the above figures would imply, because, as Richard Cooper points out, China's exports are measured in terms of gross value, not the value-added in China. Because many of China's exports involve the processing of imports, a total cessation of exports to the United States would be something less than 8.5 per cent of GDP.[15] Still, the abrupt loss of the US market would be highly disruptive in the short term for China. Similarly, China holds a powerful financial lever over the United States. As of March 2010, China held $895 billion (11.5 per cent) of the $7.8 trillion of total outstanding US Treasury securities that are privately held (as of December 2009). China is now the largest foreign holder of US Treasury securities, with Japan second at $785 billion.[16]

This economic relationship between the United States and China is beneficial to both. China sends a significant part of its manufactured goods to the United States, thereby stimulating its economy and helping it sustain high rates of growth. In turn, the United States is able to finance its deficit with Chinese purchases of US Treasury notes. Each state has a vested interest in the other doing well economically. For example, if the United States inflates the dollar, China's huge holding of dollars depreciates. If the United States experiences a severe recession, China's exports will suffer. Similarly, if China refuses to loan the United States money, interest rates in the U.S. would rise significantly, bringing on a severe recession. Both need each other, but also hold the economies of the other hostage to a significant degree. Economic interdependence benefits both states, but it also creates dependencies that are hurtful, although not impossible, to break. These dependencies in turn give each state a high stake in keeping their relations on a peaceful keel and in working through the conflicts that dependency inevitably creates.

While economic interdependence has peace-inducing effects now on the Sino-American relationship, it could be a waning asset in that regard over the medium to longer term. That is, if the United States succeeds in cutting its deficit, it will need to borrow less from the outside world, including from the Chinese, and if China succeeds in moving to a domestic market-driven, rather than an export market-driven economy, it will need to rely less for its economic growth on the US market. As a consequence, the level of economic interdependence between the two could diminish over the years. This is all the more reason why a secure and stable security relation between the two is essential. Nonetheless, for now and for the next several years, the high level of economic interdependence between the United States and China, though not without its problems for both states, gives each a strong stake in the economic well being of the other and thereby serves as a peace-inducing factor.

American and Chinese Interests in East Asia

A third way to assess the future of Sino-American relations is to consider the nature of U.S. and Chinese interests in East Asia. What interests do the two states share in common? What interests are opposed?

The United States has six overarching interests in East Asia. They are: (1) preservation of Sino-American mutual assured destruction; (2) stability in the Taiwan Straits and a peaceful resolution of the Taiwan issue; (3) the denuclearisation of the Korean peninsula; (4) the preservation of the U.S.-Japan alliance and the maintenance of Japan's non-nuclear status; (5) the peaceful settlement of China's maritime disputes with its neighbours and the preservation of freedom of commercial navigation in the South China Sea; and (6) the preservation of economic openness in East Asia.

First, it is in America's interest that decision makers of all nuclear-armed states, including China's, run scared, not safe, vis-à-vis one another. Nuclear-armed decision makers must believe that neither side will gain any advantage by a first strike against the adversary's nuclear forces because that might embolden one or both of them to take greater risks during a crisis and tip the crisis towards war—and even all-out war—rather than towards de-escalation. It is not in America's interest (or China's) for American decision makers to believe that they can disarm China's nuclear forces (nor, for that matter, for China's leaders to believe they can disarm America's). Therefore, just as it served America's best interests for the Soviet Union to have a secure second strike force once the Soviet Union had acquired nuclear weapons, so, too, is it in America's interest to have China achieve one. The existence of mutual assured destruction (MAD) between the United States and China is the bedrock for stability in their security relations with one another.

Second, the United States has a strong interest in the peaceful resolution of the Taiwan issue. The US position on what Taiwan's ultimate status should be is deliberately not crystal clear, although it has become progressively clearer in recent years. The United States does not favour a two China policy, or a one China and one Taiwan policy, nor does it support Taiwan's unilateral declaration of independence. This position comes awfully close to a US de facto acceptance of Taiwan as part of China, but the US government has not explicitly said so in order to keep open the option of Taiwanese independence should both the mainland and Taiwan agree, which, of course, the mainland would never agree to. The United States position on *how* Taiwan's status is to be settled, however, is crystal clear: it will not allow China to use force to bring Taiwan to heel. Rather, the United States is committed to the peaceful resolution of

Taiwan's status. Because of this commitment, the United States must have sufficient military power in the region to deter China from using force to settle Taiwan's status, or to protect Taiwan should the mainland use force against a Taiwanese government that has not provoked the mainland by moving towards, or actually declaring, independence. (The U.S. stance on protecting a Taiwan that provokes a mainland attack by moving towards, or declaring independence, is not clear, but it should be: any Taiwan government that acts in such a fashion is on its own.)

Third, it is in America's interest to denuclearise North Korea, and the primary reason for doing so is clear. The threat from North Korea is not so much its use of nuclear weapons against its neighbours (South Korea and Japan), although that can never be ruled out should it be attacked by the United States. Rather the more serious threat is its selling of fissile materials to terrorist groups that want to acquire them for purposes of either blackmail or destruction. The best way to prevent such transfers of fissile materials to terrorist groups from North Korea is to remove its fissile material and its enrichment facilities.

Fourth, it is crucial to America's position in East Asia, its general world position, and its global non-proliferation policy that the US-Japan alliance remains solid and that Japan remains a non-nuclear state. The alliance and Japan's non-nuclear status are tightly linked: Japan eschews nuclear weapons, primarily or in part, depending on the analyst consulted, because of the nuclear umbrella that the United States extends over it.[17] Were Japan to acquire nuclear weapons, this would be a clear political statement that it puts little or no credence in the US-Japan alliance. After all, why would Japan obtain nuclear weapons if it believed fully in the credibility of the U.S. nuclear guarantee? Without the US umbrella, Japan would either have to acquire its own nuclear weapons, or else forego them and thereby be at a political-military disadvantage vis-à-vis the nuclear-armed North Korea and China.

Fifth, the United States has a strong interest in the peaceful resolution of China's maritime disputes with its neighbours and in the preservation of freedom of navigation through the South China Sea. The United States has no interest in seeing any of its East and South East Asian regional allies, trading partners, and friends becoming embroiled in military hostilities with China over the extent of its territorial seas and exclusive economic zones, possession of islands, or ownership of seabed resources. It especially has to avoid a situation whereby China and Japan engage in the large-scale use of force to resolve their disputes over ownership of the Senkaku/Diaoyu Islands or over control of the East China Sea and its gas fields. America's interest lies in seeing all three offshore disputes solved through negotiation and compromise, not war.

It also has a strong interest in preserving freedom of navigation for commercial vessels through the South China Sea. The United States is a trading nation and has a vested interest in freedom of the seas, and since World War II it has been the provider of that collective security to the world. For both commercial and strategic reasons, it cannot tolerate China eventually moving to prevent freedom of commercial navigation though the South China Sea because it contains crucial sea lanes of communication (SLOCs) through which Middle East oil and other vital resources flow not only to China, but to Japan, Korea, and other states in the region.

Sixth, finally, the United States has a vested interest in preserving economic openness in East Asia, for two reasons. The first has to do with its own prosperity; the second, with the prosperity and political relations of the states in the region. Asia and the Pacific Rim have become of central economic importance to the United States. The Pacific Rim now constitutes the most important region—economically—to the United States, having over taken North America, which was traditionally first, when measured in terms of total dollar volume of exports and imports of both merchandise goods and services.[18] In 2009, the United States sent 16 per cent of its exports to, and received 28 per cent of its imports from the Asia Pacific region, while it sent 21.5 per cent of its exports to Canada and Mexico and received 21 per cent of its imports from them. (Comparable figures from Europe were 16.7 per cent and 17 per cent.)[19] America's trade with East Asia benefits the United States in innumerable ways, though there remains much controversy over its trading deficit with the region, especially with China and Japan, and with the loss of manufacturing jobs to China. Whatever the resolution of these controversies, on balance, America's trade with Asia is a net plus for the American economy and maintaining as free and open a trading regime with East Asia as possible continues to be in America's economic interest.

An open economic order is also of benefit to the states in the region. Regional trade within East and South East Asia has grown significantly over the last ten years, in part because of the stabilising military presence of the United States. Peace is good for trade, and trade benefits the rise of middle classes within countries, which is a net benefit ultimately for democracy. Increasing economic interdependence within the region is also good for peaceful relations, even if it cannot develop them on its own.

If we consider these six American interests, we see that there is much overlap, though certainly not completely so, between what the United States wants and what China wants. China and the United States share a broad agreement on many goals, even if they do not wholly agree on the means to attain them or on the priority that each state gives to these goals.

China wants to have a secure second strike capability vis-à-vis the United States. So, too, does the United States want, or should want, that for China. China prefers stability in the Taiwan Strait, a peaceful resolution of Taiwan's status, and no unilateral moves towards independence; so, too, does the United States. China favours denuclearisation of the Korean peninsula; so, too, does the United States. China favours preventing Japan from going nuclear; so, too, does the United States. Even in light of its recent aggressiveness, China still appears to favour the peaceful settlement of its maritime disputes with its neighbours; it clearly benefits from economic openness with the United States and among states within the region; and it does not want to see economic closure with either. So, too, does the United States. At this general level, then, America's goals for the region overlap with many of China's goals for the region.

To stress that China and the United States share many common goals for East Asia is not to make light of their serious differences. They differ on how strongly North Korea can be pushed, with the United States favouring stronger action than China. Moreover, China certainly does not want the Koreas unified if that brings US troops up to the Korean-Chinese border. The United States would like to see the North Korean regime disappear from the face of the earth. China is more ambivalent about the US-Japan alliance than is the United States and sees it as bolstering a tough Japanese foreign policy toward China. Taiwan is a sore point with China. Even though it favours the peaceful settlement of Taiwan's status, China does not appreciate U.S. meddling in what it considers its internal affairs. Finally, the United States and China disagree strongly over the status of the South China Sea, with China claiming sovereignty over it and the United States opposing that claim. So, although the United States and China have many goals in common, they differ, and that too strongly, on a few crucial ones.

The point is not that there is complete agreement on what each state wants in East Asia, but rather that there is a basic agreement between China and the United States on the fundamental goals for the region, and that makes this relationship qualitatively different from the previous rising versus dominant power conflicts of the last one hundred years. Still, there remains a fundamental tension between the United States and China that is not easy to bridge. The United States intends to maintain maritime supremacy in East Asia because its power and influence in the region depend on it. No ally will take the United States seriously if it loses its maritime supremacy to China. But China, by all accounts, appears to be on the verge of taking a serious set of steps to eventually challenge America's maritime supremacy in East Asia, as illustrated by its claims over the South China Sea, its decision to build aircraft carriers, and its development of advanced military technology, such as anti-ship missiles, to keep the United States at bay from its homeland. China is the hegemon on the East Asian mainland; the United States is the hegemon at sea. If China seriously contests America's maritime hegemony in East Asia, then there will be an intense maritime arms race over who is, or should be, number one at sea.

Thus, an analysis of U.S. and Chinese interests in East Asia yields a mixed picture. There are a few serious differences, but there are also an important number of shared interests. The course of the Sino-American relationship is not yet determined, leaving room for compromise and wise diplomacy. Whether that relationship takes on a largely conflictual, mostly cooperative, or mixed character will depend on China's ability to manage its nationalism and especially its nationalistic urban public, on how much restraint it exercises in the region as its economic and military power continues to grow, on the U.S. continuing to pursue a patient engagement strategy backed by sufficient military power in East Asia, and on the US ability to avoid overreacting to China's forays. None of this will be easy, but neither is it impossible.

Implications for Indo-American Relations

Finally, quite briefly, what are the implications of the foregoing analysis for US-India relations?

As I see it, there are three major concerns that powerfully impact US policy towards India. One is the state of US-China relations. A second is the war in Afghanistan, which causes the United States to favour good Pakistani-Indian relations and a settlement of the Kashmir conflict so that Pakistan can redirect more of its military forces to its western border to clean out the Taliban and al Qaeda sanctuaries there. A third is the U.S. desire for access to the Indian market. The first and the last concerns work in ways that cause the United States to become more akin to a suitor of India—that is, to "woo" India, so to speak, in order to obtain Indian cooperation. The second concern works in ways that causes the United States to pressure India—that is, to demand things from India. None of these three dominate the other two; all are constantly in play, although the relative importance of each will vary over time.

The effect of US-China relations on US Indian relations should be clear. If Sino-American relations are highly conflictual because China has become militarily stronger and has become assertive, the United States will want Indian cooperation to balance, and help restrain, China. If Sino-American relations are mostly cooperative and if the United States can counter China's military power on its own, then the US will come, less hat in hand, to seek Indian cooperation against China. This situation is roughly akin to, although not completely equivalent to, the relations the United States had with its NATO allies during the Cold War. When US-Soviet relations experienced periods of détente, the United States pressured its allies less; when those relations were characterised by intense hostility, which was more often the case, the United States sought to tighten the alliance and put more pressure on its allies.

Roughly the same dynamic is likely to be at work over the course of the next decade or so in the case of India. But if I had to make a bet, it would be this: the United States and India will continue to draw closer because China's power will continue to grow, and, as a consequence, so, too, will its assertiveness. Historically, the assertiveness of emerging or emergent great powers, has grown as their power has grown. Because this is the nuclear age, however, US-China relations will not escalate to war. How cooperative or conflictual the resulting peace will be—no one can predict with certainty.

Notes

1. This paper draws in part from Robert J. Art, "The United States and the Rise of China: Implications for the Long Haul," *Political Science Quarterly*, Vol. 125, No. 3 (Fall 2010), pp. 359-391.
2. See Dale C. Copeland, *The Origins of Major War* (Ithaca, NY: Cornell University Press, 2000); and Robert Gilpin, *War and Change in International Politics* (Cambridge: Cambridge University Press, 1981).
3. Hegemonic wars and hegemonic political-military competitions are conducted to determine which power will be number one and be able to set the rules of the international system.
4. These three cases do not include all the great power cases of rising-versus-dominant power dyads of the last one hundred years. Most prominently and deliberately excluded is the US-British case of the late nineteenth—early twentieth centuries. I have excluded that case because the prime reason that the British came to accept US hegemony in the Western Hemisphere was that the British Empire was overextended and necessity mandated that it begin to reach understandings with local powers—in the western hemisphere, with the United States; in East Asia, with Japan; in Africa, with France; and in Central Asia, with Russia. Overextension of empire does not apply to the US-China case. Instead, I focus on two other types of rising-versus-dominant power competitions: those that resulted in a great power hegemonic war, or those that entailed a hostile, intense, and sustained political competition for hegemony and that involved heavy reliance on military force to fight proxy wars or to engage in arms races so as to achieve political

hegemony. The US-British case does not fall into either category, although the United States did threaten the British with a naval arms race after World War I if they did not renounce the Anglo-Japanese Treaty of 1902. By focusing on the three cases I have chosen, I have, in the lingo of methodology, selected on the dependent variable—on those cases in which the outcome is either war or a sustained political-military rivalry rather than peace or peaceful accommodation, with the result that the conclusions of this analysis are to a degree biased. Nonetheless, there is still analytical merit in focusing on these two types of hegemonic competitions to see what conclusions we can draw. For a slightly different list of important power transitions, see Jacek Kugler and A.F.K. Organski, *The War Ledger* (Chicago: University of Chicago Press, 1980), p. 49. For an analysis of why the US-British case ended in peace, not war, see Stephen R. Rock, *When Peace Breaks Out: Great Power Historical Rapprochement in Historical Perspective* (Chapel Hill: The University of North Carolina Press, 1989), chap. 2. For a more general treatment of the strategies that dominant powers employ to cope with rising powers, see Randell L. Schweller, "Managing the Rise of Great Powers: History and Theory," in Robert S. Ross and Alastair Iain Johnston, eds., *Engaging China: The Management of an Emerging Power* (London: Routledge, 1999), pp. 1-32.

5. See Robert J. Art, *A Grand Strategy for America* (Ithaca: Cornell University Press, 2003), pp. 66-67.
6. In 1913, 10 per cent of Britain's total trade (imports and exports) was with Germany, and 12 per cent of Germany's total trade was with Britain. These percentages are derived from the Correlates of War data. The data was prepared by Katherine Barbieri for her doctoral dissertation "Economic Interdependence and Militarized Interstate Conflict, 1870-1985 (Binghamton University, Binghamton, NY, 1996). The dataset is at http://cow2.la.psu.edu/. I am indebted to Loren Cass for arranging the data for easy use.
7. In 1938, 4 per cent of Britain's total trade (imports and exports) was with Germany, and 6 per cent of Germany's total trade was with Britain. See note 17 for data source.
8. In 1977, for example, during the height of détente between the United States and the Soviet Union, when one would expect trade to be the highest, the United States sent 1.4 per cent of its total exports to the Soviet Union and received. 3 per cent of its total imports from the Soviet Union. By 1983, when détente had ended and U.S.-Soviet relations were hostile, the United States sent. 1 per cent of its exports to, and received.1% of its imports from, the Soviet Union. In 1977, the Soviets exported 2.5 per cent of its exports to the United States and took 8 per cent of its imports from the United States. By 1983, these Soviet figures had fallen to 1 per cent and 5.7 per cent, respectively. See International

Monetary Fund, *Direction of Trade Statistics Yearbook, 1984* (Washington, DC: International Monetary Fund, 1984), pp. 378 and 385.

9. This is close to the argument that Dale Copeland makes about the causes of hegemonic great power wars. See Dale C. Copeland, *The Origins of Major War*, chaps. 1 and 2.
10. The stability-instability paradox says either that two nuclear-armed and hostile states will start a conventional war with each other because they feel confident that it will not escalate to all-out nuclear war, or that they will not start such a war because they fear that it might escalate to all-out nuclear war. During the Cold War, the paradox produced the second, not the first, effect. Glenn Snyder was the first to formulate the paradox. See Glenn H. Snyder, "The Balance of Power and the Balance of Terror," in Paul Seabury, ed., *The Balance of Power* (San Francisco: Chandler, 1965), pp. 198-199. Also see Robert Jervis, *The Illogic of American Nuclear Strategy* (Ithaca, NY: Cornell University Press, 1984), pp. 31-33 and 148-157.
11. The exception came in November 1983 with the NATO exercise "Able Archer." This was a serious crisis from the Soviets' point of view, and they briefly believed a nuclear from the West was imminent. At the time, however, neither the United States nor NATO was aware of Soviet concerns, making this crisis hard to classify. See John Lewis Gaddis, *The Cold War: A New History* (New York: Penguin Press, 2005), pp. 227-228.
12. International Monetary Fund (IMF), *Direction of Trade Statistics Yearbook, 2009* (Washington, DC: International Monetary Fund, 2009), pp. 141, 145, 148, 541, and 542. The figure for China's total exports to the world is based on Chinese export data. The figure for China's exports to the United States, however, is based on U.S. import data, not China's export data. These two figures are not the same because, as the International Monetary Fund makes clear, there are discrepancies in bilateral trade statistics as reported by China and its industrial trading partners. I used U.S. import figures from China to measure China's exports to the United States because I deem the U.S. figures more accurate. Finally, in 2008, the United States exported to China, Hong Kong, and Macao $93 billion, or 7 per cent of its total exports. See *Direction of Trade Statistics Yearbook, 2009*, pp. 541-542.
13. China's GDP for 2008 is taken from the World Bank. See http://data.worldbank.org/indicator/NY.GDP.MKTP.CD. The GDP figure is rounded.
14. A useful benchmark for assessing China's dependence on the U.S. market is the percentage of U.S. exports to Canada, which is America's largest trading partner. In 2008, U.S. exports to Canada amounted to

$261 billion, or 20 per cent of total U.S. exports (1.3 trillion) for that year (IMF, *Direction of Trade Statistics Yearbook*, 2008, p. 541).

15. Richard N. Cooper, "Is 'Economic Power' a Useful and Operational Tool?" p. 25, at http://economics.harvard.edu/faculty/cooper/papers/Economic%2Power.pdf.
16. As of March 2010, total foreign holdings of U.S. Treasury securities were $3,884.6 trillion, of which Japan and China held $1.680 trillion. See the U.S. Treasury's International Capital Flow data at http://www.treasury.gov/tic/mfh.txt. Total U.S. treasury securities as of December 2009 amounted to $12.3 trillion. Of that total, $4.5 trillion was held by the Federal Reserve and U.S. Government accounts and the rest privately held. See United States Treasury, Table FD-1 (Summary of Federal Debt), *Treasury Bulletin*, March 2010, p. 22 at http://www.fms.treas.gov/bulletin/index.html.
17. Analysts differ on why Japan continues with its non-nuclear status. Some believe it is due to the alliance with the United States. Others believe that Japan's "nuclear allergy" is strong enough to keep Japan non-nuclear should the alliance collapse. I adhere to the first view. So, also, does Richard Samuels in his excellent book on Japanese security. See Richard J. Samuels, *Securing Japan* (Ithaca, NY: Cornell University Press, 2007), chap. 6.
18. In 2009, total U.S. exports and imports of merchandise goods and services from the Pacific Rim countries (Australia, China, Japan, Malaysia, the Philippines, Hong Kong, South Korea, Singapore, Taiwan, Indonesia, and other) equalled $787,946 billion, while the comparable figure from North America (Mexico and Canada) was $735,146 billion. North America remains the largest market for U.S. exports. Figures come from U.S. Census Bureau, U.S. Bureau of Economic Analysis, "FT900: U.S. International Trade in Goods and Services, December 2009," Exhibits 14, release date February 10, 2010, at http://www.census.gov:80/foreign-trade/Press-Release/2009pr/12/.
19. Percentage figures calculated from *ibid.*

6

Challenges of Rising China: A New Cold War or Neo-Imperialism?*

Masako Ikegami

As China's military muscle—particularly its naval power—grows to the extent of challenging the existing US dominance of the Asia Pacific, the region is getting increasingly marked by disputes and political instability. The more prominent of these are the recent maritime disputes in the South, and East China Sea and the Yellow Sea. China's firm support to North Korea's attacks on Cheonan and Yeonpyeong in South Korea might trigger off hostilities or be a prelude for a new Cold War. Also, by abusing its rapidly growing economic and military power, Hu Jintao's China is adopting a highly aggressive stance towards its neighbouring countries such as North Korea and Myanmar who are rich in natural resources yet politically vulnerable. Japan's Manchukuo policy of the 1930s is interpreted as (1) large investments in economic infrastructure for extracting natural resources, (2) military interventions for protecting economic interests, and (3) social-political absorption by means of a puppet government. China's current strategy toward its neighbouring countries can be well explained by this historical model. It has similar effects in terms of the incremental and discreet expansion of its strategic objectives initially disguised as investment for industrial infrastructure or "economic cooperation". This indicates that China is a new imperial power, notwithstanding its rhetoric of "peaceful rise" or "peaceful development".

---

* A part of this paper is based on an article originally published in Japanese, 'Chugoku ga tenkai suru Giji-Manshukoku Seisaku [China's Quasi-Manchukuo Strategy toward North Korea, Mongolia and Myanmar]', in *Chuo-Koron* (April 2010), pp. 162-173, Tokyo: Chuo-Koron-sha.

Introduction: The Implications of the Rise of China?

The rise of China poses many inquiries: Will a powerful China behave responsibly in world affairs, complying with the rules and norms of the 'international community'? Or will it defy global standards, and pursue its own interests thereby challenging the US dominated world order; especially in the matter of international trade, arms control and non-proliferation, environmental and climate issues, and the last but not least, human rights?

Records show that China is not a responsible actor when it comes to global key issues such as Iranian-and North Korean nuclear crises, RMB exchange rate manipulation in international trade, global warming and emission reduction, media and information censorship, and the brutal crackdown of dissidents and some ethnic minorities in Xinjiang and Tibet (under the garb of "anti-terrorism"), food security and food safety, the exhaustion of natural resources, and mass-manufacture of poor quality and harmful products. For instance, at the 2009 United Nations Climate Change Conference (COP15), China mobilised a group of developing countries including Sudan, Iran, Venezuela, Zimbabwe, to sabotage the attempts to establish legally binding targets for carbon emissions as proposed by the West.[1]

China is eager to dispel international apprehensions regarding the rise China by trumpeting the rhetoric of "peaceful rise" or "peaceful development". This rhetoric shows that China has learned from the experience of Germany and Japan who directly challenged the Anglo-American hegemony and triggered World War II that led to the ruin of the two countries. China will thus progressively acquire hegemony without committing the mistake of prematurely challenging US hegemony. Recently many Chinese officials and academics have been chanting the virtues of a "multi-polar system" or "new security" to emphasise the importance of economic and diplomatic cooperation rather than the 'Cold-War mentality' that revolved around alliances and power politics. However, the latest maritime disputes involving China and its neighbouring countries defy the premise of China's "peaceful rise". China's aggressive demonstration of its rapidly expanding maritime forces has caused a series of frictions and aggravated disputes in the Yellow Sea, the South and the East China Sea in 2010. The very reason that China touts "multi-polarity" is that only in a multi-polar world or international anarchy, China can gain a relative advantage over US hegemony and maximise its national interests by creating power vacuum. Its increasing self-confidence arising from its economic and military power, has prompted China to openly start challenging US hegemony. Terms such as the "US-China co-management, G2" or "China as a stakeholder" that have become prevalent in the past few years, now sound like baseless rhetoric coined by pro-China lobbyists to pave the way for raising China to 'super power' status.

Retrospectively, the US-China co-management framework, a strategic concept coined by pro-Beijing scholars, has been adopted by Beijing's diplomats to strip the US of its influence in critical regional security issues—e.g. the Korean Peninsula and the cross-Strait issue. Beijing's intention to collaborate with the United States on the North Korean nuclear issue was arbitrated by Beijing's desire to gain Washington's support on the Taiwan issue. This formula helped China avoid a direct confrontation with the United States—as China is relatively weaker—and at the same time "muddle through" issues like the North Korean nuclear crisis to protect China's strategic interests. Indeed, this pragmatic tactic functioned well—especially while the United States was preoccupied with the war on terrorism in Iraq and Afghanistan, China was able to convince the United States to "delegate" the management of the Six Party Talks almost entirely to China since 2003. The consequence was rather counter-productive: China managed to muddle through the North Korean nuclear issue without taking any responsibility to pressure or contain the North Korean regime, let alone take any initiatives or substantial actions toward tangible denuclearisation.[2] Beijing's inaction, or subtle support, allowed Pyongyang to conduct two nuclear tests in October 2006 and May 2009 to establish itself as a *de facto* nuclear weapon state, at least in Pyongyang's eyes. In spite of the provocative tests, China continues to openly support the North Korean regime, while the Six Party Talks are again stagnant since September 2007.

In 2010, China stepped up its offensive stance by claiming territories and control in the Yellow Sea, the East-and South China Sea, which caused disputes with its neighbouring countries. When a North Korean torpedo allegedly sank the South Korean warship the *Cheonan* with the loss of 46 lives, China disappointed South Korea and the United States by firmly defending North Korea against international condemnation and sanctions. As a response to the joint US-South Korean drills in the wake of the *Cheonan* incident, the Chinese navy also held artillery exercises in the Yellow Sea.[3] In March, China described the South China Sea as its "core interest" and warned the United States that it would not tolerate interference in the region.[4] Also at the ASEAN Regional Forum in Vietnam in July, China described the US secretary of state Clinton's call for conflict prevention in the South China Sea, as "in effect an attack on China" and reiterated that it is ready to protect its core interest with military means.[5]

Being prompted by the Japanese coast guard's detention of a Chinese captain (released later) who rammed Japanese coast guard vessels at the Senkaku Islands/Diaoyutai, China escalated the dispute by stopping the export of rare earth (China has a 95 per cent share of global rare earth metals production) first to Japan, and later even to the United States and Europe.[6]

China's rapid maritime expansion is commonly explained by China's aspiration for great power status, or as an extension of the traditional Chinese empire and Sino-centrism; Sir Halford Mackinder's phrase reads, "[China's power expansion well beyond their borders] might constitute the yellow peril to the world's freedom".[7] Although the peculiar characteristics of the traditional Chinese empire cannot be underestimated, the China's current and rapid expansion needs a more in-depth analysis of its operational mechanism and consequences. This paper will analyse China's ever-stronger assertiveness that gradually betrays its imperialistic nature, and conclude that China in all respects is a new hybrid imperial power of the 21st century. China's imperialistic characters derive from its nature as a Leninist state as well as its peculiar geo-political and strategic concepts that are historically imprinted on it.

China as a Leninist State: Excessive Centralisation of Power and Resources

Over the past decades, China has been pursuing market-oriented reforms gradually under one-Party rule. The concentration of power has allowed the Chinese Communist Party (CCP) elites to adopt unpopular measures and be insulated against pressures for redistribution, while its authoritarian rule provides the elites with organisational means for strengthening power. There has been a long-lasting debate as to how China has been able to achieve economic success without experiencing major political changes, and whether China should speed up its political reform. So far, there have been no clear signs of a political opening up in China, despite its sustained high growth. Nowadays, China's institutional quality is inferior to that of most former socialist countries and the largest developing countries, as it is measured by income equality, private sector development, economic freedom and the rule of law.

In reality, China's reform is unfinished, or deliberately being prolonged, since over time, the elites are gradually losing interest in undertaking egalitarian reform initiatives, except pledging to make some seemingly pro-grass roots policy adjustments to maintain their legitimacy. The post-Deng leaderships are keen on attaining the Leninist 'commanding heights' to strengthen their power. Their foremost concern is how to control the expanding economy and, consequently, the 'state' directly holds more than 65 per cent of the country's fixed assets and dominates the key industries. As a result, the whole economy is divided into a highly market-oriented part, and a state-controlled yet crony capitalistic part. The latter is vital to the ruling elites. Because of their political supremacy, the elites can use their privileges to dispense a large portion of national wealth for securing political support and implementing a development strategy. Due to the lack of restraint on the elites' distribution of power over the past decades, income inequality has turned out to be a major issue in the Chinese society.

In the post-reform period, in terms of income distribution, China deteriorated from one of the most equal cases into one of the most unequal cases in the world. According to the UNU WIDER World Income Inequality Database (2006), China's Gini coefficient dropped from 0.32 in 1980 to 0.26 in 1984, and then rose quickly—in 1990, the figure was 0.36; and in 2003, the figure was 0.45. As per World Bank figures (2004), China's Gini coefficient ranked as the 85th highest (from low to high) among the 120 countries surveyed. As for the 35 more unequal countries, almost all of them (32 out of the total 35) are Latin American and Sub-Saharan African developing countries afflicted by highly split societies, civil war, and economic stagnation. Apparently, in China, as per capita income rises, income inequality also rises at a fast pace. Over time, the worsening income distribution has been causing multiple social problems. The overall inequality can be seen as rural-urban inequality and regional inequality. In China's case, rural-urban inequality rose steadily throughout the 1990s. The urban-rural income per capita ratio (urban disposable income to rural net income) increased from 1.86 in 1985 to 3.11 in 2002 (World Bank, 2004). If state subsidies for housing, pension, medical care, and education in the urban areas are considered, the ratio will increase to about four, making China's rural-urban inequality one of the highest in the world. The disparity between inland China and coastal China has also increased, especially since the latter half of the 1990s. Compared with its inland counterpart, coastal China is more industrialised and is endowed with a geographical advantage for international trade and FDI inflows. Over the past decade, fiscal transfers had been arranged by the Chinese central government in the name of 'developing the west' and 'revitalising the northeast', but the regional gaps have not being narrowed. More far-reaching transfer programmes are not in sight, because of the central authorities' stance against excessive redistribution.

On the other hand, despite its sustained economic reform, so far China has not seen any signs of political reform. The Chinese Communist Party (CCP) elites keep a firm grip on the reform process by controlling the commanding heights. Over time, the reformers themselves have become conservative and evolved into special interest groups resistant to political change, as they are the largest stakeholders in an expanding economy. As a result, in China, economic freedom is enhanced, solely by removing the constraints of a planning economy (e.g. devolving management powers to local administrations and enterprises), while far-reaching privatisation and entrepreneurship are not taking place.

China's economic reform started with the de-collectivising of agriculture and the privatising of small state-owned enterprises, yet gradually the reform lost its momentum. Despite price liberalisation and market formation, state-monopoly of the Leninist state type and government manipulation of the individual industries (which resulted in non-contestable markets) appear to be major problems. This results in: low efficiency of the state owned enterprises despite their cheap inputs, concentration of resources in a few administrative monopolies, and policy discriminations against the private sector. Due the special interest groups' resistance against further liberalisation, the reform process has been deliberately prolonged. In recent years, major structural problems, such as, soaring land prices, stock market bubbles, and energy shortages have arisen.

Internationally, Chinese ruling elites have become more resistant to foreign pressures than before. They keep a tight control over Chinese currency (and the interest rates), which causes external tensions and internal distortions. Through currency manipulation, the elites could achieve two aims: to reduce incentives for speculative capital inflows, and to maintain a spread with the US rates (in order to gain on the Chinese foreign exchange reserves invested in US treasury bills). Domestically, low official interest rates make investment highly profitable for the monopolies, while there is little windfall going to the working class. Between 1982 and 2006, China created fewer jobs (as a percentage of the workforce) than Brazil, though the latter's annual average growth rate is only 3-4 per cent over the period. At this point, China's spectacular growth is camouflaging the fact that its working class is getting a meagre return on the resources they invested (McGregor, 2007).

Over the past decade, China's foreign exchange reserves surged to unprecedented levels (now estimated at 12 per cent of China's annual GDP), and many people view this as a testimony of China's strength. However, they fail to discover the actual use of this national wealth and its political implication for the ruling elites. In fact, China does not need such huge reserves, if the country's economy is being managed in a liberalised way. An excessive accumulation of reserves actually reflects a country's inadequacy in financial intermediation (due to the restrictions against lending to the private sector) and the lack of economic freedom. A paradox arises—China holds the world's largest foreign exchange reserves, yet many Chinese enterprises still rely on foreign capital.

China's reform logic has been sanctified as a major policy lesson for other developing countries. Nonetheless, as noted above, the presumed magical power of the 'China Model' in fact is found in the Leninist state control, which creates inequalities, distortions and contradictions. Under such a system, the ruling elites are not willing to redistribute for social equality's sake; what they instead do is mobilise resources for implanting the development strategy. In recent years, China's rise under authoritarian rule, a number of developing countries converted to the 'Beijing Consensus' (Ramo 2004), dreaming that they could achieve prosperity while keeping their regimes intact. However, the essence of the 'Beijing Consensus' is not the antithesis of the American-led neoliberal economy, but rather it is an ideological prescription: 1) only to focus on the economic scale and superficial economic growth regardless of fair distribution of wealth, 2) to maintain the authoritarian regime, and 3) to prevent real political reform and democratisation. Eventually, the fluent inflow of foreign direct investment only enriches and strengthens the CCP regime, far from liberalising the economy, let alone political liberalisation. China's economy is essentially following red capitalism (Leninist) which actually was the basis for Deng's economic reform.

It is noteworthy that China's economy predominantly controlled by the CCP is, in many respects, similar to Japan's pre-war economy that was predominated by a few *Zaibatsu* 財閥 families. *Zaibatsu* were the great family-controlled banking and industrial conglomerates involving a family, a holding company, and subsidiary companies in Imperial Japan, which led and controlled significant parts of the Japanese economy with strong political influence from the Meiji period (1868-1912) until the end of World War II.[8] By the end of World War II, the Big Four *Zaibatsu* (Mitsubishi, Mitsui, Sumitomo, Yasuda) alone had direct control of more than one third of all bank deposits, one third of foreign trade, one third of Japan's mining, chemical, metals industries and almost half of the machinery and equipment market, a significant part of the foreign *commercial merchant fleet* and 60 per cent of the commercial *stock exchange*.[9] With the economic and industrial predominance, *Zaibatsu* wielded great influence over Japanese national and foreign policies. Therefore the Supreme Commander for the Allied Powers that occupied Japan after the war (1945-1952) regarded the *Zaibatsu* as the driving force behind Imperial Japan's expansionism and dissolved the *Zaibatsu* as a major reform for democratising Japan.

Arguably China's quasi-market economy dominated by the CCP elites causing severe rural-urban inequality is comparable to Japanese pre-war economy dominated by *Zaibatsu* and power elites. In Japan in the 1930s, the severe rural-urban inequality and widening income gap led to ideological radicalisation (e.g. ultra-nationalism), which eventually triggered the 2.26 military coup d'état in 1936 that eventually ended Japan's fragile democracy in embryo, paving the way for militarism. Japan in the 1930s was characterised by rapid military build-up and economic growth (especially heavy industry),10 the impoverished rural area, widening income gap, and rural-urban inequality, which harboured ideological extremism and ultra-nationalism. This, in many respects, parallels China's current situation. It is not a coincidence that Japan in the 1930s introduced the system of a planned economy in Manchuria led by the Japanese technocrats—prominently Nobusuke Kishi, the then technocrat of the ministry of commerce and industry, and prime minister after the war—who were influenced by the Soviet's concept of planned economy, similar to 'Red Capitalism' or China's socialist market economy. Rapid industrial development in Manchuria in the 1930s was the result of effective collaboration between the Kwantung army, the government's technocrats and new Zaibatsu such as Nissan.11 The similarity is alarming, given that dynamism of Imperial Japan in Manchuria was a prelude to the Manchurian Incident in 1931, which triggered a total war between Japan and China, eventually leading to a clash of interests with the allied powers and triggered World War II.

China's Strategy toward Strategically Vital North Korea?

North Korea is virtually a failed state that survives by being entirely reliant on supply and support from China. At the United Nations Security Council following North Korean missile test-firing in April 2009 and second nuclear test in May 2009, China firmly opposed economic sanctions against North Korea. In the *Cheonan* incident in 2010, China again firmly resisted any tough measures against North Korea at the UN Security Council. In spite of the UN Security Council economic sanctions against the export of heavy arms and luxury goods to North Korea after its first nuclear test in 2006, China's export of luxury goods to North Korea increased significantly.[12] Even after the UN Security Council Resolution 1874 on sanctions to be imposed following North Korea's second nuclear test in May 2009, China's ports (e.g. Dalian, close to North Korea's port of Nampo, one of the critical trans-shipment points in China for North Korea's cargo) and airspace continuously provided channels for North Korea's arms exports and sensitive WMD-related technology/components to states such as Iran.[13] The Six-party talks framework was a trick of China to muddle through the North Korean nuclear crisis without jeopardising the North Korean regime such as economic and financial sanctions.

These records reiterate the fact that China and North Korea's 'blood-bound alliance' firmly remains real and substantial since the Korean War. The latest studies on the Korean War based on newly declassified documents from the United States (North Korean top-secret documents confiscated by the US forces during the Korean War), Russia (diplomatic correspondence between Stalin and Mao Zedong) and China, show that Mao Zedong had strategically decided to support North Korea's invasion of South Korea well before U.S. forces reached the Yalu River neighbourhood, and well before June 1950.[14] Upon the request of Kim Il-Sung, who badly needed a larger army for invading the South, Mao in the summer of 1949 generously granted Kim Il-Sung approximately 30,000 ethnic Korean PLA troops fully equipped with modern arms left by the Soviet Red army, (the PLA 166th division) who had played an important role in the decisive battles against Chiang Kai-shek's nationalist government troops in Manchuria in the early stages of the Chinese civil war. As the troops were transferred to Kim Il-Sung's command, they reappeared as the 6th division of the Korean People's army to play another decisive role in the Korean War by first attacking Kaesong, Seoul, and Taejon t the further south.[15] During the Korean War, Mao sacrificed as many as 400,000 soldiers and, most importantly, he missed the opportunity to seize Taiwan by force, but the Korean War eventually established China as a major military power in North East Asia. Having paid such a high cost for defending North Korea in the last century, China indeed cherishes the vital strategic interests of North Korea.

Thus, why would China pressurise North Korea on the nuclear issue? The North Korean refugee crisis and its already significant economic stake in North Korea are also factors for China to sustain the hermit kingdom. Kim Jong-il is well aware of North Korea's strategic value for China, and may have made a tempting proposal to Beijing: When China attacks Taiwan, North Korea could destabilise the Korean Peninsula in order to divide US forces between the Taiwan Strait and the Korean Peninsula.[16] Thus, there is a common interest between Beijing and Pyongyang to sustain the North Korean regime.

Taking advantage of North Korea's isolation, China is steadily enclosing North Korea by commanding North Korea's lifeline: China is the chief food supplier for North Korea and accounts for nearly 90 per cent of its energy imports, and 80 per cent of North Korea' consumer goods. China's trade with North Korea has continued to increase; Bilateral trade between China and North Korea reached $2.79 billion in 2008, a 41.3 per cent increase from 2007; Chinese exports of banned luxury goods averaged around $11 million per month in 2009.[17] China accounts for over 70 per cent of North Korea's trade, nearly 90 per cent of foreign direct investment in North Korea, nearly 70 per cent is for exploiting mineral resources in North Korea such as coal, iron, gold, copper, zinc, lead.[18] For instance in 2007, China bought for about $1 billion exclusive 50-year mining rights to the Musan iron ore mine, the largest open-air iron mine in Asia, which was originally developed by Mitsubishi in the 1930s.[19] Also, China has acquired the right to use North Korea's Rajin Port for 10 years possibly with another 10 years lease; The Rajing Port, originally developed by Japan in 1905 in tandem with its development of Manchuria, is a strategically important gateway to the Sea of Japan, not only for trade but also militarily.[20] It is alleged that Kim Il-song in the 1949 secret agreement leased Rajing port to the Soviet Union to be used as a submarine base.[21] Through robust infrastructure investment, China is now integrating North Korean natural resources into its north eastern industrial area, the centre of China's heavy industry. This area overlaps with the legacy of Manchuria where imperial Japan made massive investment for infrastructure and developed heavy industry and munitions manufacturing in the 1930s, without which Imperial Japan could have never contemplated a war with the Allied forces.

If the North Korean regime collapses, China will not only suffer from a massive refugee inflow, but will also face a huge loss of economic interests. North Korea is a critical interest for China in terms of its economy, military and strategy. Therefore, in spite of North Korea's second nuclear test, Chinese Premier Wen Jiabao visited Pyongyang in October 2009 for commemorating the sixtieth anniversary of diplomatic relations between China and North Korea. Reportedly, Wen, accompanied by a large delegation of high ranking officials, signed several documents including an agreement on economic and technological cooperation, and an economic cooperation package worth $50 million and the construction of a new bridge over the Yalu (Amnok) River between their two countries (estimated to cost over $150 million).[22] Further more in February 2010, Wang Jiarui, head of the international department of the Communist Party of China, visited Pyongyang to pledge a large-scale investment of about $10 billion—about 70 per cent of North Korea's gross domestic product (GDP) of some $15 billion—for infrastructure building, e.g. railroads, highways, bridges, harbours and houses, in North Korea to revive its faltering economy.[23] This mirrors pre-war Japan's heavy investment for infrastructure and industrialisation in Manchuria throughout the 1930s.

In the past, Japan fought wars with Qing China (1894-5) and Russia (1904-5) over Korea which China is now monopolising by taking advantage of North Korea's isolation. Geopolitically an important strategic point, a huge infrastructure investment and economic interests, aligned with *de facto* annexation by military intervention and the puppet regime: All these three elements together, in China's approach to North Korea resemble the 'pseudo-Manchukuo' strategy.

China's Strategy toward Strategically Vital Myanmar and Mongolia

Myanmar/Burma is another case where China applies the quasi-Manchukuo strategy for monopolising access to the natural- and strategic resources of Myanmar. The geo-political importance of Burma is evident from its war-time history when the allied forces constructed the Burma Road linking Burma (then a British colony) with China to supply logistic support to Chiang Kai-shek's Kuomintang troops in their fight against Japanese forces since 1937. The 'Chiang Kai-shek Route' became a trigger for the Pacific War with the Allies when Japan tried to cut off the lifeline to China, leading to the catastrophic Imphal operation (1944), the biggest Japanese defeat of the war on ground. Along such a strategically important route, China is currently building an international highway connecting Kunming in Yunnan province, Yangon and Mandalay in Myanmar, and Bangkok, Thailand, with funding from the Asian Development Bank. Given China's close relationship with the Myanmar military junta in recent years, it is obvious that the highway can be converted into a land route for transportation of troops and military supplies in case of a military dispute.

As China's maritime power expanded, the PLA navy extended the definition of its 'offshore defence' (1985) by claiming the 'two island chains': The first is usually described as a line through the Kurile Islands, Japan, the Ryukyu Islands, Taiwan, the Philippines, and Indonesia (Borneo to Natuna Besar); the second island chain runs along a north-south line from the Kuriles through Japan, the Bonins, the Marianas, the Carolines, and Indonesia; Together, they encompass a maritime area of approximately 1,800 nautical miles from China's coast, including most of the East China Sea and East Asian Sea Lines of Communications (SLOCs).[24] As per these, China would be able to exercise control over the Yellow Sea, the East China Sea and the South China Sea, all located within the 'first island chain' of the Pacific Ocean (*ibid.*). Some Chinese analysts even include Diego Garcia, the key US military base in the Indian Ocean, in the geo-strategic belt along the Chinese coasts (*ibid.*). Myanmar is a critical strategic point in this concept.

Currently nearly 80 per cent of China's oil imports pass through the Strait of Malacca from the Indian Ocean, and thus the "Malacca dilemma" is an issue of the strategic importance for China. The road construction and oil- and gas pipeline from China through Myanmar out to the Indian Ocean becomes a critical alternate route for China.[25] In June 2010, the China National Petroleum Company (CNPC) officially commenced construction of oil and gas pipelines between China and Myanmar; Each pipeline has an overall length of about 1,100 km, and a designed transport capacity of 22 million tonnes of oil per year, and 12 billion cubic meters of natural gas annually.[26]

In the past, Japan was the largest donor of official development assistance (ODA) to Myanmar due to its strong historical tie with Burmese military leaders back in World War II and at its independence. Japan's ODA accounted for nearly 70 per cent of the total bilateral ODA received by Myanmar between 1979 and 1990, and accounted for nearly 40 per cent of its imports.[27] However, when Japan suspended its economic assistance after the 1988 military coup, China quickly filled the vacuum to become the largest donor of economic aid for Myanmar. Besides, China since 1989 has been exporting armaments such as light- and heavy artillery, fighter jets, military helicopters, tanks, and surface-to-ship missiles, to Myanmar.[28] According to Stockholm International Peace Research Institute (SIPRI) data, Myanmar is among the largest recipients of Chinese arms, together with other states of concern such as Sudan, Nigeria, and Pakistan. Also reportedly, China has built a maritime reconnaissance and electronic intelligence station on the Great Coco Island in the Bay of Bengal, the most important Chinese electronic intelligence installation in Myanmar, as well as a base on the Small Coco Island in the Alexandra Channel; These two islands, which have been leased to China since 1994, are located at a crucial point in traffic routes between the Bay of Bengal and the Strait of Malacca.[29] This is Chinese navy's future strategy for extending its reach into the Indian Ocean.

Due to its strategic importance, it is in China's interest to keep the Myanmar junta isolated so as to maintain control over it. For China, Myanmar is critical not only as a strategically vital route, but also as a tool for strategic manipulation, against for instance India. A Burmese expert points to a post-election scenario of further ethnic armed struggle or disintegration of the Myanmar, i.e. non-Burman/Myanmar ethnic nationalities may join up with their respective ethnic groups living on the other side of international borders, e.g. the Kachin and Shan states to join with Yunnan province in China.[30]

Mongolia is another country that is strategically important for China. The Mongolian economy is rapidly becoming reliant on China. Mongolia is reliant on China for 70-80 per cent of its food and daily goods supply, as well as for development of natural resources. Mongolia's total trade with China soared 83 times in 2008 compared to 1990, reaching $2.8 billion, making China Mongolia's largest trading partner.[31] Also, China is the largest investor in Mongolia; Chinese investment in Mongolia in four years alone (2004-2008) jumped four times to $16 billion, with approximately 70 per cent for mining (*ibid.*). While Mongolian exports of raw materials to China are increasing, the manufacturing industry has not grown enough to compete with the low-cost manufacturing in China. For instance, the Mongolian cashmere industry could not compete with cheap Chinese goods, and thus could not become an export industry, a potentially a great loss for Mongolian economy.[32] It is indeed a typical case of classical imperialism that Mongolia is exploited for natural resources, while being reduced to become a market monopolised by Chinese products.

After the democratisation of Mongolia in the early 1990s, China employed a strategy to differentiate Inner Mongolia from Mongolia by enriching Inner Mongolia with rapid economic investment. The envy of their neighbour, is leading the Mongolians to increasingly favour the idea of a 'unified Mongolia' as a part of China.[33] Among Mongolian government officials, there is a growing concern over being absorbed by China. Indeed, if they so choose, Chinese forces are strong enough to occupy Mongolia in a few hours, as in the case of 'the Prague Spring' of 1968. Currently, the most contentious issues in Mongolia are 1) The Dalai Lama's visit to greet Mongolian Buddhists who share the same roots as Lama Buddhism, 2) defence cooperation with the United States, and 3) Academic and cultural exchanges with Taiwan. When the Dalai Lama visited Mongolia in 2003, China reportedly closed the border and the railway link with Mongolia, which triggered a rise in the price of food and daily goods in Mongolia in a few days. A remarkable contrast to its attitude to North Korea when it did not stop its nuclear tests or punish North Korea by closing its border and stopping food and energy supplies. In this respect, it is evident that China does not consider Mongolia to be a sovereign state.

Obviously, China's approach to North Korea, Myanmar, and Mongolia is the *de facto* encircling of strategically vital countries and areas by employing economic, political and military means in an effective and concerted manner. Such an action is only possible by a single party-state, an authoritarian regime led by the Chinese Communist Party (CCP),[34] whose power monopoly is guaranteed by the People's Liberation Army (PLA), the army by its nature serving the party and not the nation.[35]

China's Quasi-'Manchukuo Strategy' and 'Stealth Imperialism'

Now China is enclosing North Korea in order to monopolise its mineral resources to ensure the stable supply of raw materials for the heavy industry of Northeast China, formerly Manchuria. Its strategic and military-economic implications should not be overlooked. Similar military-economic risk factors can be found in China's large-scale investment and infrastructure building of railways, highways, and ports in strategically vital areas from Tibet throughout Myanmar and Pakistan.

For instance, the Qinghai-Tibet railway was built in 2006 not only to accelerate the mass settlement of Han Chinese into western ethnic minority regions, but also for greater military effectiveness. According to strategic analysts, the main purpose of the Qinghai-Tibet railway is strategic rather than economic, i.e. to give the PLA a greater ability to move heavy weaponry in response to military emergencies. For instance, the PLA is focussing on rapid assault capabilities to meet requirements of theatre wars, and has been moving heavy weaponry frequently. In the exercise dubbed "Peace Mission 2007," which it conducted with the Shanghai Cooperation Organisation (SCO) countries in August 2007, the PLA succeeded in transporting armoured vehicles over a distance of 10,000 km.[36] The railroad also opens up previously unrealised strategic, tactical and conventional possibilities for the PLA to direct military firepower toward South Asia and beyond; the PLA will have excellent hiding places for its new rail-mobile ICBM, the DF-31A, and each missile train will be able to carry up to thirty nuclear warheads capable of destroying any strategic targets in Japan and many in the western United States.[37]

The Chinese operations parallel Japan's strategy toward Manchuria in the 1930s in many respects. Retrospectively, Japan's Manchukuo policy was a very sophisticated and strategic route to 'stealth imperialism' for a relatively weak latecomer imperial state who was trying to expand its own interests incrementally and discreetly by avoiding direct confrontations or frictions with existing imperial powers such as Great Britain. China's current quasi-Manchukuo policy is also an effective measure for 'latecomer imperial' China to steadily consolidate its strategic foothold for expanding its vested interests by avoiding immediate collisions with other existing powers such as the United States and Japan.

Confronting with New Imperial China?

If the United States and the Soviet Union were the "new empires" during the Cold War era, China's current economic, political, and military expansion is literally in the manner of an orthodox imperial state, an empire that spreads authoritarian ideology by expanding its 'satellite states' worldwide. This is all aimed at sustaining PRC's own CCP single-party authoritarian regime that has otherwise lost its legitimacy and credibility among the Chinese people due to enormous corruption and malfunctioning of domestic social-economic institutions. Inevitably the CCP regime increasingly relies on coercive measures such as ever tightening control and censorship of information and media, propaganda, indoctrination of the people through education and controlled media, and repression against journalists, intellectuals, or any criticism among the people. Under the risk of regime malfunction (if not yet breakdown), China's behaviour seems to be deviating from the shared code of conduct of the international community. Beijing's desperate efforts to sabotage the ceremony for the 2010 Nobel Peace Prize Laureate Liu Xiaobo, and its abrupt suspension of the export of rare earth metals were a disgrace.

China's rapid maritime expansion and aggressive stance to claim its territoriality by threats or actual use of force is evidence that China does not abide by the Western originated Westphalian system, the basic international system of mutual respect for sovereignty attributed to territoriality. Under the Westphalian system, borders are basically stabilised by the international rule of law and mutual recognition of sovereignty, not by force or power. However, instead of adapting to the modern concept of 'sovereignty', China follows its traditional concept of 'strategic frontier'(戦略的辺疆), a non-Westphalian notion based on geopolitical landscape (版図) (Hiramatsu 2006). This concept is potentially aggressive and expansionistic, because strategic frontiers can be expanded corresponding to its national power and forces.[38] China's brutal annexation of Tibet and Xinjiang (literally 'New Territory'), and over-commitment to the Korean War to support North Korea by sending as many as one million soldiers, are concrete examples of China's adaptation of the 'strategic frontier' concept. This may explain Johnston's finding that during the Cold War China was "more dispute-prone than most of the other major powers except for the United States", and "China tended to resort to higher levels of violence in disputes especially over territory, than did other major powers and India", despite the fact that China was a far weaker military power compared to the United States in that period.[39]

On the one hand, China's political and economic institutions still retain characteristics of Leninism, which makes China virtually the successor of the Soviet Union, in spite of its careful introduction of market economy. At the same time, China is learning the use of soft power from the US such as extensive utilisation of mass media, and cultural and academic activities for sophisticated propaganda. In this regard, China is the new hybrid empire of the 21st century.

If the Cold War of the 20th century was interpreted as a confrontation between the two empires, the United States and the Soviet Union, there might well be another new Cold War in the 21st century, between the United States and China. Indeed, Hu Jintao's proactive diplomacy in Africa, Central Asia, and Latin America and elsewhere, assimilating block-formation, are countering the other coalition led by the United States. As China, out of its strategic calculation, incorporates rogue states, failed states, and other authoritarian regimes, it will inevitably lead to block-formation (for details, see Ikegami 2009). As the Soviet Empire needed the satellite states of the Eastern block to maintain the empire, possibly the PRC is also building a block of satellite states in order to maintain its empire. In other words, as the Soviet Empire collapsed when it lost its vital satellite states following the Velvet Revolution, the PRC regime might be also jeopardized if its vital satellite states such as North Korea and Myanmar junta collapse. This explains why China is trying to protect and sustain the North Korean and Myanmar regimes by all means, even at the cost of frictions with the international community.

Or worse, the high tension of the territorial disputes in the Asia Pacific might escalate into an armed conflict, if China takes aggressive stance by employing the quasi-Manchukuo strategy of 'stealth imperialism' to expand its strategic frontiers. Do we want to see another age of empires in this 21st century? China might well review carefully Japan's pre-war Manchukuo strategy that triggered the catastrophic war which ruined Imperial Japan in the end. This guides us to ponder and envision a new world order free of the coercive system of empire.

The 2011 Egyptian Revolution followed by a series of demonstrations against authoritarian regimes in the Middle East is evidence that democracy is not the prerogative of the West. Indeed, in my view, the very essence of democracy is to prevent abuse of power, the most dangerous aspect of human nature, by a system of checks-and-balances and transparency. This is a universal value, regardless of the West or East, North or South. Even in China, democracy is in the interest of the Chinese people who are suffering from the CCP power elites' abuse of power in the form of corruption, coercion, and repression. China still has the great power aspiration of the age of empires, when the world community is obviously rejecting imperialistic values of the previous centuries. Some day, if or when China accomplishes two critical reforms: liberalises its political system and transforms its military into a national army instead of serving the CCP, then China could finally become a truly responsible and respectable great state.

Notes

1. UK Climate Change secretary Ed Miliband accused China for the fiasco of the Tokyo, conference saying "China vetoed" ('Ed Miliband: China tried to hijack Copenhagen climate deal', *Guardian*, 20 December 2009); 'Eyewitness: How China sabotaged climate talks' ABC News, 23 December 2009 <http://www.abc.net.au/news/stories/2009/12/23/2779498.htm?section =justin> accessed on 15 October 2010.
2. For further details on this argument, see Masako Ikegami (2008) 'Time for Conflict Prevention Across the Taiwan Strait', *China Brief*, Vol. III, Issue 7, The Jamestown Foundation (28 March 2008).
3. 'China begins military exercises in Yellow Sea', BBC, 1 September 2010, available at <http://www.bbc.co.uk/news/world-asia-pacific-11149923>.
4. Andrew Jacobs, 'China Warns U.S. to Stay Out of Islands Dispute', *New York Times*, 26 July 2010.
5. 'Foreign Minister Yang Jiechi Refutes Fallacies On the South China Sea Issue', PRC Ministry of Foreign Affairs, 26 July 2010 <http://www.mfa.gov.cn/eng/zxxx/t719460.htm>.
6. 'Amid Tension, China Blocks Vital Exports to Japan', *New York Times*, 22 September 2010; 'China to cut rare Earths minerals export by 30 percent', Homeland Security Newswire, 20 October 2010, available at http://homelandsecuritynewswire.com/china-cut-rare-earths-minerals-export-30-percent; 'China Just Banned Export Of Rare Earth Metals To The US', *Business Insider*, 19 October 2010 available at <http://www.businessinsider.com/chinas-has-just-started-ban-of-rare-earth-metal-exports-to-the-us-2010-10>.
7. Robert D. Kaplan 'The Geography of Chinese Power: How far can Beijing reach on land and at sea?', *Foreign Affairs* (May/June 2010), pp. 22-41, p. 22.
8. Hideaki Miyajima (1994) 'The Transformation of Zaibatsu to Post-war Corporate Groups', *Journal of the Japanese and International Economies* 8, pp. 293-328 (1994).
9. Peirson M. Hall, 'The Japanese Zaibatsu', 28 March 1946, *Vital Speeches of the Day*, 4/15/46, Vol. 12 Issue 13, pp. 394-400; 'Zaibatsu' in the Columbia Electronic Encyclopedia.
10. Takahusa Nakamura (1986) *Showa Keizai-shi* [Economic History of the Showa Era], Tokyo: Iwanami.
11. Kobayashi, Hideo (2008) *Manshu no Rekishi [History of Manchuria]*, Tokyo: Kodan-sha.
12. Marcus Noland, "The (Non-) Impact of UN Sanctions on North Korea", Asia Policy, No. 7 (January 2009), 61-88; Noland, "North Korean Missile Test: Remedial Action", Asia Pacific Bulletin, No. 33 (6 April 2009). According to the U.S. Congressional Research Service, Chinese

exports of banned luxury goods averaged around $11 million per month in 2009 (Dick K. Nanto, *et al. China-North Korea Relations*, CRS Report for Congress R41043, 22 January 2010).

13. Shirley A. Kan (2010) *China and Proliferation of Weapons of Mass Destruction and Missiles: Policy Issues*, August 2010, CRS Report for Congress RL31555, Washington DC: Congressional Research Service.
14. Zhu, Jianrong (2004) *Mo Takuto no Chosen Sensou [Mao Zedong's Korean War]*, Tokyo: Iwanami.
15. Hagiwara, Ryo (1997; 2003) *Chosen-Senso: Kin Nissei to Macarther no Inbo [The Korean War: A Conspiracy of Kim Il Sung and MacArthur]*. Bungei-Shunju, Tokyo.
16. See, Tsang, Steve (ed.) (2006) *If China Attacks Taiwan*. London/New York: Routledge, p. 2. Also see <http://club.backchina.com/main/viewthread.php?tid=529596&extra=page%3D1>.
17. Jayshree Bajoria, 'The China-North Korea Relationship', Council on Foreign Relations, 7 October 2010 <http://www.cfr.org/publication/11097/chinanorth_korea_relationship.html>.
18. Ting-I Tsai, 'North Korea-China trade hotter than kimchi', *Asia Times*, 6 October 2007; 'China, N Korea trade boom despite rocket tensions', Agence France-Presse (AFP), 5 April 2009; 'China's economic influence over N Korea grows', *The Business Times*, 18 May 2009; Nanto, Manyin and Dumbaugh (2010) 'China-North Korea Relations', CRS Report for Congress, 22 January 2010.
19. *US Geological Survey Minerals Yearbook* 2009, 14.2, US Department of the Interior. Reportedly Pyongyang terminated the contract in 2009.
20. Kim Young-yun, '[Viewpoint] North Korean port is slipping away', *Korea Joongang Daily*, 19 March 2010; Jung Sung-ki, 'China to Use N. Korea's Rajin Port for 10 Years', *Korea Times,* 3 August 2010; Kim So-hyun, 'China to invest in North Korea', *The Korea Herald*, 29 March 2010.
21. Abe, Keiji and M. Kimura (2003) *Kita-Chosen no Gunji-ka [Militarization of North Korea]*, Tokyo: Chisen-shokan, p. 206.
22. Jayshree Bajoria, 'The China-North Korea Relationship', *ibid.*, p. 11.
23. Jung Sung-ki, 'China Plans $10 Billion Investment in North Korea', *The Korea Times*, 15 February 2010.
24. 'People's Liberation Navy-Offshore Defense', *Global Security* <http://www.globalsecurity.org/military/world/china/plan-doctrine-offshore.htm >.
25. You Ji, 'Dealing with the Malacca Strait Dilemma: China's efforts to enhance energy transportation security', EAI background Brief No. 329, 12 April 2007, < http://www.eai.nus.edu.sg/BB329.pdf>.
26. 'China starts building oil, gas pipelines to Myanmar', *China Daily,* 4 June 2010.
27. Toshihiro Kudo "China and Japan's Economic Relations with Myanmar:

Strengthened vs. Estranged", Mitsuhiro Kagami (ed.) *A China-Japan Comparison of Economic Relationships with the Mekong River Basin Countries*, Institute of Developing Economies, Japan External Trade Organization, BRC Research Report, No. 1 (2009).

28. Daniel Byman and Roger Cliff (1999) *China's Arms Sales: Motivations and Implications*, RAND.
29. 'Coco Islands' available at <http://www.globalsecurity.org/intell/world/china/coco.htm>.
30. Lian H. Sakhong (2010) *In Defence of Identity*, Bangkok: Orchid Press.
31. 'Current Situation of Mongolia-China Trade and Economic Cooperation and Further Trends', J. Shishmishig, Representative of the Ministry of Foreign Affairs and Trade of Mongolia, Joint Dissemination Workshop on the Study of Zamyn-Uud, Erlian Border Development Programme, 14 April 2009, Hohhot, IMAR, PRC.
32. Migeddorj Batchimeg, "Future Challenges for the PRC and Mongolia: A Mongolian perspective", *China Brief*, Volume: 5 Issue: 10 (2005).
33. Remarks of Mongolian officials at a conference in Ulan Bator, April 2009.
34. For a good guidance on this issue, see for example, Ming Xia 'The Communist Party of China and the "Party-State"', *New York Times College* available at <http://www.nytimes.com/ref/college/coll-china-politics-002.html>.
35. This nature and role of the PLA is evident in the historical 1989 Tiananmen Square massacre. It was Deng Xiaoping, then Chairman of the Central Military Commission of the CCP, being backed by the military and CCP conservatives, who brutally repressed the pro-democracy demonstrators and purged General Secretary Zhao Ziyang, then the top of the CCP and true founder of economic opening and reform. For details see, Zhao Ziyang (Author); Bao Pu, Renee Chiang, Adi Ignatius (eds) (2009) *Prisoner of the State: The Secret Journal of Premier Zhao Ziyang*, New York: Simon & Schuster. This is a stark contrast to the Egyptian military that played a decisive role in the 2011 Egyptian Revolution when the military eventually sided with the civilian demonstrators against Mubarak's authoritarian regime. This means the Egyptian military transformed to a modern national army, presumably through its long-term collaboration with the United States, while the Chinese PLA remains as a party army, and not *national* army.
36. *East Asian Strategic Review 2009*, Tokyo: The National Institute for Defense Studies, p. 140.
37. William C. Triplett, II, 'The Dragon in the Indian Ocean', *China Brief*, vol. 3, issue 4 (February 25, 2003).
38. Hiramatsu, Shigeo (2006) *Chugoku wa Nihon wo Heigo suru [China's Annexation of Japan]*, Tokyo: Kodan-sha International.
39. Alastair Iain Johnston (1998) 'China's Militarized Interstate Dispute

Behaviour 1949-1992: A First Cut at the Data', *The China Quarterly*, No. 153 (Mar., 1998), pp. 1-30.

References

Abe, Keiji and M. Kimura (2003) *Kita-Chosen no Gunji-Kogyo-ka [Military Industrialization of North Korea]*, Tokyo: Chisen-shokan.

Alastair Iain Johnston (1998) 'China's Militarized Interstate Dispute Behaviour 1949-1992: A First Cut at the Data', *The China Quarterly*, No. 153 (March 1998), pp. 1-30.

Choucri, Nazli, Robert North, and Susumu Yamakage (1992) *The Chellenge of Japan before World War II and After: A Study of National Growth and Expansion*, London/New York: Routledge.

Daniel Byman and Roger Cliff (1999) *China's Arms Sales: Motivations and Implications*, RAND.

East Asian Strategic Review 2009, Tokyo: The National Institute for Defense Studies.

Hagiwara, Ryo (1997; 2003) *Chosen-Senso: Kin Nissei to Macarther no Inbo [The Korean War: A Conspiracy of Kim Il Sung and MacArthur].* Bungei-Shunju, Tokyo.

Hiramatsu, Shigeo (2006) *Chugoku wa Nihon wo Heigo suru [China's Annexation of Japan],* Tokyo: Kodan-sha International.

Ikegami, Masako (2010) 'Chugoku ga tenkai suru Giji-Manshukoku Seisaku [China's Quasi-Manchukuo Strategy toward North Korea, Mongolia and Myanmar]', *Chuo-Koron* (April 2010), Tokyo: Chuo-Koron-sha, pp. 162-173.

Ikegami, Masako (2009) 'China's Grand Strategy of 'Peaceful Rise': A Prelude to a New Cold War?', Hsin-huang Michael Hsiao and Cheng-yi Lin (eds) *The Rise of Chin: Beijing's Strategies and Implications for the Asia-Pacific*, London: Routledge, pp. 21-54.

Ikegami, Masako (2008) 'Time for Conflict Prevention across the Taiwan Strait', *China Brief,* Volume 8, Issue 7 (March 28, 2008), Washington, D.C.: Jamestown Foundation.

Ikegami, Masako and Chin, Hai-Iau (2007) 'Democratization and economic growth: the relevance of Taiwan's experience', *Stockholm Journal of East Asian Studies*, vol. 17, pp. 9-29.

International Campaign for Tibet (ICT) (2006) *Crossing the Line: China's Railway to Lhasa*, Tibet, Washington, D.C./Amsterdam/Berlin.

Jung Chang and Jon Halliday (2005=2006) *Mao: The Unknown Story*, London: Vintage Books.

Kaplan, Robert D. 'The Geography of Chinese Power: How far can Beijing reach on land and at sea?', *Foreign Affairs* (May/June 2010), pp. 22-41.

Kato, Yoko (2002) *Sensou no Kin-Gendai-Shi [Japanese Modern and Contemporary War History],* Tokyo: Kodan-shinsho.

Kobayashi, Hideo (2008) *Manshu no Rekishi [History of Manchuria],* Tokyo: Kodan-sha.

Kudo, Toshihiro (2009) "China and Japan's Economic Relations with Myanmar: Strengthened vs. Estranged", Mitsuhiro Kagami (ed.) *A China-Japan Comparison of Economic Relationships with the Mekong River Basin Countries*, Institute of Developing Economies, Japan External Trade Organization, BRC Research Report, No. 1 (2009).

McGregor, Richard 'Untamed: China's leaders struggle to restrain a surging but unbalanced economy', *Financial Times*, 22 May 2007.

Miyajima, Hideaki (1994) 'The Transformation of Zaibatsu to Post-war Corporate Groups', *Journal of the Japanese and International Economies* 8, pp. 293-328 (1994).

Migeddorj Batchimeg, "Future Challenges for the PRC and Mongolia: A Mongolian perspective", *China Brief*, Volume: 5 Issue: 10 (2005)

Military Power of the People's Republic of China 2009, Office of the Secretary of Defence: Washington, D.C.

'Modern Japan in Archives', the Japanese National Diet Library, available at http://www.ndl.go.jp/modern/

Nakamura, Takahusa (1986) *Showa Keizai-shi [Economic History of the Showa Era]*, Tokyo: Iwanami.

Noland, Marcus, "The (Non-) Impact of UN Sanctions on North Korea", *Asia Policy*, No. 7 (January 2009), pp. 61-88.

Noland, Marcus, "North Korean Missile Test: Remedial Action", *Asia Pacific Bulletin*, No. 33 (6 April 2009)

Ramo, Joshua Cooper (2004) *The Beijing Consensus*, London: The Foreign Policy Centre.

Romerstein, Herbert and Eric Briendel (2001) *The Venona Secrets: Exposing Soviet Espionage and America's Traitors*, Washington, DC: Regnery Publishing, Inc.

Sakhong, Lian H. (2010) *In Defence of Identity*, Bangkok: Orchid Press.

Shirley A. Kan, *China and Proliferation of Weapons of Mass Destruction and Missiles: Policy Issues*, US Congressional Research Service, January 2009 and January 2010.

Tsang, Steve (ed.) (2006) *If China Attacks Taiwan*. London/New York: Routledge.

"Undermining Democracy: 21st century Authoritarians", Freedom House/ Radio Free Asia/Radio Free Europe: Washington, D.C. June 2009.

William C. Triplett, II, 'The Dragon in the Indian Ocean', *China Brief*, vol. 3, issue 4 (25 February 2003).

You Ji, 'Dealing with the Malacca Strait Dilemma: China's efforts to enhance energy transportation security', *EAI Background Brief* No. 329, 12 (April 2007)

Young, Louise (1998=1999) *Japan's Total Empire: Manchuria and the Culture of Wartime Imperialism*, Berkely/Los Angels/London: University of California Press.

Zhao Ziyang (Author); Bao Pu, Renee Chiang, Adi Ignatius (eds) (2009) *Prisoner of the State: The Secret Journal of Premier Zhao Ziyang*, New York: Simon & Schuster.

Zhu, Jianrong (2004) *Mo Takuto no Chosen Sensou [Mao Zedong's Korean War]*, Tokyo: Iwanami.

7

Realist and Pragmatic Elements in China's Grand Strategy: Assessment of China's Relations with Japan, Taiwan and India

Mumin Chen

Introduction

Although its leaders have reiterated that China's rise would be "peaceful" and no regional alliances have been formed to balance China, realist scholars still predict that hegemony is the fate of a rising power, and China is not exempt from this. Yet one finds that a pragmatic foreign policy over the past decades has been successful in maintaining good relations with neighbours and has avoided challenging the existing international order. The existence of both realist and pragmatic elements in China's foreign policy indicates that China's strategic behaviour is not entirely aggressive or passive, but a combination of both. Based on this hypothesis, this paper analyses China's strategic behaviour in recent years in the light of its policies towards three neighbours: Japan, Taiwan and India. In all three cases China has disputes, political or territorial, and each of these countries has the potential to enter into an anti-China alliance with the US. Recent developments suggest that the Chinese leaders tend to be more aggressive with a neighbour when the latter demonstrates a potential to challenge China's strategic interests. Yet the Chinese government sometimes adopts a two-pronged strategy of antagonistic behaviour and benevolent utterances.

One China, Different Interpretations

Ever since Deng Xiaoping initiated economic reforms in the late 1970s, the world has witnessed China's transformation from a politically isolated and ideologically radical society to a global economic giant. David Kang notes:

> China is unlikely to replace the US as the most technologically advanced and militarily dominant country in the world within the foreseeable future ... by virtue of its population, geography, economic growth, and military power, China is already a major actor in East Asia, and by some measures the largest and the most powerful in the region.[1]

Academically, China's rise is viewed both pessimistically and optimistically. The pessimists focus on China's domestic situation, particularly whether the Chinese Communist Party (CCP) government is capable of maintaining social stability while maintaining a high rate of economic growth. They argue that the collapse of the authoritarian regime is inevitable, as increasing domestic crises and social unrest are undermining the CCP's ruling base. Pei Minxin and Gordon Chang are prominent advocates of this perception. Pei argues that China's developmental strategy of combining market reform and preserving a one-party state has created contradictions as market-oriented economic policies in China are implemented in a context of exclusionary politics and predatory practices, which expose the Communist party government as an increasingly self-serving ruling elite, and not as a proletarian party serving the interests of the working people.[2]

A more optimistic view of China's development emphasises the peaceful transition of the society during the reforms. These scholars argue that the CCP has devoted tremendous effort to managing domestic crises while maintaining a high degree of social stability. China observers such as Andrew Nathan, Yang Dali and Jagannath Panda apparently belong to this group. Panda points out that China's progress is visible and noticeable in the institutionalisation of the leadership system and the military, the accommodative character of the relationship between the CCP and the military, increasing government accountability, better constitutional rights of the people and the CCP's adaptability to new social conditions. "These changes have moved alongside China's progress in achieving a decent socio-legal system respecting the rule of law", he concludes.[3]

The other focus of debate is China's foreign relations: whether China's rise will alter the power structure of East Asia or whether it will increase the chances of confrontation between China and the other great powers in the region. Realist theorists predict that no great revisionist power can ever rise peacefully, and that fear of China's dominance in the region will force its neighbours to counterbalance its rising influence. In other words, as long as balance of power continues to be the principal dynamic in international politics, the growing Chinese influence will intensify China's clashes with its neighbours. John Mearsheimer represents this perspective: "China will try to dominate the Asia-Pacific region much as the United States dominates the Western Hemisphere.... China will want to make sure it is so powerful that in Asia no state has the wherewithal to threaten it".

Consequently, it is expected that a powerful China will push the: "United States out of the Asia-Pacific region, much as the United States pushed the European great powers out of the Western Hemisphere in the 19th century".[4]

David Kang presents an opposite view that perhaps except Taiwan, no other country in East Asia worries about China's rise because "most East Asian states view China's return to being the gravitational centre of East Asia as inevitable and have begun to adjust their policies to reflect this expectation."[5] Kang's arguments are based on his study of the region's history where in China had always been accepted by its neighbours as the hegemonic power; no other country in East Asia today is "arming itself against China nor seeking military alliance with which to contain China."[6]

Predicting China's political future is difficult. On the one hand, even if China meets all conditions of collapse, no one can tell when or how the Communist government will dissolve, as was the case with the former Soviet Union or more violently via civil war as it happened in Yugoslavia. Nevertheless we can attempt to assess China's relations with neighbouring countries by analysing its attitudes towards certain unresolved disputes. But before that, the following section will discuss Chinese foreign and strategic policies: whether they demonstrate more realism or pragmatism.

Realism or Pragmatism

Realist scholars believe that China cannot rise peacefully simply because the realist theory holds that a rising power will challenge the interests of a status-quo power, thus making confrontation inevitable; and realism remains the dominant perspective among those who analyse China's strategic orientation. It needs to be noted, however, that Chinese scholars and policy makers were not familiar with Western international relations theories until the early 1990s, when the works by American realist scholars such as Hans Morgenthau, Kenneth Waltz and Robert Gilpin were first introduced into China. Yet the scholarly community and policy circles in China quickly became faithful supporters of realism: everyone seemed to accept the realist premise that great powers dominate world order; and that it is natural for all great powers, including China, to maximise their influence on the world stage.

Therefore, it is fair to say that Chinese leaders today are card-carrying realists. They are convinced that the world is governed by great powers and not by international laws or principles of justice. As the most populous country and armed with a nuclear arsenal, China is already considered an important player in world politics. Its accumulated economic strength has made China stronger than ever. One may predict that China will protect its interests and expand its influence in Asia and elsewhere, as well as demonstrate its military capabilities when it deems necessary. However, Chinese foreign policy in the past decades has shown a high degree of pragmatism and flexibility. This docile aspect was developed during China's long-international isolation from the 1950s to 1970s, and immediately after the Tiananmen incident of 1989. The bitter experiences of the isolation period compelled the leaders to adopt practical positions while handling foreign policy issues.

When the debate on whether China would "rise", gained traction in China's strategic community in the late 1990s, Chinese leaders responded by creating the term "peaceful rise" (*heping jueqi*). It revealed a shift in Chinese foreign policy from "keeping a low profile" (or hiding its light, *taoguang yanghui*), a principle laid down by Deng Xiaoping. The idea of a peaceful rise largely demonstrates the mainstream strategic thought of the ruling elites under Hu Jintao's leadership.

Given this interpretation, one can predict that China's grand strategy will be neither categorically aggressive nor passive, but will adopt a middle way or some degree between the two poles. The leaders are attempting to strike a balance between the goal of seeking hegemony, the realist prediction of a rising power's fate, and creating a stable and peaceful external environment, an idea derived from the tradition of pragmatism. Therefore, China will not be a benevolent great power as long as the territorial or sovereignty disputes with neighbours remain unresolved. One can hardly imagine Chinese leaders allowing Taiwan to become independent or surrendering claims over the Spratly Islands in the South China Sea. Yet Chinese leaders will endeavour to prevent such disputes from escalating into military confrontation because that will greatly undermine China's image internationally and will force its neighbours into an anti-China alliance.

But when and how will China's policies be more assertive? On what conditions will Chinese leaders show more goodwill to China's neighbours? Apparently, it is impossible to formulate a comprehensive framework to elucidate China's relations with all its neighbours, but perhaps we gain some clarity by focussing on the leaders' assessments of the external security environment and their decisions for dealing with security-related issues. For instance: How does the current Chinese government react when faced with a diplomatic crisis? How do Chinese leaders evaluate the degree of threat posed by China's neighbours? Do China's decision-makers pay more attention to domestic or external factors when they formulate foreign policy?

To answer these questions, we need to interpret China's national interests as well as strategic behaviour. Certain scholars have attempted to do so. John Garver and Fei-ling Wang have recently characterised China's policies towards India and Japan as an "anti-encirclement strategy" (*fan baowei zhanlve*), where China had used "a combination of carrots and sticks in an effort to persuade those neighbouring states not to enter into anti-China alignments with the United States and with one another."[7] Their assessment is plausible, but their data premises are outdated. For instance, their judgement on the shift in China's policy towards Japan was based on a series of debates among Chinese strategic scholars between 2005 and 2006. More recent developments have altered China's policy directions.

Using more accurate and timely information about China's strategic intentions, I select three cases for analysis: Japan, Taiwan and India. China's relations with these three neighbours over the past year do not match David Kang's predictions of a "benevolent" China. They all have disputes with China, politically or territorially. More importantly, they are—or have the potential to become—allies of the United States. If China exhibits more aggression or antagonism, all three can enter into an anti-China alliance with the US.

China-Japan Relations: Disputes over Senkaku Islands Intensified

China and Japan, the two largest powers in East Asia, are now competing for leadership. Kent Calder argues: "As with the United Kingdom and Germany a century ago, the contest for regional leadership between China and Japan today is creating new security dilemmas, promoting concerns over Chinese ambitions in Japan and fears of renewed Japan militarism in China".[8]

In 2005, GlobeScan, an international polling firm in Canada, conducted a survey in 33 countries for BBC World Service and for determining people's impressions of other countries. The result showed that Japan was the country most widely viewed as having a "positive" influence. The only exceptions to this were the opinions in China and

South Korea, where 71 and 54 per cent of those polled said that Japan had a negative influence.[9] In a similar survey conducted in China in 2002, only 5.9 per cent of the Chinese considered Japan as a friendly nation, while 43.3 per cent had a negative view of Japan.[10]

Over the past decade China-Japan relations have been overshadowed by contentious issues like including Japan's war responsibilities, history textbook controversies, Chinese resentment over Japanese politicians making pilgrimages to the Yasukuni Shrine, and trade-related disputes. In 2005, the textbook issue escalated into a series of large-scale anti-Japanese demonstrations in several Chinese cities: demonstrators attacked and damaged Japanese consulates, supermarkets and restaurants, prompting the Japanese government to demand an apology and compensation for damages. The crisis lasted several weeks until mid-April when the government declared all "unauthorised marches" to be illegal.

Among the issues debated, the sovereignty over Diaoyu/Senkaku Islands is perhaps the only one that is most likely to mire the two countries into a military conflict. Diaoyu/Senkaku is a group of eight islands located north east of Taiwan and is currently under Japanese jurisdiction; it is also claimed by Taipei and Beijing. The status of these uninhabited islands has been a source of diplomatic disputes among the three countries for decades. The Chinese government claims that the islands have been part of its territory since ancient times and have served as important fishing grounds for Taiwanese and Chinese fishing boats. The Japanese government argues that the islands have been under Japanese administration since 1971, when the US government transferred the administrative power of the islands, along with Okinawa, to Japan. The Japanese government argues that neither the Taiwanese nor Chinese government raised their claims between 1945 and 1971 when the islands were under US trusteeship. In other words, the Japanese government denies any dispute over the islands because they are already part of Japan. Nevertheless, for years the Chinese government has pestered Japan to "return" the islands on the grounds that they are part of Taiwan and Taiwan is part of China. The controversy has another dimension that makes it even more difficult for either Japan or China to compromise: the jurisdiction rights and interests within the maritime boundaries in the East China Sea boast of vast natural gas and petroleum reserves.

The dispute suddenly escalated in September 2010, when a Chinese fishing boat collided with Japanese coast guard patrol vessels near the disputed islands. The Japanese coast guard detained the boat and its crew for several days, which resulted in the most severe diplomatic crisis between the two countries in recent years. Anti-Japan protests broke out in several Chinese cities, with several Japanese-owned companies and supermarkets being vandalised. Further, the Chinese government cancelled ministerial-level meetings between the two governments and also detained four Japanese employees of Fujita, a construction company, for allegedly videotaping military sites.

It was also reported that the Chinese government halted export of rare earth (*xitu*) minerals to Japan. Chinese Premier Wen Jiabao unprecedentedly threatened to take further action if the captain of the Chinese boat was not released; he was handed over two days later. During this time, several anti-China protests also erupted in Japan. These actions reflect the rising anti-China sentiment in Japanese society as well as people's disappointment with the current DPJ government's policy towards China.

China and Japan are now each other's biggest trade partners, but their bilateral relationship is hardly amiable. Many in China have serious concerns about a potential Japanese military resurgence; and people in Japan worry about China's military modernisation and growing nationalism. But why did China suddenly harden its position towards Japan on the Diaoyu/Senkaku issue in September 2010? One explanation is that a new generation of 'young turks' in the military, and their growing influence exerted great pressure on the foreign ministry, forcing them to take a more assertive position towards Japan. John Pomfret of the *Washington Post* explains:

> Emboldened by China's economic expansion, these officials are taking advantage of a weakened leadership at the top of the Communist Party to assert their interests in ways that would have been impossible even a decade ago. In previous crises, China's Foreign Ministry has acted as a calming influence, but this time ... the military led the charge.[11]

China-Taiwan Relations: Closer Economic Ties but Far from Political Reconciliation

Taiwan's political future constitutes another challenge to China's foreign relations. The confrontation between China and Taiwan grew out of the prolonged civil war between the KMT (Kuomintang, the Nationalist Party government) and the CCP in the late 1940s. In 1949, when the Communists established a new "People's Government" in Beijing, the KMT regime fled to Taiwan and proclaimed itself—the Republic of China—as the only legitimate government of both the mainland and Taiwan. The People's

Republic considered Taiwan as a renegade province and the recovery of the island as the ultimate and sacred goal of Chinese nationalism. During the next half century, the confrontation grew into a stalemate: neither side was militarily and politically capable of crossing the Taiwan Strait to "reunify" the country.

The deadlock began easing slightly in the 1980s when Beijing abandoned the policy of military liberation and began to call for "peaceful unification". In 1987 Taipei lifted the ban on civilian travel to the PRC; limited numbers of PRC citizens were permitted to visit Taiwan three years later. Since then millions of Taiwanese people have visited China and many of them have seized upon China's economic opening-up to turn in a profit. In 2003 China became the largest trade partner of Taiwan ($46.32 billion), while Taiwan became China's second-largest importing country, next only to Japan. In terms of capital flows and foreign direct investment (FDI), Taiwanese firms have invested $50.5 billion in China since 1979, which accounts for more than 47 per cent of Taiwan's approved outward investment.

Taiwan-China relations suddenly deteriorated in the new millennium, as the DPP government in Taipei adopted a series of policies to exercise Taiwan's autonomy as a truly sovereign state. In the presidential election of 2004, incumbent president Chen Shui-bian of the DPP proposed a referendum to demonstrate Taiwan's courage against the "military threat from China". The referendum, finally held on March 20, alongside the presidential election, attracted only 45 per cent of eligible voters, thus failing to pass the threshold for validation. After the referendum crisis of 2004, the DPP government continued to use Taiwan's sovereignty issue as a tool to bolster popular support, but these policies were strongly resisted both by Beijing and Washington.

When the KMT won the presidential election of 2008, the new president Ma Ying-Jeou promised to end nearly six decades of political rivalry with China and seek greater economic cooperation with the mainland. Ma based the new policy on two principles: "maintaining the status quo across the Taiwan Strait" and "one China, respective interpretations". The latter principle—referred to by the media as the "1992 Consensus" (under which both sides agreed to disagree over how to define "One China")—was particularly opposed by the previous DPP government. Beijing's reactions to the KMT's strategic reversion were positive, partly because leaders from both sides had established a certain level of mutual trust in previous years. Chinese leaders also recognised the value of this olive branch.

Over the next two years, the KMT government adopted several unprecedented policies, including resuming semi-official talks with

Beijing, establishing regular cross-strait flights, and allowing Chinese tourists to visit Taiwan. In 2010, the two sides signed the Economic Cooperation Framework Agreement (ECFA), for establishing a closer economic partnership and integrating Taiwan's economy into the booming Chinese market. The KMT government claimed the ECFA would, "lay down basic rules for the normalisation of cross-strait economics and trade" and would "improve cross-strait relations to an extent that will induce foreign governments to enter more actively into FTA (Free Trade Agreement) talks with Taiwan."[12] In spite of vehement opposition by the DPP, the ECFA was signed in September. In terms of its objective of inducing foreign governments to sign FTAs with Taiwan, however, only Singapore has demonstrated an interest.

Will the KMT-CCP reconciliation lead the two parties across the Taiwan Strait to lasting peace? President Ma once openly advocated signing a "Cross-Strait Peace Agreement" (*liangan heping xieyi*) with Beijing, claiming it was time to end six decades of political confrontation. Some Western observers predicted that a peace agreement would "greatly reduce the chance of a crisis that could draw the US and China into a military conflict" and also "provide a positive example that might apply to other cases of long-term political or ethnic conflict."[13] Yet by the end of 2010, neither side had proposed bringing the issue to the negotiation table. According to Taiwanese media reports, Ma realised that the public might not agree with his pro-Beijing stance. The KMT government's aversion to further cross-strait policies may reveal KMT leaders' frustration at Beijing's refusal to review its basic policy towards Taiwan.[14]

China-India Relations: Strategic Competition along Two Frontlines

It is difficult to assess the temperature of the India-China relationship, as cooperation and conflict keep occurring. On the one hand, the two governments reiterate that bilateral relations are as good as ever. When India's external affairs minister S.M. Krishna visited Beijing in April 2010, he signed an agreement with his Chinese counterpart Yang Jiechi to establish a hotline between the leaders of the two countries. This is the first time in recent years that India has established a dedicated hotline facility with any country.[15] In December, Premier Wen Jiabao visited Delhi and expressed confidence that the two countries would reach an "important strategic consensus" and take bilateral ties to an "even higher level".[16]

On the other hand, the India-China relationship is still marred by a series of quarrels ranging from undemarcated borders along the Himalayas to strategic competition in the Indian Ocean. The territorial dispute that led to a war in 1962 remains the top issue hindering either side from taking a truly conciliatory path. Reports about Chinese troops crossing the borders have constantly appeared in the Indian media, despite denials by both governments. Media in both societies continues to portray the other society in negative ways, and facts are not highly valued. As BBC News reported: "If you have tuned into one of the most hawkish India television channels or are reading the views of the many experts on India and China, it might seem like the two countries are at each other's throats."[17]

From a strategic perspective, two developments in India-China relations deserve special attention and analysis. The first is China's attitude and policy regarding India-controlled Kashmir. Evidence shows that Beijing has gradually readjusted its position on the Kashmir issue by terming the area "a separate territory from India". Indian media first noticed this when the Chinese government issued a different form of visa to passport holders from Jammu and Kashmir in 2009, and then published a map which depicted the Indian-controlled area as a separate territory. In August 2010, Indian media reported that China had denied a visa to General B.S. Jaswal of the Northern Area Command of the Indian army, who was to go to China to participate in a high-level official meeting. As General Jaswal was the field commander of Indian ground forces in Jammu and Kashmir, this denial caused considerable agitation in the Indian society. At the same time, Western and Indian media began reporting that China would deploy troops in Pakistan-controlled Kashmir. The original report appeared in the *New York Times* in August 2010, indicating that "Pakistan is handing over de facto control of the strategic Gilgit-Baltistan region in the northwest corner of disputed Kashmir to China", and that "an estimated 7,000 to 11,000 soldiers of the People's Liberation Army" would be deployed in the region to work on the railroad as well as other infrastructure works.[18] Both Beijing and Islamabad denied these reports, but it did not reduce the suspicion in New Delhi.

The other frontline of the emerging China-India strategic competition is in the Indian Ocean. When the concept of "String of Pearls" was first touted years ago, very few strategic analysts took it seriously because historically China was a continental power and had little interest in developing a blue-water navy.[19] Within years it became one of the most discussed topics among the strategic community in India and elsewhere.[20] The key issue that emerged from the discussions was—energy security: China is located in a strategically vulnerable position where the world's most salient energy shipping lanes, the oil pipelines on land or oil tankers from the sea, are beyond its control. China's economic growth makes the country increasingly dependent on foreign oil, and this dependence is likely to grow in the near future. Consequently, Chinese strategic analysts no longer avoid talking about formulating a strategy for the Indian Ocean, and some even openly talk about it in the foreign media. Hu Zhiyong, a young Chinese strategic scholar based in Shanghai, published an article in the *United Morning News* (*Lianhe Zaobao*) of Singapore in September 2010, calling for the building of strategic partnerships with countries along the Indian Ocean, including building Chinese naval bases in India's front yard.[21]

Interpretations of China's Strategic Behaviour

It would be seen from the three cases considered here that China's strategic behaviour in recent years has both cooperative and conflictive elements. Beijing's stance towards Japan has become acutely bombastic. Although it is still too early to say whether this change is a result of the power struggles between hardliners and pragmatists leaders within its political system, China has shown unabashed vitriol and vigour in dealing with Japan. On India-China relations, one can easily sense the growing strategic competition between the two rising powers on land and sea, yet Chinese leaders have refrained from using provocative language with New Delhi. With Taiwan, China has shown more benevolence by inducing Taiwan to sign ECFA and promising to sign a peace agreement, but has remained unwilling to alter its basic position on the island's political status.

These three cases also provide some insight for us to interpret the reactions of the current Chinese government when facing a diplomatic crisis, and how China's leaders evaluate the degree of "threat" posed by its neighbours. Based on China's domestic political circumstances and its past behaviour in the international community, three factors can be identified:

- The first is the degree of threat to China's military security, or whether the neighbour *constitutes any challenge to China's military capabilities*.
- The second factor is whether an assertive response to the crisis will affect China's image in the international community, or whether a tough position will *undermine the prestige of China* internationally.
- The third factor is whether a current policy gains legitimacy from society because the degree of *support from domestic society* affects the attitudes of the leadership.

The following table provides a description of these factors:

	Positive	*Negative*
"Clear and present danger"	The neighbour's military capabilities pose a challenge to China's national security.	The neighbour's military capabilities do NOT constitute a challenge to China's national security.
"Undermine the country's prestige"	China's assertive response to the issue will NOT harm the country's image in the international community.	China's assertive response to the issue will raise concerns of a "China threat" in the region.
"Support from domestic society"	The public will support the government's assertive position towards the crisis.	The public will NOT support the government if it adopts tough measures.

If there are more positive signs in the three factors, we conclude that Chinese leaders feelings towards the neighbouring country are less than friendly China's strategic behaviour leans towards aggressiveness, as predicted by realists. In contrast, if the case shows more negative signs, Chinese leaders are more likely to adopt a benevolent stance, thus fostering more cooperative relations with the neighbour. In the Senkaua/Diaoyu dispute, Chinese leaders knew well that if they did not react quickly and vigorously, they would be strongly criticised by their domestic constituency. Indeed, policymakers in Beijing will not hesitate to label Japan a competitor if not a rival in the military arena. The Chinese public is overwhelmingly supportive of a tougher position towards Japan, particularly on the Senkaku/Diaoyu dispute; though most Chinese do not understand why these islands should belong to China. Moreover, since clashes over disputed territories between countries are quite common in international relations, China's tough stance is less likely to prompt other countries to regard China as an aggressive power.

China-Taiwan relations demonstrate an opposite trend, in which the confrontation between the two sides has been barely visible. Chinese leaders do not consider Taiwan a threat to China's national security unless the latter acquires some offensive capabilities. Since the KMT government adopted a conciliatory policy towards Beijing in 2008, and eagerly built closer economic ties with the growing Chinese market, the Chinese public as well as government leaders seldom mentioned using military force against Taiwan.

It is more difficult to assess India-China relations, as we do not know if military leaders in China perceive India as a threat and if China's new strategic deployments in South Asia will raise concerns over a "China threat" in the region. Yet as discussed above, China has strengthened its strategic posture in South Asia in recent years, and if the dispute with India intensifies, the public will support the government in dealing with India in a tougher way. In other words, China will increase its influence in the region, even if Chinese leaders do not see India as a threat. The following table shows a comparison among the three cases.

	China-Japan Relations	*China-Taiwan relations*	*China-India Relations*
Major Developments	Fishing boat dispute; Growing hostility in both societies	Signing of ECFA; Peace agreement stagnant	China's hardening position on Kash-mir; Strategic com-petition between the two sides in the Indian Ocean
"Clear and present danger"	Positive	Negative	Negative/Positive
"Undermine the country's pres-tige"	Positive	Negative	Positive/Negative
"Support from domestic socie-ty"	Positive	Negative	Positive

Conclusion

An observation of China's relations with these three neighbours also reveals some interesting aspects. First, Chinese leaders tend to treat a neighbour in a more aggressive way when the later demonstrates the potential of challenging China's strategic interest. This explains why Beijing deals with Japan and India differently. Second, the Chinese government sometimes adopts a two-way strategy in dealing with its neighbour: it behaves assertively but shows benevolence in words. This trend is obvious when Chinese leaders interact with their counterparts from Taiwan and India. Finally, although China's political system remains authoritarian by nature, the leaders have become more responsive to the growing nationalistic sentiments in society. Nationalism is certainly a tool utilised by CCP leaders to strengthen their legitimacy, yet it is also a force that constrains the government from adopting some radical policies. The strategies adopted by the government towards Japan, Taiwan and India largely reflect the public's views towards these three societies. Therefore influencing the Chinese public, particularly the 450 million internet users, is perhaps a better way for the neighbouring regimes to coax China into a more cooperative foreign policy.

Notes

1. David Kang, *China Rising: Peace, Power, and Order in East Asia* (New York: Columbia University Press, 2007), p. 12.
2. Pei Minxin, *China's Trapped Transition: The Limits of Developmental Autocracy* (Cambridge, MA: Harvard University Press, 2008), pp. 7-8.
3. Jagannath Panda, *China's Path to Power: Party, Military and the Politics of State transition* (Delhi: Pentagon Security International, 2010), p. 8.
4. John Measheimer, "The Gathering Storm: China's Challenge to US Power in Asia", *The Chinese Journal of International Politics*, vol. 3, 2010, p. 389.
5. David Kang, *China Rising ...*, n. 1, p. 50.
6. *Ibid.*, p. 4.
7. John W. Garver and Fei-ling Wang, "China's Anti-Encirclement Struggle", *Asian Security*, 6(3) (2010): 238-61.
8. Kent E. Calder, "China and Japan's Simmering Rivalry", *Foreign Affairs*, 85(2) (March/April 2006): 129.
9. Figures obtained from GlobeScan, http://www.globescan.com/news_archives/bbc06-3/index.html
10. "Japan's Diplomatic Failure toward China", *People's Daily*, 10 December 2002.
11. John Pomfret, "Dispute with Japan highlights China's foreign-policy power struggle", *The Washington Post*, 24 September 2010, http://www.washingtonpost.com/wp-dyn/content/article/2010/09/23/AR201009 2306861.html
12. *ECFA Policy Explanation*, Mainland Affairs Council, http://www.mac.gov.tw/english/english/macpolicy/ecfa_exp.pdf
13. Phillip C. Saunders and Scott L Kastner, "Bridge over Troubled Water: Envisioning a China-Taiwan Peace Agreement", *International Security*, 33(4) (Spring 2009): 87.
14. Ko Shu-ling, "KMT slow pace of Cross Strait ties," *Taipei Times*, December 27, 2010, p. 3.
15. "Krishna in Beijing: Hotline to bring India, China 'closer' ", *Times of India*, http://timesofindia.indiatimes.com/india/Krishna-in-Beijing-Hotline-to-bring-India-China-closer/articleshow/5770884.cms#ixzz1AmwHpk1p
16. "India a great neighbor, visit to take ties to higher level: Wen", *The Hindu*, http://www.thehindu.com/news/national/article956128.ece
17. Amit Baruah, "China bashing in the Indian media", *BBC News*, 16 September 2009, http://news.bbc.co.uk/2/hi/8258715.stm
18. Selig Harrison, "China's Discreet Hold on Pakistan's Northern Borderlands", *New York Times*, 26 August 2010, http://www.nytimes.com/2010/08/27/opinion/27iht-edharrison.html?_r=1

19. The phrase "String of Pearls" was first used in a report by the US department of defence. The report described the strategy as including a new naval base under construction at the Pakistani port of Gwadar, naval bases in Myanmar, a military agreement with Cambodia, strengthening ties with Bangladesh and even a plan to build a canal in Thailand to bypass the Strait of Malacca.
20. Two great discussions about how India looked at China's String of Pearls strategy are: Gurpreet S. Khurana, "China's String of Pearls in the Indian Ocean and its Security Implications", *Strategic Analysis*, 32(1) (2008): 1-39; and Iskander Rehman, "China's String of Pearls and India's Tactical Advantage", IDSA comment, 8 June 2010, http://www.idsa.in/idsacommentsChinasStringofPearlsandIndiasEnduringTacticalAdvantage_irehman_080610;
21. Hu Zhiyong, "China's Indian Ocean Strategy and Its Content" (*Zhongguuo de Yinduyang Zhanvue ji qi Neihan*), *United Morning News* (*Liahe Zhaobao*), 13 September, 2010.

Section Three

REGIONAL SNAPSHOTS

8

Russia and Geopolitics of East Asia

Alexander Lukin

General Framework

Russia's role in Asia and the Pacific has been largely overlooked since the end of the Cold War. Other countries have been preoccupied with other issues, and have not fully observed or appreciated the major changes in Russian policy and the new opportunities they present. In the past 15 years, Russia and East Asia have undergone tremendous change. While Russia was passing through various stages of market reforms and democratisation (with a re-assessment of its objectives and roles on both the regional and global stages), China's economy has grown rapidly, whilst Japan has struggled with a long recession. On the international front, the WTO has come into existence; APEC has developed into an influential organisation; the region was shaken by the 1997 Asian financial crisis; followed by Russia's 1998 crisis, and the current world recession. All of these brought about a new set of international conditions, and Russia has embarked on new policy of opening up towards the region.

The evolution of Russian policy towards East Asia since the emergence of the Russian Federation in place of the former Soviet Union can be traced in terms of four major factors:

1. The objective of greater integration into the world economy;
2. New principles and approaches in Russian diplomacy, including an emphasis on multilateral approaches in dealing with issues and problems;
3. Recognition of the distinct interests and orientation of Russia's Asian regions;
4. A more pragmatic and dynamic pursuit of economic and strategic objectives.

Immediately after the collapse of the Soviet Union, Russian foreign policy was mostly focussed on relations with the West and in entering the "common European house", initially disregarding relations with its closer neighbours. While official representatives of the Russian foreign ministry did not articulate their lack of interest in fostering relations with the East, they stressed Russia's Western connection and their desire to make Russia a good citizen of the Western community. However, the Russian government soon had to change its attitude. Speaking at the Chinese Association of People's Diplomacy on January 27, 1994, Foreign Minister Andrei Kozyrev said: "The realisation of Russian interests not through confrontation but through cooperation with the outside world allowed us in many respects to rediscover for ourselves a whole number of Asian states."[1] If in 1993 the Asian Pacific region was officially rated the sixth among Russia's international priorities (after relations with the CIS, arms control and international security, economic reform, and relations with the U.S. and Europe), in 1996 Foreign Minister Yevgenii Primakov promoted it to the third position (after the CIS and Eastern Europe).

There were practical reasons for this "rediscovery". For the new Russia, achieving greater integration into the world economy is a principal means of transition to a market-based, democratic system. In East Asia this has meant increased trade, investment and interaction, with neighbouring countries such as China, Japan, the two Koreas and Mongolia. The defining concept was "from the system of security towards a system of economic cooperation in East Asia."

Russian policy in East Asia has developed within the framework of Russia's "New Diplomacy" which has evolved to reflect the interests of the new state after December 1991. Based on the acceptance of market principles for economic development, a pragmatic, non-ideological approach for formulating foreign policy, and a broad re-definition of national interests, the new foreign policy promotes Russia's integration into the world economy together with strengthening international and regional stability and security, both of which are essential if Russia is to address its internal political and economic issues. Multilateralism is an important element in the new foreign policy with increased emphasis on membership and active participation in international organisations and other multilateral mechanisms (including international economic institutions and regional organisations) to address issues and problems.

In East Asia, Russia has applied its New Diplomacy in the following directions:

1. Constructive support for stability and security, to ensure a stable external environment.

2. Virtual non-involvement with the Asia-Pacific economy pre-1992 has been replaced by a wide-ranging policy of promoting international trade and investment in East Asia and the Pacific.
3. Russia has sought membership in all existing regional cooperative structures and forums of regional integration.
4. The development of strong, mutually beneficial bilateral relations with key countries in the region has been another major priority. Dialogue with China has been wide-ranging and intensive with a number of important agreements laying the basis for what is already emerging as a strong economic partnership. Efforts have also been made to strengthen relations with Japan, although with less concrete results to date. And the dialogue with the two Korean states represents a historic initiate for restoring active Russian diplomacy with both strategic and economic aims in the Korean Peninsula.

The new Russian approach to Asia was summarised by President Vladimir Putin in an article published just before an APEC Bangkok summit in October 2003. Expressing satisfaction with the APEC members' decision to accept Russia as a member in 1998, he stated:

Russia is a reliable political and economic partner. This will become more obvious and accepted. That not a single serious global or interregional problem can be solved without Russia's active and equitable participation, moreover, contrary to its interests, is a geopolitical reality now. This is why Russia's course is secured in its foreign policy conception on the active development of interaction in all vectors of the Eastern and Asia-Pacific regions. Kipling's well-known postulate, which seemed to be unshakeable in the past of "West is West and East is East" is outdated. These vectors are equal for Russia.[2]

Russia has good reasons to pursue this new foreign policy activism. Over the last few years, its social and economic development has progressed apace. Governance has been rationalised and the supremacy of the federal constitution and laws among all members of the federation has been secured. The economic results have also been positive: until recently GDP and industrial output grew by 7-8 per cent annually. Inflation dropped to between 12-14 per cent, investment grew, and Russia managed to repay $17 billion of its foreign debt without undue pressure on its state budget. The Russian economy has shown consistent growth over the years.

Currently East Asia is one of the main strategic areas for Russian diplomacy because developments here have global implications which can ensure to a lasting peace and stability in the region and create a solid basis for meeting modern challenges which transcend state borders. To make these positive processes irreversible, regional states as well as the international community need to step up their efforts for resolving the most urgent regional problems.

One aspect that currently needs to be addressed is the need to develop a normative basis of relations between the countries of East Asia. Russia is ready to contribute to this process. It is also interested in supporting collective efforts for fighting international terrorism, drug trafficking, proliferation of WMDs and international crime. To this end, it is essential to create multilateral structures for regional security because attempts at achieving security based exclusively on a bipolar system of blocs undermines the possibility of securing a real peace in the East Asia. Security in East Asia should not be narrowly defined as military security but understood as a complex set of conditions that can promote economic growth and internal stability in countries of the region as well as ensure their access to international markets, new technologies and investment.

Regional Cooperation

Russia has contributed significantly to regional cooperation. Both Russia and Japan are members of the international anti-terrorist coalition and have built up high level contacts between the defence and law enforcement agencies of the two countries which would have been unthinkable during the Cold War. The two countries coordinate their foreign policy in international organisations, especially the ARF. Maintaining and strengthening stability in East Asia has become an important focus of Russian-Japanese relations. Russia especially values the participation and cooperation of Tokyo in the six-party talks on the North Korean WMD problem.[3] Other areas of Russian-Japanese cooperation include developing energy security in the region and Eurasian transport corridors. This cooperation can significantly strengthen Russia's integration into regional economy.

Russian-Japanese trade and economic cooperation is also fast increasing. The Russian-Japanese inter-governmental committee on trade and economic cooperation has been meeting regularly since 1994. In 2007 Japan became Russia's 6th biggest trading partner overtaking the US, Britain and France. In 2007 the growth was 64 per cent, more than Russia's trade with any other country of the region. As a Japanese trading partner Russia became as important as Britain and France, and above India and Vietnam. In the peak year of 2008 the trade volume reached $29 billion. While, in 2009 because of the world economic crisis bilateral trade dropped by almost 50 per cent, it grew very fast in 2010 almost reaching pre-crisis levels ($24 billion). Although the crisis in Japan, caused by natural disasters could have a negative impact on the Russian-Japanese trade and economic cooperation it will possibly ease the territorial dispute since the Japanese government will have to turn its attention to internal matters for the long term.

Russia and DPRK

The Russian approach to the crisis over North Korean WMD should be understood in terms of Moscow's perspective on the Korean Peninsula. Russia's neighbour, the Democratic People's Republic of Korea (DPRK), is not an ordinary country. Putin's leadership, has, however, concluded that it is necessary to maintain normal relations with Pyongyang. Improving relations with Pyongyang was a manifestation of the new pragmatism of Moscow's foreign policy. It is quite clear that the collapse of the North Korean regime is inevitable and few doubt this in the Kremlin. It may take five, ten or fifteen years but it will disappear sooner or later from the world political map and a new, united Korea will emerge as Russia's neighbour. It will become a major country, in terms of its population and economic strength. South Korea which is more populous and developed would obviously be dominant so it would not be reasonable for Russia to pursue a strategy of long term relations with Pyongyang.

What is important, both for Koreans and their neighbours, is when and how the reunification will occur. Few wish the North Korean regime—which has stockpiled advanced weapons—to collapse abruptly with its population seeking refuge in neighbouring countries. This scenario must surely be Seoul's worst nightmare—its leaders know full well that even Germany which is far more developed experienced serious difficulties in uniting with its Eastern part (which was also much more developed and prosperous than North Korea). It would be much more acceptable to Seoul if a reformist leadership came to power in the North and implemented gradual economic and political reforms.

In the interest of promoting gradual reforms in the DPRK, Moscow's traditional ties with Pyongyang are an important asset as the latter has top level contacts with a very limited number of countries. It is in Russia's interest, in cooperation with China and South Korea, to demonstrate to the North Korean regime the advantages of the market economy and to encourage reformist tendencies within it hinting that the Korean Communists would be able to last longer or even to become an integral part of the new political system by introducing market reforms. The example of China or Vietnam where the ruling Communists have managed to stay in power by improving the living standards of the population in the wake of economic reforms would prove to be even more persuasive.

The problems of the Korean peninsular will be resolved sooner or later. However, this will most probably unfold gradually as an Oriental scenario rather than as an action-packed American-style Western. During this time it would be important for Russia to maintain its dialogue with Pyongyang based on its own long-term economic and political interests and in coordination with its East Asian neighbours who share Russia's interests and responsibilities in the region.

Russia and South Korea

Political relations between Moscow and Seoul have been growing steadily since the late 1980s when the Soviet leader Mikhail Gorbachev took the first decisive steps towards normalisation of relations. Over ten years of Russian-South Korean relations, a system of government level contacts have been established and regular summits have been held. Speaking to the MBC and KBS TV channels on February 26, 2001 President Putin commented: "I believe there are no disputes between the two countries. The level of political relations is very high. In the international arena we often hold identical positions."[4]

The economic cooperation between the two countries is also growing fast, although the economic crises in both Russia and South Korea in 1998-1999 dealt a serious blow to this. Since that time favourable economic conditions in both countries have been catalyzing new bilateral economic cooperation.

Bilateral trade grew more that 50 per cent from $9,8 billion in 2006 to $15 billion in 2007. Russia became 8th Korea's trading partner (12th in 2006). In 2008 trade volumes touched $18 billion. In 2009 trade dropped to about $10,5 billion but in 2010 showed signs of improvement reaching about $17 billion.

Russia and Korean WMD Crisis

Russia is very serious about cooperating with other countries on enforcing the non-proliferation regime. Moscow has announced on many occasions that North Korea should renounce all nuclear and weapons of mass destruction programmes. However, Russia also believes that Washington should assume its share of responsibility for the failed 1994 agreement with Pyongyang and reach a compromise with North Korea that avoids hostilities—a position Moscow shares with Beijing.

Russia has two fundamental interests concerning the Korean Peninsula First, Russia does not want weapons of mass destruction anywhere in the world, least of all near its borders. Second, Russia does not want a war in Korea.

Russia has made two proposals for stabilising the Korean peninsula. Firstly, by negotiating a package deal whereby North Korea dismantles its WMDs in return for security and energy guarantees. The second Russian proposal involved the holding of multilateral talks. Russia specifically supported the Japanese proposals from the early 1990s for six-party talks. It is no coincidence that Pyongyang announced its wish to participate in the six-party talks via its ambassador in Moscow during a meeting with the Russian deputy foreign minister Yurii Fedotov in July 2003. Russia has played an active role in the talks and persuaded Pyongyang to continue with them after the first round failed to bring specific results and was an active participant at all subsequent round of talks.

Russia and China

Russia's relations with China have reached their highest point since the process of normalisation began in the second half of the 1980s. In an interview with Chinese and Russian media ahead of his visit to the PRC, in July 2000, President Putin reaffirmed that China is Russia's strategic partner and noted that both countries have similar international perspectives in: "Our aspiration to maintain and strengthen a multipolar world, our joint efforts in preserving strategic equilibrium and balance in the world, and putting in place conditions for a peaceful, steady, and effective advancement of our states."[1]

In April 1996, during Yeltsin's visit to China, the two sides announced their desire to develop: "a strategic partnership directed toward the twenty-first century." As the former Russian president explained, the purpose of the partnership was to promote a multipolar world order against hegemony by any one power at a time when "there are absolutely no controversial issues between Russia and China."[5] Since then, the "strategic partnership" has become the official policy of both sides. One indication of the new level of bilateral relations was the Russian-Chinese Good Neighbourly Treaty of Friendship and Cooperation, signed by Russian President Putin and PRC President Jiang Zemin at a summit meeting in Moscow in mid-July 2001.

The treaty aroused considerable interest and stirred up a lively controversy both in Russia and the world at large. Some maintained that the two former Communist giants were advancing towards a new alliance aimed against the United States and the West, while others contended that the document was purely rhetorical and did not have any practical value.

Both views are superficial. Indeed, the Russian-Chinese treaty does not establish an alliance. Its text does not contain obligations of joint defence against aggression. The restrained, cautious wording of the 2001 treaty and the avoidance of any ideological assertions or clearly unrealistic obligations shows that the leadership of both countries is taking a long-term view, and seeking to avoid mistakes of the past.

The treaty, reflecting Russian-Chinese cooperation in general, has two aspects: international and bilateral. The virtually total synchronicity of views on international problems is a major driving force behind the advancement of bilateral relations. The Beijing and Moscow leadership has repeatedly stated that Russian-Chinese rapprochement is not aimed against third parties, including the United States, given that the United States and the West as a whole are not seen as an enemy either by China or by Russia. On the contrary, both countries are greatly interested in economic and political cooperation with the West: it is a major factor in the development of both states and fully in line, therefore, with their strategic objectives. However, it is also true, that Russian-Chinese rapprochement is driven, to a certain extent, by a number of negative (from the perspective of Moscow and Beijing) international trends initiated by Washington.

The treaty is also very important for charting the course of bilateral relations. Of significance for China is the leeway it accords to China for choosing its own development model which is far cry from Russian attempts in the early 1990s to preach at China about human rights and political liberalisation. Such attempts were ironic to say the least given Russia's far from perfect human rights record and political system that defied definition.

The treaty also addresses important bilateral issues that could escalate in the future. Noteworthy is Article 8 which prohibits the use of Russian or Chinese territory by third countries, organisations or groups that could pose a threat to either country's sovereignty, security, and territorial integrity. Clearly, this refers to the fight against separatist movements supported by international terrorist organisations or third countries and is of vital interest to both states. The wording, however, may be too unclear. For example could a Russian organisation or a division of an international organisation championing the independence of Tibet operate in Russia? The Russian federation's constitution does not forbid this, but under the treaty (which takes precedence over internal laws), China now has the right to demand such a ban. Russia is certainly not interested in the disintegration or destabilisation of China, but as a sovereign and democratic state it can hardly allow the activities of its NGOs to be regulated from abroad.

Article 6 which recognises the existing state borders and the need to preserve the status quo on sections that have yet to be demarcated is very important to Russia. This provision addresses the fears of Russian Sinophobes that China purportedly intends to claim tracts of Russian territory or follow the policy of establishing Chinese settlements in Russia's far eastern region resulting in de facto occupation. Although there has been a lot of fuss over the issue, no one has as yet produced any evidence of official Chinese plans to settle Russian territories. On the contrary, Chinese leaders constantly urge their citizens in Russia to obey Russian laws, and accept all measures proposed by Moscow to straighten out the border regime.

Cooperation with China is very important for Russia for several reasons. China shares Russia's views with regard to a future world order which they both see as multipolar. In practical terms this means that both countries want a world which is not dominated by a single force but which has several important power centres that cooperate with each other on the basis of the international law and the UN Charter. This vision can be explained by simple example: Russia and China, like some other countries, are strong enough to have their own interests and approaches to regional and international issues. The world where a single power dictates its vision to everybody else cannot satisfy them.

At the same time both Russia and China are satisfied with the existing system of international law under the aegis of the UN Security Council. Their permanent membership in the Security Council makes their say equal to that of the US while in all other respects the US is still a much more powerful country which feels constrained by international law and tries to undermine it.

Both Russia and China are critical of the external criticism of their political systems, and support each other in fighting separatist tendencies. Therefore Russia and China support and coordinate their positions on most international problems such as the Iranian and North Korean nuclear issues.

Economic cooperation with China is very important for both countries. It can help Russia to develop its vast Siberian and far eastern regions while China uses trade and investment from Russia to acquire important natural resources and to develop its north eastern provinces.

Russia and China have established an elaborate mechanism of "strategic partnership and cooperation". Its essence is very close relations without entering into an alliance. Their heads of state hold formal meetings every year (although they also meet four or five times on the sidelines of various international forums). While the top leaders usually discuss political and strategic questions, the prime ministers at their annual meetings set the agenda for economic cooperation. The Intergovernmental Committee for the Preparation of the Regular Meetings of Heads of Government chaired by deputy premiers with its sub-committees oversees almost every area of economic and trade (see Table 1). It has been in place since 1997. In 2000 an intergovernmental committee on humanitarian cooperation was formed which since then has been taken the lead in promoting cooperation in areas such as education, culture, medicine, sports, travel, mass media, film industry, archives, youth, etc.

Despite some problems, bilateral trade has been steadily growing over last two decades. In 2007 China became Russia's second most important trading partner. In 2008 bilateral trade reached $56,8 billion. In 2009 it dropped about 32 per cent to $38,8 billion, but grew again significantly in 2010 to $55,4 billion. Russia is worried about the growing trade deficit since 2007 (before 2007 Russia had a positive trade balance), the small share of machinery in the Russian exports which dropped to 1-4 per cent depending on the year and the massive illegal trade (because of the corrupt customs authorities). Some of these questions are being effectively dealt with by the two governments and others are closely connected to the Russian economic system in general and can not be resolved unless serious reforms are put in place.

Russian-Chinese cooperation on regional security has been developing both on a bilateral basis and within the Shanghai Cooperation Organization (SCO). The SCO evolved from the Shanghai process and originally comprised five states: Russia, China, Kazakhstan, Kyrgyzstan, and Tajikistan. The process started with border negotiations between the USSR and the PRC. Once the border issues were resolved, the Shanghai Five was not dissolved and emerged instead as a regional organisation because member states also share other interests. These include, above all, joint efforts in fighting terrorism and Islamic extremism as well as multilateral cooperation in the political, economic and cultural spheres.

The establishment of the SCO was officially announced at the Shanghai summit in June 2001. It included members of the former Shanghai Five plus Uzbekistan.[6] The SCO member states promptly responded with a special statement condemning terrorist acts and emphasising that they were already fighting terrorism, extremism, and separatism. They readily share their experience and resources, assisting

Table 1: Russia and the Shanghai Cooperation Organisation

Cooperation Mechanism
(1997 intergovernmental treaty)

- Regular meeting of Heads of Government (annually, coordinates trade, economic, humanitarian cooperation)
- Intergovernmental committee for the preparation of the regular meetings of the heads of government (headed by vice premiers)
- Subcommittees (headed by ministers)
  - Trade and Economy
  - Energy
  - Nuclear Energy
  - Transport
  - Science and Technology
  - Space
  - Banking
  - Civil Aviation Technology
  - Communications and Information Technology
  - Environment protection
- Fishery Committee

the antiterrorist coalition. Cooperation amongst the defence and law enforcement agencies of the member states has reached a high level and joint military exercises targeting terrorist forces have become a norm. However, the SCO is a young organisation and its full potential has yet to be developed. Its secretariat was officially opened in January 2004 in Beijing. On the same day the SCO Regional Antiterrorist Structure (RAS) which coordinates the organisation's antiterrorist activities began operations in the capital of Uzbekistan, Tashkent.

The international interest in the SCO is hardly accidental. Thanks to its clear-cut position on international issues, the SCO has the potential to emerge not only as a dynamic Asian regional cooperation organisation, but also as a body that attracts states who favour multipolarity and oppose the trend towards a unipolar world.

Russia and ASEAN

Official relations between Russia and ASEAN were established more than twenty years ago and some progress has been made since then. Russia is interested in the growing political influence and economic potential of ASEAN and in its strengthening cooperation with it. For Russia ASEAN is a reliable and promising partner, a neighbour of the Russian far eastern region which can play an important role in its economic development.

In recent years the Russia-ASEAN interaction has been growing. The main task here is implementing and broadening the agreements reached at the first ASEAN-Russia Summit held in Kuala Lumpur on December 13, 2005. The agreements included the leaders' political declaration, the Comprehensive Programme of Action to Promote Cooperation between the ASEAN and the Russian Federation in 2005-2015, and the Agreement on Economic and Development Cooperation.

In 2007 the Russia-ASEAN Dialogue Partnership Financial Fund was launched. In 2007 Russia has made its first contribution of $500 thousand to the Fund and later decided to contribute to it on annual basis. This money is being used to finance cooperation projects. Russia-ASEAN cooperation for fighting new threats and challenges is also developing. The working group on fighting terrorism and trans-national crime is to become a mechanism for information exchange between Russian and ASEAN member states law enforcement agencies and special services. The two sides are working on a cultural agreement.

At the same time, Russia recognises that the level of economic cooperation with ASEAN countries is currently too low despite the existing potential. In 2008 the trade turnover was just $7 billion which is only 0,3 per cent of the total volume of ASEAN trade. The PRC-ASEAN trade in the same year was $190. Therefore Russia is determined to increase trade with ASEAN countries. Among the promising areas are: energy security, including developing alternative energy sources, energy saving and "clean" technologies, non-military nuclear energy. Russia also would like to participate in joint infrastructure projects, such as building the Trans-ASEAN natural gas pipeline or creating an ASEAN unified energy system. There are also projects to train tour operators and to establish an ASEAN Research Centre at Moscow State Institute of International Relations, the University of the Russian Foreign Ministry. Russian and ASEAN deputy foreign ministers meet annually as does the ASEAN-Russia joint cooperation committee.

At the same time Russian experts highlight several problems in Moscow's relations with ASEAN in both political and economic areas. There years after its signing it is clear that the intergovernmental economic agreement is not really working. Generally, the ASEAN approach is rather passive. One possible explanation for this is that ASEAN is trying to go slow with regard to its relations with Russia so as not impact its worsening relations with the US. One example of this deterioration is that a US-ASEAN summit is still to be held. Singapore and Indonesia who have the closest relations with the US are also the most sceptical about Russia's growing role in the region. This supports the above explanation. This scepticism can be seen not only on the issue of Russia-ASEAN relations, but also over Russia's membership of the East Asian Summit (EAS). Russia wishes to become a full member of the EAS. It has clearly showed its interest on many occasions since 2005 when President Vladimir Putin participated in the first EAS as a guest.

This negative approach to Russia's participation in regional cooperation in East Asia seems to be a consequence of an outdated image of Russia as one of two world superpowers. This is a remnant of Cold War thinking because of which ASEAN as a grouping non-aligned with either of the two world superpowers should try to maintain a balance in its attitude to them. That is why Russia is welcome in the ARF (of which the US is also a member), but not in the EAS.

However, the Cold War is long over and contemporary Russia is very different both from the Soviet Union and the United States. Unlike the Soviet Union, Russia is no longer a world superpower; it is no longer the power centre of the two world systems seeking global dominance. It seeks genuine pragmatic cooperation with as many partners as possible and aims not at global dominance but its own development. But, unlike the US, Russia is an integral part of East Asia. It's Siberian and far eastern regions which constitute two thirds of Russia's vast territory and where most of Russia's natural resources are located are situated in East Asia. This makes Russia a natural participant in regional cooperation, which Russia needs, to fulfil its strategic task of developing the Russian Far East. This new reality should be understood by other regional actors which may also have a stake in seeking cooperation with Russia.

Another new geopolitical factor which encourages the ASEAN countries to seek cooperation with Russia is growing China's economic and political power. Russia has very good relations with China and is actively trying to use its neighbour's potential to promote development of its Asian regions. Most other countries of the region are also developing constructive cooperation with China. However, creating closer ties between Russia and ASEAN countries may help both sides to diversify their political and economic attachments so that they do not become too one-sided. This element has been understood by some countries of the region, such as Japan, South Korea and India which are actively strengthening their economic and political ties with Russia.

Notes

1. Andrei Kozyrev, "'Aziatskim'" putem—k sisteme bezopasnosti v Azii" [Via an "Asian" Way To a Security System in Asia] *Segodnia,* February 4, 1994, p. 3.
2. Vladimir Putin, Russia and APEC: topical issues and prospects for cooperation, October 17, 2003. http://www.president.kremlin.ru/eng/speeches/2003/10/17/1548_54203.shtml
3. Interview of Deputy Foreign Minister A.P. Losyukov to journal *Business Match,* No. 36, 2003, http://www.ln.mid.ru/ns-rasia.nsf/3a0108443c964002432569e7004199c0/432569d80021985fc3256e0400 46895a?Open Document
4. Interv'yu presidenta Rossiyskoy Federatsii V.V. Putina yuzhnkoreyskim telekanalam "Em-Bi-Si" i "Key-Bi-Es" (Interview of the President of the Russian Federation Vladimir to the South-Korean TV channels MBC and KBS), http://194.226.80.159/events/168.html
5. Vladimir Mikheev, "Boris Yeltsin vypolnit v Kitae osoboe zadanie 'semerki,'" [In Beijing Boris Yeltsin Will Carry Out a Special Instruction of the "Big Seven"] *Izvestiia,* April 26, 1996, p. 1.
6. http://www.ln.mid.ru/ns-rasia.nsf/3a0108443c964002432569e7004199c0/d7ad36556e7f330843256bd1005e561d?OpenDocument

9

Iranian-American Relations and the Future Security Architecture in Western Asia[1]

Arshin Adib-Moghaddam

Introduction

Asia is emerging as an important centre of gravity in a world order that is becoming increasingly multi-polar. Yet on the west of the continent, there exists an arc of crisis stretching from Palestine, over Afghanistan to Pakistan. Iran and the United States have emerged as central players in this 'greater west Asian crisis' and this paper attempts to evaluate the prospects of dialogue and reconciliation between these two actors. Starting with an analysis of Iran's strategic preferences with a particular emphasis on the country's self-perception in international politics before and after the Islamic revolution of 1979, the paper carves out both the areas of mutual interest between Iran and the United States and areas of potential conflict. Improving relations between Iran and the United States, it is argued, is central to any viable security architecture in western Asia and beyond.

There has been a lot of movement in the relations between the United States and Iran since the election of President Obama in November 2008, some of which merits critical reflection in order to gauge its impact on the security situation in western Asia and beyond. In his first sit-down interview—symbolically given to the Arabic satellite network Al-Arabiya—Obama directly addressed the leadership of Iran, asking them to 'unclench their fist' and to open up a new chapter in Iran's relations with the United States. On the occasion of the Iranian *nowrouz* (new year) celebrations in March 2009, he reiterated his willingness to talk to Iranian leaders setting a markedly different tone than that of his predecessor George W. Bush.[2]

On the other side, the current Iranian President Mahmoud Ahmadinejad became the first Iranian leader in three decades to officially congratulate a US president-elect, a gesture acknowledged by Obama at a news conference in January 2009.[3] In parenthesis one should add that when the reformist President Mohammad Khatami talked about dialogue and détente with the United States in the late 1990s, he was castigated by the Iranian right-wing and blamed for sacrificing Iran's revolutionary ideals. There have been no such complaints about Ahmadinejad's far more proactive overtures to the United States. There are two reasons for this: First Obama's cautious rhetoric set the stage for a new phase in the relationship between the two countries. And second, the Supreme Jurisprudent of Iran, Ayatollah Ali Khamenei, indicated before the elections in the United States that Iran would consider talking to any president other than George W. Bush.[4] Since then, however, this relationship has not moved beyond the stage of a rhetorical rapprochement which continues to be hampered by intermittent accusations and suspicions on both sides. But there is a lot of political activity going on behind the scenes—including backdoor messages via third parties (e.g. Turkey and Brazil on the nuclear dossier) and a good dose of 'veiled' or clandestine diplomacy that extends into the 5 plus 1 negotiations over the nuclear energy programme of Iran.

This paper charts a course through the obstacles and opportunities that characterise Iranian-American relations in the light of current trajectories. It is argued that a strategic understanding of these requires an analysis of the grand strategic preferences of both countries. In many ways the nuclear issue is only a side-show to this. Some of the central questions are: How compatible are the strategic preferences of Iran and the United States? How do both states perceive their respective roles in world politics in general and west Asian politics in particular? Which norms, institutions, interests and ideologies inform their foreign policies? How mutually exclusive are they? In order to address these questions I will start by sketching what I have called the 'foreign policy culture' of both countries.

I have defined the term foreign policy culture in depth elsewhere.[5] The term refers to the socially constructed perception of elites who are involved in the foreign policy decision-making process of a particular country. More specifically, the term denotes an integrated system of cultural artefacts (imageries, discourses, ideologies, norms, institutions etc.) which define the pervasive and embedded grand strategic foreign policy preferences, based on the processing of indigenous and exogenous socialisation, affecting the mental disposition of agents vis-à-vis their environment and giving content to the interest to be pursued. The concept thus appreciates that different countries and regions approach the key issue of war, peace and strategy from the deeply embedded perspectives that are intrinsic to the distinct political cultures of the agent(s). To put it simply: By constituting the perception of decision-makers, a particular foreign policy culture shapes the broad contours of a country's foreign policy agenda, defined in terms of its grand strategic preferences.

US Grand Strategy after the End of the Cold War

Based on this formulation, what are some of the central tenets of the contemporary foreign policy culture of the United States, especially since the end of the Cold War? For four decades, the Cold War context provided the US and the Soviet Union with a central ideological and discursive basis for a shared identity. In the case of America what the country was (the leader of the free world, democratic, respectful of human rights, Christian) was easily differentiated from what it was not (communist, atheist etc.). The global recognition of this role was at its pinnacle after the fall of the Berlin Wall and the adoption of capitalist and democratic systems by the former Warsaw Pact states. David Campbell in his *Politics Without Principle: Sovereignty, Ethics, and the Narratives of the Gulf War* provides a succinct account of these changes:

> Following the demise of the Soviet Union, and the strategic proclamation that uncertainty and unpredictability are the new enemies, US foreign policy discourse [was] casting around for a new set of principles to guide action and aid judgement about the place of the United States in the world. Yet, for all the novelty of the situation, the debate associated with these developments is structured along traditional lines, as a battle between the advocates of "isolationism" and the proponents of "intervention."... In the absence of few actual champions of isolationism, the argument countering intervention is largely a mythical construction of an impossible position on the part of those who intend to disparage it. ... [T]here is little contention among those putatively at odds with one another that the United States must retain an internationalist policy.[6]

This was the ideological background out of which the 'new world order' emerged. The rhetoric about a new world order served to reconfirm the claim of the US to international prominence, by positioning its post-Cold War mission in the tradition of the country's achievements in the past.[7] The invasion of Kuwait by Saddam Hussein in 1990 was very conducive to this new narrative. The invasion was pasted onto a larger effort to construct the image of a post-Cold War world order that was presumably more dangerous and unpredictable than the familiar symmetries of bipolarity had been. On the one side 'liberal' analysts were caught in the euphoria of a new era of global peace and prosperity. On the other side the right wing, which reconstituted itself under the neo-conservative banner, warned of new risks and challenges, criticising the decline in US military spending and the neglect of President Reagan's missile defence system by the Clinton administration.

By attacking Kuwait, Saddam Hussein unconsciously provided the US with the opportunity to invent new 'others' that the US felt destined to combat. Hence, rather suddenly, there emerged, new terms in the foreign policy discourse of the country (e.g. rogue states, Islamic fundamentalists, terrorist fanatics). More specifically, the narrative of the new world order has had three impacts on the self-perception of the US and by extension the country's foreign policy culture: First, the international legitimation of the US role during 'Operation Desert Storm' (Second Gulf War between 1990-1991) re-invoked the euphoria of US predominance in the absence of serious competition on the global level. Second, and inter-related with what I just said, from the perspective of the 'neo-conservative' political and academic elite, the demise of the Soviet Union and the quick success in the Second Gulf War confirmed the country's self-perception as the unchallenged superpower in international politics. And third, the victory over Iraq in Operation Desert Storm and the demise of communism were presented as the inevitable triumph of superior values, norms and people, reinvigorating the 'America first syndrome' in influential political circles in the United States. As former National Security Advisor Anthony Lake put it:

> We no longer have to fear Soviet efforts to gain a foothold in the Persian Gulf by taking advantage of our support for one of these states to build relations with the other. The strategic importance of both Iraq and Iran has therefore been reduced dramatically, and their ability to play the superpowers off each other has been eliminated.[8]

So in strategic terms, the US-centric perception of world politics assumes that unipolarity (the systemic dominance by one pole, i.e. the United States) has eliminated the Cold War balance of power calculations, allowing for a redefinition of world politics in keeping with US interests. This brings us to a second question: How did these modifications in America's foreign policy culture affect the country's foreign policies vis-à-vis Iran and western Asia in general? I believe there are at least four policies, which exemplify the modified, rather more aggressive posture I am alluding to: First, the dual containment policy with regard to Iraq and Iran in the 1990s rather than the more passive balance of power rationale which had dominated US policy towards the region during the Cold War—especially during the 1960s and 1970s. Second, laws such as the Iran-Libya Sanctions Act (ILSA) signed by President Clinton in 1996.[9] The ILSA which had an extra-territorial connotation, provided for sanctions to be imposed against any person or corporation investing $40 million or more in the Iranian (or at that time the Libyan) petroleum sector. In legal terms then, the ILSA extended the jurisdiction of domestic US law to the international environment. The president was empowered to punish *any* foreign unit making a significant investment in the oil sector of these countries. And third, the intensified bombardment of Iraqi targets, beginning in September 1996 and escalating into a 7 hour bombing campaign in December 1998 by US American and UK air forces codenamed 'Desert Fox', which was largely opposed by China, France, Russia, Germany and other countries. And finally, of course the invasion of Iraq in 2003, an invasion that was opposed by most members of the international community and that was not sufficiently legitimated via the United Nations. In this sense it was the first war waged under the pre-emption doctrine, and in many ways a unilateral war.

The Iranian Counter-discourse: Global Shifts and Regional Realities

So far I have argued that the end of the Cold War led to the emergence of a rather more aggressive US foreign policy culture which has exacerbated existing sources of conflict in western Asia and beyond. Now this idea of unipolarity clashes in many ways with Iran's foreign policy culture and the grand strategic preferences of the Islamic Republic. Iranian foreign policy norms emphasise the establishment of a 'multipolar world order' (or *donyay-e chan ghodbi* in Persian). This narrative is based on the premise that the systemic shift in international politics towards US unipolarity would not be in the interest of developing nations, and by extension Iran. Indeed, Iran increasingly sees itself as an important pole within this emerging strategic configuration. This attitude is exemplified by following quote which I have taken from a 'strategic paper' written by Kayhan Barzegar who is a Senior Research Fellow at the Centre for Strategic Research in Tehran and who is also affiliated with Harvard's Belfer Center for Science and International Affairs:

> From the perspective of the governing elites in Iran, any direct talks with the U.S. in an unequal condition will endanger Iran's national security and interests, and as past experiences show,

> Iran would ultimately lose in such talks. Meanwhile, Iran's strategic value, along with the legitimacy of its role among the friendly political factions and states in the region, is based on playing an independent role and avoiding direct engagement with the U.S. on the regional issues. Instead, Iran should resolve its strategic differences with the U.S. through strength in the region. This policy will give Iran more importance, and subsequently better serve the interests of its friends in the region. ... With an "active diplomacy", and in the peak of its regional role, Iran should take advantage of this momentous time and, for the sake of "sustainable security" and subsequently "sustainable development," resolve its issues with America.[10]

We find in this paragraph at least three points that link up with Iran's foreign policy culture: an emphasis on independence (when the author talks about the perception of Iran in the region), an emphasis on legitimacy within the region, and an emphasis on Iran's special status within the region and the wider Islamic world. These are symptoms of a larger cultural-strategic context which indicative of the components of the contemporary foreign policy culture of the Iranian state, the development of which will be the topic of the next paragraphs.

Whereas before the revolution the Iranian monarchy rationalised its alliance with the United States and its western cultural policies via an identity discourse emphasising Iran's pre-Islamic heritage and its Persian identity, the political culture of Iran underwent a radical change during the 1960s. The protest movement against the Pahlavi state, which emerged more vociferously during this period, did not reflect dissatisfaction only with domestic issues. The historical claim accentuated by the Iranian movement, which was inherent in the revolutionary-internationalist ethos, transcended the nation-state, creating the dynamism that propelled the revolution to spiral out of the Iranian context. Opposition activists and intellectuals not only protested against the institution of monarchy, but also they demanded the redefining of the country's identity and a change of direction in its relations with the world. As Ayatollah Mottahari' one of the leaders of the 'clerical left', argued:

> If it is decided that [the] basis in determining the limits of the Iranian nation is the Aryan factor, the ultimate end of that is proclivity toward the Western world. But this proclivity in our national and political mission involves submissions and consequences, the most serious being a severance with neighbouring Islamic nations that are not Aryan and an attachment to Europe and the West. ... [I]f we [would choose

> as] the foundation of our nation our intellectual, behavioural and social heritage over the past fourteen centuries, [however,] we would have a different mission and other costs ... Therein, Arab, Turk, Indian, Indonesian and [Chinese] would become our friends, even kinsmen.[11]

In order to legitimise the monarchy, the Pahlavi rulers stressed the ancient, pre-Islamic Persian heritage of Iran. Moreover, both Mohammad Reza Shah (r. 1941-1979) and his father, Reza Shah Pahlavi (r. 1925-1941) nurtured the idea of 'Persianism.' They embedded the Iranian self in the romantic discourse of a superior 'Aryan' nation (*mellat-e aryan*), married to its Indo-European heritage because of common linguistic roots and hence different from the 'Arab-Semitic other.'[12] Golnar Mehran, a professor at Al-Zahra University in Tehran, argues in a similar vein: 'The self presented during the late Pahlavi period was mostly Aryan, taking pride in Iran's "superior" civilisation, language, and culture, and filled with a sense of supremacy toward neighbouring nations and cultures.' 'What was important', she elaborates, 'was Iraniyat. Thus, government sponsored schoolbooks, especially history and Persian-language textbooks, instilled a sense of Iranianness, Iranian spirit, and Iranian identity among the young.' Conversely, in the textbooks of the Islamic Republic, there is a 'shift to the Irano-Islamic identity, as opposed to the solely Iranian one. The *national-religious* identity', in short, 'replaced the exclusive emphasis on *national* identity.'[13]

At the heart of the revolutionary process was a relentless battle for the ideational fabric of modern Persia. If the Pahlavi state attempted to externalise the 'Arab-Semitic other' from the Iranian-Aryan self in order to position Iran more firmly in the 'western' camp, oppositional intellectuals constructed the narrative of 'west-toxification' to protest this 'westernisation' of Iran. The particular strength of Jalal Al-e Ahmad's influential book, which was published in the fall of 1962 under the Persian title *Gharbzadegi*, (west-toxification) was its focus on the dissonance, the articulation of an increasing gap between what was considered to be the authentic self of Iran and the 'distortions' created by 'western' modernity. Employing a medical analogy, Al-e Ahmad deprecated the decadent, mediocre and inauthentic character of Pahlavi Iran. If left untreated, he argued, the spread of the disease-like present would lead to the demise of the country's cultural, political and economic independence, because society was becoming susceptible to 'western' penetration.[14] Moving beyond the Iranian context, Al-e Ahmad saw the struggle against *Gharbzadegi* (west-toxification, occidentosis or westitis) in terms of a conflict between the 'Occidental west' and the 'Oriental east.'[15] Resorting to the metaphor of 'the machine,' he argued that while the 'west' had learned to master the 'technology of modernity,' the 'east' had been kept in a state of political and economic dependency. The subjugation and power was dramatised in order to alert the 'eastern mind' about the creeping intrusion of 'west-toxification' and its corrupting symptoms on societies programmed to be subservient to their imperialist masters:

> So the time is now past when we divided the world into two 'blocs', the two blocs of East and West, or communist and non-communist. Although the first articles of most of the constitutions of the governments of the world still contain that huge twentieth-century sham, the flirtation between the United States and Soviet Russia (the two supposedly unrivalled leaders of those blocs) over the Suez Canal and Cuba showed how the owners of two neighbouring villages can sit down together comfortably at the same table ... Behind the scenes at every riot, *coup d'état*, or uprising in Zanzibar, Syria, or Uruguay, one must look to see what plot by what colonialists company or government backing it, lies hidden. ... These days any schoolchild not only sees the expansionist aims of mechanised industry on both sides of the dispute at work behind the scenes in the Second World War, but also sees the things that were happening in Cuba, the Congo, the Suez Canal, and Algeria were disputes over sugar, diamonds, and oil. The bloodshed in Cyprus, Zanzibar, Aden and Vietnam was for achieving a bridgehead to protect trade routes, which are the first determinant of the policy of governments.[16]

The second dominant narrative that had a determining impact on Iran's shifting self-perception and its relationship to the 'west' emerged from the writings of Ali Shariati. With reference to the anti-dependency theory of Al-e Ahmad and the Islamic-reformist writings of Mehdi Bazargan, Ayatollah Mottahari and others, Shariati developed a comparably critical position toward imperialism and the cultural, political and socio-economic dependence on the 'west.' During his education at the Sorbonne in Paris, Shariati had been in contact with figures of the French left whose political outlook and intellectual paradigms influenced his later writings. Those included Catholic Islamologist Louis Massignon to whom he was a research assistant between 1960 and 1962, the Jewish-Russian émigré George Gurvitch who was his professor of sociology, Islamologist Jacques Berque whose class on the 'Sociology of Islam' Shariati audited in 1963-1964, Frantz Fanon whose seminal *The Wretched of the Earth* he translated (in collaboration with others) into Persian, and Jean-Paul Sartre whose attempt to reconcile existentialism with Marxism and humanism had an important influence on Shariati's own attempt to synthesise social scientific concepts with Shia-Islamic political thought.[17]

In one of his main publications titled *Bazgasht beh-khish* (Return to Oneself), which appeared as series of articles in the Iranian daily *Kayhan* between April 22 and June 22, 1976, Shariati juxtaposed what he considered the true and authentic identity of Iran as a nation, whose inner structure is defined by Imam Hussein's epic *shahadat* (martyrdom), with the inauthentic fallen present under the Pahlavi Shahs, whose inner structure made Iran susceptible to the corrupting influences of 'western' culture. For Shariati, the former symbolised was rectitude and wholeness, the latter disintegration and inadequacy:

> In the nineteenth century I would have felt as an Iranian that I was attached to the great civilisation of the fourth, fifth, sixth, seventh and eighth Islamic centuries ... I would have felt attached to a culture that was more than two thousand years old, which has created in various ways new intellectuality, literature and art in the world of humanity. ... I would have felt attached to an Islam that created the most beautiful spirits and the most sublime faces of humanity. And I would have been able to feel, as a human, a human personality in relation to the world and all the people. So how could they change such an 'I' into a tool whose only worth is to consume new merchandise?
>
> They must empty him of his personality. They negate the 'I' that he feels within himself. And they compel him to believe that he is attached to a weaker civilisation, culture and way of life. He must believe that European civilisation, Western civilisation and race are superior. An African must believe that he has been wild in order to create the temptation in him to become civilised. ... Even their medical doctors and biologists have proved (!!) that the Westerner's brain has an additional grey layer which is not found in Eastern man or negroes, and which assists in the intelligence and the sensitivity of Western man. They have also proved that the Western man's brain has an additional part (which the Easterner lacks) that is the reason for his greater talent and intelligence.[18]

The romantic dimension of Shariati's world-view, that is the emotionally charged assertion of the self and the value of the individual together with the sense of the infinite and transcendental, can be attributed to his interest in Sufism (or 'Islamic mysticism') and the role he allocated to it in politics. While the trajectory of his intellectual thought makes it difficult to discern any genuine Sufi tendencies, nonetheless one might argue that he presented elements of Sufism as a revolutionary and libertarian programme, suitable for challenging the status quo in Pahlavi Iran. As Ali Rahnema argues in his perceptive political biography of Shariati:

> In a way Shariati argued that an individual's gnostic experience was an educational process which paved the way for the meaningful dedication of one's life to the cause of the people. By the time the Sufi wayfarer is free of all worldly chains including his love for life and ready to be accepted by Him, he has acquired all the attributes of a true warrior for the cause of God. ... Thus Shariati replaces the Sufi concept of self-annihilation and subsequent assimilation or living in God with self-annihilation and subsequent assimilation of living in 'the people.' This is certainly a novel interpretation. According to it, Che Guevara becomes an armed and socially responsible reincarnation of Hallaj and 'Ayn al-Quzat Hamadani [two Persian Sufis executed on charges of heresy]. They are both selfless martyrs of love.[19]

The anti-imperialism advocated by Al-e Ahmad and Shariati, translated into reality by Iran's revolutionary generation and constitutive of the Islamic state, became a dominant institution in Iran. The country's foreign policy culture, and Iran's perception of international affairs and its role in world politics were being transformed. This driving agency, which commanded its own reality, soon broke down the boundaries between political idiom and political action. Consistent with the concepts of the Muslim left, the encroachment of the Islamic world by 'corrupting' 'western' concepts was deemed poisonous for the evolution of a just society and the emergence of the ultimate *Homo Islamicus*. In theory, regaining authenticity—in Shariati's terminology returning to the self (*Bazgasht beh-khish*)—and retaining independence, required detachment from the bipolar international system that was perceived as 'dangerous for humanity.'[20] Many scholars of Iran's modern intellectual history agree, when they explain that the question of authenticity has been central to the country's political culture, and that 'this has translated into a rejection of the apish imitation of the west on the grounds that mimicry and submission are fraudulent and counterfeit states of being. As a result', it is argued correctly, 'precarious policies (i.e., hostage taking, export of revolution, the death sentence against Salman Rushdie) should not come as a surprise.'[21]

The Islamic Republic, as many Iranians saw it during the revolutionary period, existed in a severed, schismatic form: on the one side of the Atlantic the so called 'great Satan', the United States, on the other the atheist Leviathan, the Soviet Union (and in between other, lesser enemies such as apartheid South Africa, Israel and Saddam Hussein). Having completed their diagnosis of the Iranian and Islamic civilisation, and their prognosis that revolutionary action will bring about the renewal of both, Iran's revolutionary generation was injected with the idea that radical independence from both superpowers will start a process, at the end of which the existing world order would be transformed. Hence the emergence of the *na sharghi na gharbi, jomhur-e eslami* norm (neither eastern nor western, only the Islamic Republic). Iran self-consciously repositioned itself within the 'third world's', Asia, the Non Alignment Movement and the Arab-Islamic world. In short the country's elites moved Iran away from the western context. Today, this shift is reflected in Iran's active participation in 'non-western' institutions such as the G77, the Developing 8, the Shanghai Cooperation Organisation (SCO), the Organisation of the Islamic Conference (OIC), the Non Alignment Movement (NAM) and the G-15.

To open a parenthesis, here I am not claiming that there is a consensus among the different factions of Iranian politics on every foreign policy decision. That would be to oversimplify the differences between the wide spectrum political institutions in Iran. After all, there are at least eight institutions involved in Iran's foreign policy process: the office of the leader, the foreign ministry, the office of the president, the head of the Expediency Council, the Supreme National Security Council, the Parliament (primarily through its National Security and Foreign Policy commissions), the Strategic Council for Foreign Relations which was established in June 2006 to oversee Ahmadinejad's performance and the Revolutionary Guards. There is no doubt that these institutions follow different agendas. But there appears to be a culturally constituted consensus about the country's role in international affairs that is strong enough to transcend the factions of—and fractions in—Iranian politics. This foreign policy culture refers to a higher level of abstraction than the day-to-day affairs of the state. It functions as the guardian of an (invented) identity, represents a web of shared ideals, images, norms, institutions, and provides for the foreign policy elites a coherent, if systematically abstract, over-all orientation for the conduct of international affairs. Pro-Palestinian sentiments, anti-Zionism and anti-imperialism, Islamic communitarianism, 'third-worldism' (recently reinvigorated by Iran's close relationship with leftist governments in Latin America especially Nicaragua, Bolivia, Cuba and Venezuela and receptive states in sub-

Saharan Africa), and cultural and political independence have functioned as the ideational points of fixation reconstituting the Iranian self during the revolutionary process of the 1960s and 1970s and are not easy to discard. They have acquired the status of cognitively objectified and formally codified social institutions reabsorbed by Iran's contemporary elite, one that is infused with the penetrating ideational force of this cultural reality. Despite the domestic power struggles in Iran then, the shared interests of reformers and conservatives meet where their competition ends: at the junction of Iran's foreign policy culture and—by implication—the grand strategic preferences of the state.

The composition of Iran's contemporary foreign policy culture shows both residual elements of the revolutionary utopias and signs of an emergent 'counter-culture' that signals loyalty to the country's commitment to a multi-polar world order, yet less 'raucous' methods to achieve that goal. Ultimately, the strategic preferences of the country continue to oscillate around at least five grand strategic preferences: First, maximising economic independence manifesting in the idea of economic autarky, relative distance from the World Bank and IMF and indeed mastering the full nuclear fuel cycle, an emotive issue that ties up with the economic nationalism that prevailed after the revolution in 1979. Secondly, support for Palestinian movements, especially HAMAS. Thirdly, the challenge to Israeli and US claims to regional hegemony exemplified by support for the Lebanese Hezbollah and opposition to US military bases in central Asia, the Persian Gulf and elsewhere in Iran's geo-strategic neighbourhood. Fourthly, Islamic communitarianism, i.e. strengthening of pan-Islamic cooperation primarily through the Organisation of the Islamic Conference (OIC). And finally, 'third world empowerment' primarily via the Non-Alignment-Movement recently highlighted by Iran's close relationship with Hugo Chavez in Venezuela, the Castros in Cuba, Evo Morales in Bolivia, and Daniel Ortega in Nicaragua.

There is no suggestion here that these strategic preferences are absolute. For instance, Iran's commitment to 'Islamic communitarianism' does not mean that the country pursues a 'fanatically' pan-Islamic agenda. But that the Muslim world is its primary geo-strategic neighbourhood and therefore 'Islam' recurs as an ideational plank in the foreign policy discourse of the Iranian state. So it is entirely true to argue that successive Iranian governments have been rather silent about the situation in Chechnya and the heavy handed approach of the Russian military there. The Iranian reaction has been comparably muted in view of China's suppression of the Muslim minorities in its western Xinjiang province. Moreover, Iran has repeatedly supported Armenia in its disputes with

Azerbaijan and it has had a comparable proclivity toward India in the country's disputes with Pakistan. And as indicated Iran has by far better relations with 'leftist' Latin American states than it has with some of the countries in its own region. But in the self-understanding of the Islamic Republic, the country is considered a 'natural' leader in the Muslim world and the political elites in the country use an Islamicised discourse to reiterate that claim whenever possible. Therefore, strategic preferences such as the ideal of closer relations between Muslim majority countries continue to be a priority, but Iranian foreign policy elites know that such priorities have to be repeatedly compromised for achieving the ultimate goal of positioning Iran as a major regional power with global ambitions.

So I am referring here to ideal-types which do not determine each and every foreign policy decision of the Iranian state. Rather they are indicative of the country's transnational interests which are a product of the country's pre- and post-revolutionary political cultures. As such they can be linked to the discourse of pre-revolutionary intellectuals such as Ali Shariati and Jalal al-e Ahmad and the early policies of the Islamic Republic: to the *na sharghi na gharbi, jomhur-e eslami* norm (neither eastern nor western, only the Islamic Republic); to the country's decision to end its membership of Cold War institutions such as the Central Treaty Organisation (CENTO); and to Ayatollah Khomeini's decision to prolong the war with Iraq which was seen as a part of an international conspiracy to subdue the revolution. The costs of these changes to the country's international affairs were accepted, even if it meant isolation, and Iran being labelled a 'rogue' or 'outlaw' state by prominent members of the international community. It must follow quite logically that sanctions or indeed bombing the country into submission would not change these very ingrained ideational precepts of the Iranian state. Indeed, any confrontation merely serves the political agenda of the crisis-ridden right-wing in both Iran and the United States. Both Iranian and US American 'neo-conservatives' can only function effectively within the context of conflict.

Conclusion: Signposts of Engagement

In conclusion the foreign policy cultures of Iran and the United States and the corresponding strategic preferences clash along at least four binaries: a) unipolarity versus multi-polarity; b) America's pro-Israeli policies versus Iran's subversion of Israeli power within the region and beyond; c) America's efforts to contain populist Islamic movements such as HAMAS and Hezbollah versus Iran's support for them; d) and the United States' opposition to populist leftist movements especially in Latin America which clashes with Iran's close cooperation with them. So, on the one side we have Iran with its self-perception as an ideological superpower poised to export the revolutionary call for empowerment and cultural authenticity to receptive agents in the international environment. On the other side, the political elites governing the United States (including the Obama administration of course) firmly believe in the Americo-centric configuration of world politics. These self-perceptions are in many ways mutually exclusive, but that does not mean that the US and Iran need to be perennial enemies. What we can hope for in the short run is what I have called a 'Cold Peace' elsewhere.[22] At the current juncture this 'Cold Peace' can be achieved within three interrelated contexts and along three mutual interests: In Iraq where both the US and Iran have an interest in the stability of the al-Maliki government and the unity of the Iraqi nation-state, which has already yielded some low level diplomatic engagement throughout 2007. In Afghanistan, an equally important strategic theatre, where both states have an interest in preventing the resurgence of the Taliban and to support the government of Hamid Karzai politically and economically. And generally on a global scale, both the US and Iran are enemies of al-Qaeda type movements that are virulently anti-American and anti-Shia/anti-Iranian. These are some components for a positive-sum game between the two countries within which both can benefit from a dialogue and an acknowledgement of each other's interests within a context of mutual respect and engagement. To my mind, such engagement could not only be the cornerstone of a viable security architecture in western Asia, it would feed into the stabilisation of world politics on a global scale.

Notes

1. This abbreviated paper is based on a lecture delivered at the IDSA, New Delhi on 17th February 2011. The argument is elaborated in Arshin Adib-Moghaddam *Iran in World Politics: the Question of the Islamic Republic*, New York: Columbia University Press, 2008 and *The International Politics of the Persian Gulf: A cultural genealogy*, London: Routledge, 2006.
2. On the necessity to reach out to Iran see further William Luers, Thomas R. Pickering, Jim Walsh, 'How to deal with Iran', *The New York Review of Books*, Vol. 56, No. 2, 12th February 2009.
3. See further, Ali-Reza Eshraghi, 'Our friend in Tehran', *The New York Times,* 22nd February 2009.
4. 'Iran Leader suggests US ties possible in future', *Reuters*, 3rd January 2008.

5. Arshin Adib-Moghaddam, *Iran in World Politics: the question of the Islamic Republic*, New York: Columbia University Press, 2008.
6. David Campbell, *Politics Without Principle: Sovereignty, Ethics, and the Narratives of the Gulf War*, London: Lynne Rienner, 1993, pp. 94-95.
7. See further Arshin Adib-Moghaddam, *The International Politics of the Persian Gulf: A cultural genealogy,* London: Routledge, 2009.
8. Anthony Lake, 'Confronting Backlash States', *Foreign Affairs*, Vol. 73, No. 2, 1994, p. 48.
10. Kayhan Barzegar, 'Only a Strong Iran Will Talk', Centre for Strategic Research, Tehran, February 2009. Available at <http://www.csr.ir/departments.aspx?lng=en&abtid=06&&depid=74&semid=1588> [accessed 25 February 2009].
11. Morteza Mottahari, *Islam and Iran*, Beirut: Dar al-Ta'aruf, no date, p. 22, quoted in Wajih Kawtharani, 'Mutual Awareness between Arabs and Iranians,' in Khair el-Din Haseeb (ed.), *Arab-Iranian Relations*, Beirut: Centre for Arab Unity Studies, 1998, p. 74. For a comprehensive analysis of Mottahari's ideas see Mahmood T. Davari, *The Political Thought of Ayatullah Murtaza Mutahhari: An Iranian theoretician of the Islamic state*, London: Routledge, 2005.
12. Indeed, one of the many titles of Mohammad Reza Shah included *Aryamehr*, which means 'light of Aryans' in Persian. His father, Reza Khan, who established the Pahlavi dynasty, promoted the name 'Iran' (or 'Land of Aryans') instead of Persia and supported the elimination of Arabic terms from the Persian language.
13. Golnar Mehran, 'The Presentation of the "Self" and the "Other" in Postrevolutionary Iranian School Textbooks', in Nikki R. Keddie and Rudi Matthee (eds.), *Iran and the Surrounding World: Interactions in Culture and Cultural Politics*, Seattle: University of Washington Press, 2002, p. 247, emphasis added.
14. Jalal Al-e Ahmad, *Plagued by the West (Gharbzadegi)*, New York: Caravan, 1982, p. 10 ff.
15. See further Arshin Adib-Moghaddam, *A Metahistory of the Clash of Civilisations*, New York: Columbia University Press, 2011.
16. Jalal Al-e Ahmad, 'The Outline of a disease', in Lloyd Ridgeon (ed.), *Religion and Politics in Modern Iran: A Reader*, London: I.B. Tauris, 2005, p. 168.
17. See Ali Rahnema, *An Islamic Utopian: A Political Biography of Ali Shariati*, London, I.B. Tauris, 2000, pp. 119-128.
18. Ali Shariati, 'Civilisation and Modernisation', in Ridgeon (ed.), *Religion and Politics in Modern Iran*, 2005, pp. 192-193.
19. Rahnema, *An Islamic Utopian*, p. 159.
20. Ayatollah Rouhollah Khomeini, Sermon delivered on 5 November 1982

in *Kayhan* (Tehran), 6 November 1982 and quoted in Rajaee, *Islamic Values and World View*, p. 75.

21. Mehrzad Boroujerdi, 'Iranian Islam and the Faustian Bargain of Western Modernity', *Journal of Peace Research*, Vol. 34, No. 1, February 1997, p. 4.
22. 'We may be near a "cold peace" between the US and Iran', *The Daily Star*, Beirut, 15th August 2008, 'How to make peace with Iran', *Opendemocracy*, 14th May 2010.

10

Security for Western Asia: Building a Cooperation Framework among Turkey, Iran and the Arabs[1]

Paul Salem

Introduction

The 2003 toppling of the Iraqi regime and the implosion of state power that ensued changed the geostrategic outlines of the eastern part of the Middle East. For most of the twentieth century, Iraq defined the north-eastern frontier of the Arab world and served as a buffer to Iranian and Turkish power. When the Iraqi state collapsed, the country was transformed from a buffer to an arena of regional and international conflict, fuelled by sectarian and ethnic divisions and the lure of oil resources. For most of the twentieth century, both Iran and Turkey were oriented West rather than South. In recent decades they have regained a strong interest in their Arab and Muslim neighbourhood to the South.

These developments have created a new web of interests, risks, and interactions among states of this "Eastern Middle East" sub-region consisting of: Turkey, Iran, Syria, Iraq, Saudi Arabia, and the rest of the GCC countries.[2] Turkey and Iran have developed strong relations with each other, and with their several Arab neighbours. Several Arab states have reached out to Turkey, Iran, or both; others have sought to keep them at bay. Trade and energy cooperation have flourished along some axes, but tensions and proxy conflicts have escalated along others, especially between Iran and Saudi Arabia. Elements of positive interdependence—trade, political contacts, and security cooperation—have increased, but aspects of negative interdependence—risk perception and proxy conflict—have increased as well.

Despite the growing interdependence, and the high energy and security stakes, this sub-region remains very disordered and unstable. There are strong rivalries and tensions, but not even the faintest prospect of any overarching economic, political, or security framework. The states in the region and outside players, primarily the United States, have an interest in the stability of this sub-region. This paper will explore the patterns of interests and threats among these players and examine the logic of increased sub-regional cooperation. Establishing a full-fledged sub-regional cooperation framework is currently premature—particularly in light of the ongoing standoff over the Iranian nuclear issue—but it is an important long-term goal that should inform thinking and policy in, and toward, this sub-region.

The Global Trend Toward Increased Regional Cooperation

Regional and sub-regional cooperation in economics, politics, and security has increased steadily since the Cold War ended. Some of this cooperation has been formalised in legal and institutional frameworks; in other cases, the cooperation is informal and ad hoc. During the Cold War, small states could gain security and economic sustenance by forming alliances with a superpower. When the Cold War ended these states ebbed in global relevance, and security increasingly had to be sought through regional cooperation, trade, and security agreements; 118 third-world countries formed the Non-Aligned Movement in an attempt to resist this polarising Cold War dynamic. With the end of the Cold War and the gradual decline of US dominance; power—especially economic—has become more decentralised. Europe, Russia, and China are emerging powers, and regions such as South East Asia, Southern Africa, and South America have their own economic influence and regional relations networks.[3]

During the Cold War, Israel, Turkey, Iran, and the Arab states were part of a highly charged confrontation. Although the end of the Cold War saw a glimmer of progress on the Arab-Israeli front—a form of regional cooperation—the Cold War logic of zero-sum confrontation endures as the dominant pattern of regional relations in the Middle East. The Cold War-era alignments have changed, but most states of the Middle East continue to play by the patterns of that time: investing in confrontation and downplaying regional and sub-regional cooperation.

Economics has played a role in increasing regional and sub-regional cooperation. "The rising share of trade in GDP and the acceleration in the broader process of global integration through investment, movement of people and increased communication of all kinds implies both greater export opportunities and increased international competition."[4] Policy makers are eager to promote and protect comparative advantages through regional and sub-regional trade agreements. This trend toward regional economic cooperation accelerated with the 1995 establishment of the WTO, and accelerated further with the 2008 impasse in the Doha round.

Regional frameworks have proliferated into a rich, overlapping global patchwork with a variety of political, security, and economic arrangements. Some do not go beyond occasional conferences; others have reached full institutionalisation, with permanent secretariats, political bodies, and sectoral institutions. Most countries are members of several such frameworks, reflecting the complex webs of state interests in today's world. Regional frameworks have become a necessary element of the decentralising global order.

In the eastern Middle East, however, no sub-regional cooperation framework has emerged. Political and economic interchanges have flourished between Turkey and the Arab countries; Turkey and Iran; and Iran and several Arab countries. Yet at the same time tensions have existed and sometimes escalated. A sub-regional cooperation framework could manage these relations, enhancing the positive and mitigating the negative. Global patterns suggest that the development of a sub-regional framework to manage and enhance economic, political, and security interests is overdue. Such a framework would not supplant existing frameworks such as the Arab League, the GCC, the D-8, and other groupings, but would add a necessary layer for managing a particular set of interests and interrelations.

Attempts at Regional and Sub-Regional Cooperation

Past attempts at regional and sub-regional cooperation have had mixed results. In 1937, Turkey joined Iran, Iraq, and Afghanistan in Tehran to sign the Treaty of Saadabad, a non-aggression pact designed to boost the signatories' security and stability. Turkey and Iran had already signed a Treaty of Friendship in 1932. The Saadabad Treaty did not develop into an important set of relationships and was overtaken by the events of the Second World War.

The Arab states, including Iraq, established the League of Arab States as the war ended in 1945, indicating their preference for an Arab regional framework. Arab nationalism, and even attempts at Arab unity, would become a hegemonic discourse in the Arab world after the establishment of Israel in 1948 and the rise of Nasser in Egypt. The wave of Arab nationalism led to actual unity between Egypt and Syria from 1958 until 1961 and the rise of the Baath party, which controlled Syria and Iraq. The Arab nationalist discourse painted Turkey and Iran in hostile terms.

Cold War efforts by the United Kingdom and United States to establish a Middle East Treaty Organisation (METO or, more popularly, the Baghdad Pact) met resistance in the Arab world. METO was established in 1955 as a defensive organisation and brought together Turkey, Iraq, Iran, Pakistan, and Great Britain (which was still the main maritime power in the Gulf and governed the Trucial States, which would later become the United Arab Emirates, along the Arab side of the Gulf). It was modelled on the North Atlantic Treaty Organisation (NATO) and was intended to contain Soviet influence. It was discredited by Nasser and Arab nationalists—and leftists in the Arab world—as a Western imperialist design. Its only Arab member, Iraq, withdrew in 1959, and the name was changed to the Central Treaty Organisation (CENTO).

The Arab League, founded in 1945 and currently consisting of 22 members, itself has remained a weak institution, enfeebled by rivalries among Arab leaders, the waning power of Egypt—which continues to insist on monopolising CENTO's leadership—and the absence of convergence among Arab states relating to political, economic, and foreign policy choices.

The Arab countries avoided regional frameworks other than the Arab League, although they did pursue intra-Arab sub-regional frameworks. Saudi Arabia, Kuwait, the United Arab Emirates, Qatar, Bahrain, and Oman formed the Gulf Cooperation Council (GCC) in 1981; it has made some progress in enhancing political, security, and economic cooperation. The limited success of the GCC drew partly on the success of the United Arab Emirates, a federal state established in 1971 by the seven emirates of the southern Gulf: Abu Dhabi, Dubai, Sharjah, Ajman, Umm al-Quwain, Ras al-Khaimah and Fujairah. They had been under British control as the Trucial States.

In other parts of the Arab world, Morocco, Algeria, Tunisia, Libya, and Mauritania established the Arab Maghreb Union in 1989 for sub-regional coordination and cooperation. The institution survives, but it has remained weak because of tensions between Algeria and Morocco over Western Sahara and tensions between the Libyan leader, Muammar Qaddafi, and his Maghrebi counterparts. Iraq, Egypt, Jordan, and Yemen established an Arab Cooperation Council in 1989, but that body collapsed when Iraq invaded Kuwait in August 1990. Egypt strongly opposed the invasion; Jordan and Yemen went along with it.

Many Arab states avoided regional frameworks other than the Arab framework but did establish strong alliances with the United States or Soviet Union during the Cold War. As that era ended, the United States' Arab allies maintained and strengthened their security relations with the

United States, especially on the Arab side of the Gulf, where the United States is the main guarantor of security for the GCC countries.

The first Gulf War, which was triggered by the Iraqi invasion of Kuwait in August 1990, was a leading indicator of the breakdown of order in the eastern Middle East. The invasion of Kuwait was preceded by the long Iraq-Iran war and was followed by the deployment of hundreds of thousands of US combat troops to the region. The United States felt compelled to intervene directly to restore order, and felt so compelled, again in 2003.

Immediately after the first Gulf War, the states of the GCC announced the Damascus Declaration, a regional security cooperation initiative that provided for stationing Syrian and Egyptian troops in the GCC to preserve security there. The militaries of both nations had helped the US-led coalition forces to liberate Kuwait. However, GCC states quietly but quickly moved away from implementing the declaration; Egyptian and Syrian troops returned home, and the GCC built new security and military basing and assistance agreements with the United States and some European countries.

In October 1991, only eight months after US troops had prevailed in the first Gulf War, US diplomats were deployed to ease tensions between the Arabs and Israelis through the Madrid Peace Process. The United States promoted a new form of regional cooperation that would be built on the foundations of peace between Israel and the Arab countries. This new approach would be embodied in cooperative arrangements over security, economics, water, environment, and refugees.

The process included working groups to address these five issues but it ground to a halt after the 1995 assassination of Israeli Prime Minister Yitzhak Rabin and the breakdown of the peace process itself. This working group framework made significant progress in several areas, and contains many useful lessons. However, some critics in the Arab world linked the process to Israeli ambitions of becoming the economic, transport, communication, education, investment, and technological hub of the Middle East. (These ambitions were expressed to a degree in The New Middle East, a 1993 book by Shimon Peres, then Israel's foreign minister.) As with the Baghdad Pact, the Madrid experiment in regional cooperation was linked to a suspected external motive—in this case, Israeli regional dominance—that many in the Arab world viewed as hostile. In their eyes, the arrangement could not be beneficial.

The next initiative for regional cooperation came in the wake of the 2003 invasion of Iraq. The United States recognised that it would need regional cooperation to manage an increasingly deteriorating situation in occupied Iraq, so it organised a Neighbours of Iraq conference. This included strong US allies such as Saudi Arabia and Kuwait, opponents such as Iran and Syria, and more neutral countries such as Turkey. The leaders and ministers of these countries have met repeatedly; often with the participation of Egypt, the United States and other G-8 members, and the United Nations. They have also set up working groups on security, energy, and refugees. Iraq has welcomed this mechanism, but regional states see the process as necessary for helping the United States manage and clean up the mess it created in Iraq. The 'Neighbours Conference' is not an autonomous mechanism for building sustainable, broad-ranging sub-regional cooperation.

This history of attempts at regional and sub-regional cooperation illustrates that the relevant states recognise the need for such cooperation, but that they resist impositions from outside and any framework needs to originate within the region. A framework will work only if states in the region view it as integral to preserving their security and promoting their economic and political interests.

Attitudes toward a Sub-Regional Framework

Official statements and policies suggest that Turkey and Iran favour the setting up of an organisation for sub-regional cooperation, as would Iraq, Syria, Qatar, Bahrain, and Oman. Saudi Arabia and Egypt have been more reluctant. Even among those states that favour more cooperation, interpretations of such cooperation are varied.

With the rise of the AKP, Turkey's interest in cooperative relations with Iran, Iraq, and the Arab east has become policy. Turkey learned about the benefits of regional frameworks partially through its proximity to the European Union, and it has seen benefits simply by working to accede to the EU. Turkey is a long-standing member of NATO and sees no contradiction between being part of that organisation and promoting its regional interests. Since 1990 Turkey has realised that sustaining its export-led growth policy requires stability and access to all of its proximate markets.

Turkey has been active in promoting political and security frameworks in the Balkans, the Caucasus and the Black Sea area. In 1985 Turkey became a founding member of the Economic Cooperation Organisation, which includes Iran and Pakistan, as well as Afghanistan, Azerbaijan, Kazakhstan, Kyrgyzstan, Tajikistan, Turkmenistan, and Uzbekistan. In 1997 it became a founding member of the D-8, an economic cooperation grouping of eight large Muslim countries: Turkey, Iran, Egypt, Pakistan, Bangladesh, Indonesia, Malaysia, and Nigeria. Turkey has developed excellent political and economic relations with Iran and its

Arab neighbours, signing dozens of trade and cooperation agreements with them. Although it has not put forward a specific proposal, Ankara would welcome the establishment of a sub-regional cooperation framework to help manage and strengthen relations with Iran and its Arab neighbours.

Iran has also indicated the need for a sub-regional cooperation framework, although the Iranian approach differs in tone and content from the Turkish. Iran proposes a cooperation framework among countries that border the Gulf, which would exclude Turkey, and any American presence. In December 2007, Iranian President Mahmoud Ahmadinejad proposed a regional pact on trade and security in his address to the GCC meeting in Doha, Qatar. The speech was not well received: It failed to make any mention the of the three gulf islands claimed by both the UAE and Iran, and it was seen to be proposing a framework where it could dominate. Iranian proposals also called for a region "free of foreign influence." However, Arab Gulf states fear that if US forces were to leave the Gulf, the balance of military power would shift overwhelmingly toward Iran.

Succesive governments in post-Saddam Baghdad have urged more sub-regional cooperation. They are aware that Iraq has been infilterated by regional and external powers, and they recognise that Iraq's stability in the immediate future will depend on more cooperation and less conflict among its immediate neighbours. The Iraqi leadership includes Shiites, Sunnis, Kurds, and even Turkmen—each a living link to regional capitals in Iran, Saudi Arabia, and Turkey.

Syria has a strong interest in seeing Turkey, Iran, and the Arab world cooperate. Iran is Syria's main strategic ally in its confrontation with Israel to regain the occupied Golan Heights. Damascus has developed strong relations with Turkey, and is strengthening its relations with Saudi Arabia and the GCC, to meet its broader economic and political interests.

Qatar and Oman are both comfortable with better relations with Iran and have called for increased cooperation across the Gulf.

Saudi Arabia has shied away from supporting the establishment of regional frameworks that would include the eastern Arab states, Iran, and Turkey. Riyadh fears that a permanent sub-regional framework might only legitimise Iranian claims for influence in Iraq and the Gulf. Saudi Arabia prefers to prop up the Arab League, through which it can ensure that Iraq and the Arab side of the Gulf are zones of exclusively Arab concern, and work through the Islamic Conference Organisation. That body represents a Sunni-majority Muslim world in which Saudi Arabia has a leadership position. It has helped form the Arab Quartet with the

UAE, Jordan, and Egypt to coordinate policy in the face of Iranian and Syrian influence.

Saudi Arabia has been much more sanguine about Turkey, encouraging its engagement in the Arab Levant in order to help counterbalance Iranian influence there. It has also participated in impromptu "Sunni Summits" in which leaders of Saudi Arabia, Turkey, Egypt, and Pakistan, among others, have met to discuss political and security developments, presumably to remind Iran of its minority sectarian status in Muslim world affairs.

In fact, Turkey is key for improving Arab-Iranian relations. The GCC fears that Tehran would dominate any Arab-Iranian cooperation, but including Turkey creates a three-way sub-regional framework that helps to assuage the council's fears. Such a move would inject a strong dose of pragmatism into the highly charged and often ideological GCC-Iran relationship.

Web of Positive and Negative Interdependence

Iran, Turkey, and the states of the Arab East are entwined in a complex web of economic, political, and security issues. Some effects of this are positive and benefit the parties; other effects are negative and are perceived as threats that create tension and proxy conflict. Numerous bilateral relationships have sprung up in recent years, but no overall cooperation framework has emerged. Numerous bilateral tensions have escalated as well—primarily between Iran and Saudi Arabia—and the two sides have sought to avoid war.

Iran, Turkey, and the states of the Arab East (taken as a group) are of comparable size. They each have populations of around 70 to 80 million, with GDPs based on purchasing power parity (PPP) of between $800 billion and $1 trillion.[5] Iran and the Arab states have large oil and gas resources. The Turkish economy has achieved large scale and sustainable growth by boosting productivity and developing regional and European export markets. This comparability in size and economic weight indicates that a regional framework among them would not be unduly skewed towards any one direction.

Growing Economic Interaction

Trade among Turkey, Iran, and the Arab East has grown significantly in recent years. Trade between Turkey and the Arab East is most significant, at about $33 billion. Trade between Iran and the Arab world is around $20 billion, and trade between Iran and Turkey stands at about $10 billion.[6]

Turkey has economic agreements with Iran, Iraq, Syria, Saudi Arabia, and the GCC. Iran has agreements with Iraq, Kuwait, Syria, Bahrain, Qatar, Oman, and Turkey, and has large investments in Dubai which were hit hard by the recent financial collapse there.

Generally, rentier oil economies, like those of Iran, Iraq and the GCC, are less prone than less rentier economies, like Turkey or Syria, to regional interdependence because of their low reliance on regional labour, investment and trade. But in this case, energy transport imperatives help counteract this tendency. Oil and gas fields and transport routes underpin many of the interrelations in the region. The GCC, Iran, and Iraq have numerous common interests as oil exporters. Turkey and Syria have conjoined interests as energy transporters—which, in the long run, should provide impetus for more sub-regional cooperation.

Iran and Turkey are interdependent: Iran needs Turkish pipelines to transport its oil and gas to Europe, and Turkey needs Iranian energy to satisfy its own needs, earn premiums, and boost its strategic importance to Europe by providing an alternative to Russian oil and gas. Iran has begun work on a "Persian Pipeline" to transfer gas from its South Pars gas field in the Gulf to the Turkish border so that Turkey can transport it to European markets. In addition, Turkey has explored with Iran the possibility of participation in the planned Nabucco gas pipeline, which also would feed European demand.[7]

Iran and the Arab Gulf states are interdependent: Both need a stable Gulf and open Straits of Hormuz to deliver their oil and gas to global markets and the rapidly growing economies in India and China. Iraq and Syria are interdependent: Iraq would benefit from a boosting of overland pipeline capacity through Syria to the eastern Mediterrannean, and Syria would benefit from being factored into the energy transport grid that delivers energy to Europe and global markets. Any conflict that would disrupt energy exports hurts the interests of all players in the region.

Security of energy installations and transport facilities are a primary concern for the GCC, Iraq, and Iran. However, Iran is concerned that ist more distant enemies, such as Israel and the United States, might at some point target its energy installations, and the GCC countries are concerned that an attack on Iran might trigger retaliation on their own installations. The threat of sabotage and attack from al-Qaeda and other radical organisations remains real. All the oil exporters have a stake in avoiding regional escalation and clamping down on radical groups that threaten their resources.

Both Iran and Iraq have an urgent need for outside financing and knowhow for developing their energy resources. Iran has been hit hard

Selected Oil and Gas Pipeline Infrastructure in the Middle East

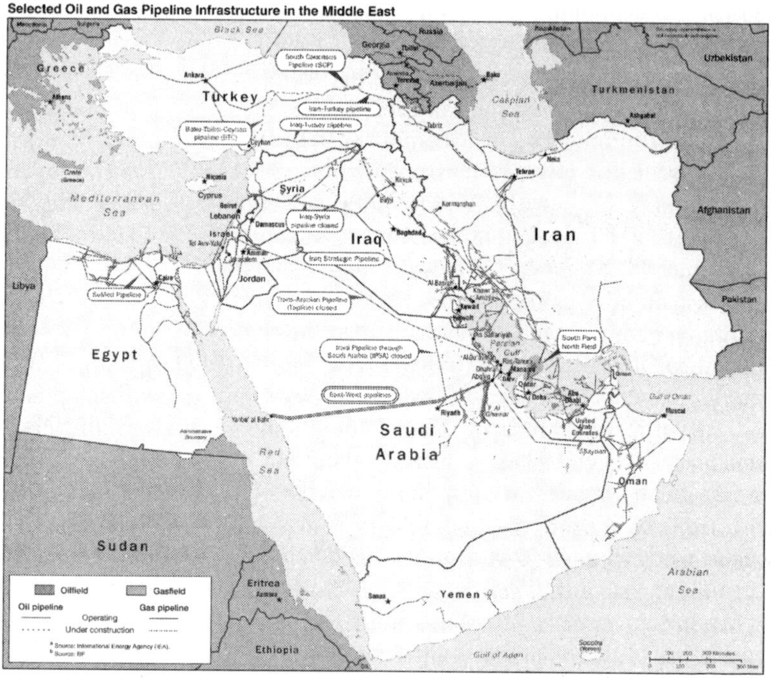

Source: U.S. Government, 2003.

by three decades of international isolation and sanctions. Iraq's isolation during Saddam's rule, and its instability since his removal, devastated its oil excavation, production, refining, and transport capacities.

In terms of transport, both Turkey and Syria are eager to develop their roles as key overland conduits for Gulf oil and gas to the eastern Mediterannean and Europe. The development and consolidation of a participatory energy transport grid would help bind the region's interests in the way, that perhaps the ECSC did in post-World War II Europe. Additionally, Turkey, and its Arab neighbous are expanding their road and rail links to enhance trade, tourism, and pilgrimage. A high-speed railway project has been launched to link Istanbul with Mecca through Syria and Jordan by 2012, thus reversing the century old destruction of the famous Hejaz Railway during the British-backed Arab Revolt during World War I, and indicating a return to Turkish-Arab integration.[8] Turkey and Iran are also upgrading their rail links from Istanbul to Tehran, and from there on to Islamabad.

Oil prices are likely to remain high for the foreseeable future. This might be a boon for the oil exporters in some senses, but it also means that the states around the Gulf will remain rentier economies with little pressure to democratise and institutionalise; although cash-rich regimes can resist internal and external pressures for change. Iran and Saudi Arabia could afford to maintain hard-line ideological and religious positions. In Iraq, oil riches would enable the government to buy support and postpone real problem solving instead of moving toward more institutionalisation and participation.

Saudi Arabia, Iran, and Iraq are all OPEC members and have cooperated on production and pricing; recently, Saudi Arabia and Iran have differed over these issues, with Iran arguing for higher pricing. Iran and Qatar share common gas market interests despite occasional differences over the exploitation of their common gas field.

Oil and gas are precious resources in the Middle East, but so is water. Syria and Iraq lack significant sources of surface water other than the Tigris and Euphrates rivers; they depend on Turkey for the river flows, which are essential for Syrian and Iraqi agriculture. Kuwait recently signed a strategic water agreement with Iran to reduce Kuwait's dependence on desalination, and Iran hopes to expand that cooperation to other GCC states.

Turkey's diversified economy differs from many others in the region. It has followed an export-led growth strategy, so Ankara recognises the need for an open regional marketplace. The Iranian and Arab economies are largely oil-based rentier economies, but Turkey lacked such resources. As of 1980, and after years of an import-subsitution approach to growth, it adopted an export-led growth model that has borne strong results and depends upon maintaining and increasing markets for these exports. This motivates Turkish foreign policy to seek stability and good relations—what Turkish foreign minister Ahmet Davutoglu calls "zero problems"—in all of its neigbouring regions: Europe, the Balkans, Caucasus, Black Sea area, Iran, the Arab Levant, and the GCC. In recent years Turkey has signed numerous energy, trade, investment, and economic cooperation agreements with Iran, Iraq, the GCC, Syria, and Jordan. Turkey would see a sub-regional framework as integral to its economic interests.

The oil exporters are able to rely heavily on oil revenues, but those with large populations (e.g. Iran, Saudi Arabia, and Iraq) realise that oil-based growth might increase revenues but does not create significant numbers of jobs. These large oil-exporting countries have high unemployment rates and large populations of youths—strong motivation for diversifying their economies. The GCC countries and Iran have been aware of this for some time. For the most part, however, the oil-exporting countries around the Gulf have not diversified enough, although all have recognised the need and are trying to move policy in that direction. Regional stability and increased intra-regional trade and investment will help these nations grow economies—diversified enough—to satisfy the demand for employment in the years to come.

Syria's oil resources are dwindling, and it has been attracted by the Turkish economic model. Syria and Turkey recently removed all restrictions on movement across their border, and Syria has also been eager for GCC and Iranian investment. As an almost landlocked country, Iraq also has a vested interest in a more ordered sub-regional environment so that it can secure multiple export routes for its energy resources and benefit from investment by its Arab, Turkish, and Iranian neighbours. Turkish investment in northern Iraq is significant; investment in central and southern Iraq awaits a stabilisation of the situation there.

Across the waters of the Gulf, trade has flourished. Iran and Qatar share a gas field; Dubai has emerged as a major hub of Iranian banking, business, and investment; and Bahrain and Oman have significant economic relations with Iran. Improving relations across the Gulf would only enhance economic benefits on both sides.

Security Interdependence

The positive economic interdependence is shaken by a political and security interdependence that most states consider negative. Iran perceives its security to be threatened by its Arab neighbours to the south, who backed Iraq in its devastating war against Iran. It suspects that some of these neighbours have encouraged the United States to threaten or even attack Iran.

The Arab Gulf states, for their part, perceive their own security as being threatened by Iran, which they fear would overpower them in the Gulf if the United States left the region. They worry that Iran has gained undue influence in Iraq, and that the empowerment of Shiite groups in Iraq, Lebanon, and Yemen, bolstered by Iranian encouragement, could encourage rebellion by Shiite groups in eastern Saudi Arabia and Bahrain. The Arab world has been a comfortably Sunni-dominated world since the defeat of the Fatimid empire in the twelfth century. The projection of Iranian power and the rise of Shiite power in the Arab world shakes the dominant Arab-Sunni elite's fundamental conception of security. Iran's ongoing nuclear programme only exacerbates Arab insecurities.

Turkey's regional security issues are centred on the Kurdish question. Syrian support for Kurdish separatists almost led to war betweeen the two countries in the late 1990s, but that conflict has been fundamentally resolved. Turkey still has concerns about potential Kurdish moves for independence in northern Iraq, but so far Ankara has developed cooperative relations with the Kurdish government there, encouraging investment in the region. There has also been cooperation between Turkey and the Kurdistan Regional Government of Iraq against armed groups of the PKK. Indeed, the Kurdish issue unites, rather than divides, Ankara, Tehran, Baghdad, and Damascus. Beyond that issue, Turkey shares the regional and global concern about Iran's nuclear programme—not because it would feel particularly targetted or threatened, but because Turkish officials feel that it would push the region into an unnecessary arms race following which Turkey and key Arab states would feel pressured to develop their own nuclear capacities.

Despite pervasive patterns of tension in this sub-region, various patterns of security cooperation have managed to emerge. Turkey has security agreements with Iran, Iraq, and Syria on border security, combating terrorism, and intelligence cooperation and it is seeking security cooperation with the GCC states. Iran has security agreements with Oman, Qatar, and Kuwait that deal with maritime security, smuggling, and other crime and security matters; it is seeking security agreements with Saudi Arabia. Iran and Syria have a mutual defence pact. Nevertheless, GCC-Iranian tensions remain high.

Security tensions often pose obstacles to cooperation, but in the long run they should be perceived as the primary reason for it. Europe's regional cooperation frameworks were created to counteract centuries of internecine warfare in Europe and particularly to respond to two world wars that had killed millions in Europe and devastated the continent. The ECSC established in 1951 had peace as one of its strategic goals. The Conference for Security and Cooperation in Europe (CSCE, now the Organisation for Security and Cooperation in Europe, or OSCE) was established in 1973 to counteract the security threats posed by the Cold War. The current high threat perception across the Arab-Iranian divide is a primary reason for the eventual establishment of a cooperative and joint consultation platform in the eastern Middle East.

Trends in the Sub-Region

Risks to Stability

Two crises loom large over the immediate future of this sub-region: the Iranian nuclear issue and the fate of Iraq.

If the Iranian nuclear programme proceeds apace without a modicum of international confidence and transparency, it will lead at least to a tightening of US and international sanctions—and possibly to a military attack by Israel or the United States. The further isolation of Iran would raise tensions in the region. Iran would likely apply counterpressure by using its influence in Iraq, Afghanistan, Lebanon, Syria, and Palestine to exact a price from Israel and the United States, and might also ratchet up support for dissastisfied groups in Yemen, Bahrain, and eastern Saudi Arabia. Iran's response to a military attack might be less predictable: Its officials have threatened that if attacked they could close the Straits of Hormuz and activate sleeper cells in the Gulf. They have also warned that an attack on Iran would affect the whole Gulf. Iranian officials occasionally have raised an old Iranian claim on the Kingdom of Bahrain, and they have hinted that the monarchs of the Gulf might fall to popular rebellions.[9]

But these negative outcomes could be avoided if Iran arrive at some form of agreement with the international community. Iran probably will insist on maintaining its uranium enrichment programme, but it might agree to mechanisms that enhance international confidence that this programme is not being diverted to military purposes. Iran will not consider a narrow deal on its nuclear programme, but it might be interested in a wide ranging understanding with the international community. That agreement would include: lifting US and international sanctions; freeing Iranian assets in US banks; halting US support for insurgency and covert activity in Iran; admitting Iran into the regional and international community and acknowledging its position as a regional power; encouraging investment, especially in the oil and gas sectors; cooperating on drug and human trafficking; cooperating on combating terrorist groups; and cooperating in Afghanistan, Iraq, and Gulf security.

Iran would also have to meet certain conditions including: agreeing to the credible transparency of its nuclear programme; adopting a more responsible tone in its foreign policy; reducing its support for nonstate actors in Iraq, Lebanon, Palestine, and Yemen; supporting the governments of Iraq and Afghanistan; abandoning its denial of the Holocaust and toning down its anti-Israeli rhetoric; and moving closer to the Arab (including Syrian) position of a land-for-peace approach to Israel. Progress by Iran and the international community on these matters would reduce tensions in the region, making it easier to increase intra-regional contacts and work toward building a sub-regional framework. As of this writing, this progress appears unlikely.

The second major security consideration in this region is the fate of Iraq itself. The country regained some stability in 2008 and 2009 after the mayhem from 2004 to 2007. This improvement was largely due to the US troop surge and a change in US military strategy that co-opted Sunni groups and tribes through the Sahwa and Sons of Iraq movements. The surge and accompanying strategies were designed to buy time for Iraqi leaders to move ahead on national cross-communal reconciliation, tackle difficult issues such as the fate of Kirkuk and the distribution of oil revenues, and build up the Iraqi national army and police forces. Some progress has been made on the oil issue and on building up the security forces, but communal relations remain precarious; Kirkuk remains an unresolved time bomb.

It is unclear whether Iraq will hold together as US troops gradually withdraw. Communal tensions could overpower the state, leading Iraq toward civil war and dismemberment. Any dissolution of the current Iraqi state would draw in regional powers. Iran would move to support the Shiite community, and Saudi Arabia and other Arab and GCC countries would support Sunni groups. Turkey might feel compelled to act in order to avoid the declaration of an independent Kurdish state in northern Iraq.

Such developments would perhaps escalate; igniting Sunni-Shiite tensions in Lebanon, Yemen, Bahrain, or eastern Saudi Arabia. Kurdish separatists in Turkey, Syria, and Iran would be emboldened. If Iraq were to remain intact, stable and with a growing economy, it could spread stability in the region and drive sub-regional cooperation.

Multiple Scenarios

Looking ahead, there are three likely scenarios for the eastern Middle East. The first, and most likely, extends the status quo: Tensions among the states of the region remain unresolved and are managed on a crisis-by-crisis basis. There would be no sub-regional discussion of common interests nor a framework for sub-regional consulation and cooperation. This likely would mean long-term instability in the Gulf and the eastern Middle East. Iran complicates matters significantly, and not only because of its nuclear issues. This is the most likely scenario, given the nuclear standoff with Iran, and the fact that a number of major players still don't see the value of such a sub-regional fraemwork. Saudi Arabia is reluctant to acknowledge Iran's role in any framework that includes the Arab east; Iran is relucant to join a regional framework that includes a continued US military presence; and the US cannot envision including Iran in any sub-regional framework even if the nuclear issue is resolved.

The second scenario: A dramatic deterioriation in the security situation, is less likely but a distinct possibility. It could be triggered by an impasse in the Iranian nuclear issue leading to an Israeli or US military attack on Iran. The Iranian response to such an attack is unpredictable—which is probably just how Iranian officials want things. It is reasonably certain that such an attack would not cripple Iran but would strengthen hardliners in the regime. It could lead to military and covert responses that could destabilise much of the Gulf and the Arab east, disrupt oil flows, implode Iraq and/or spread Sunni-Shiite strife in the Levant and Arab Peninsula. This worst-case scenario would also cause Iran to rush toward building or securing nuclear weapons, escalating tensions with Israel and the United States and spurring Turkey and key Arab states to secure or acquire nuclear weapons of their own.

The third scenario: Progress toward sub-regional accommodation and cooperation, is the least likely but the best way forward for the region. It would occur only if some progress is made for defusing the Iranian nuclear crisis. This scenario would be characterised by states in the region committing to invest in their common security and economic interests. Iran would have to move from paranoid defiance to responsible cooperation; Saudi Arabia would have to move from denial to constructive engagement with Iraq and Iran; and Iraq would have to build the basics of political and economic stability. Turkey and Syria would be ready to move toward a sub-regional cooperation framework. The United States would have to be convinced that the status quo—regional disorder, multiple confrontations, and mulitple military deployments—does not serve its interests in ensuring the secure and free flow of oil from the Gulf.

Moving Forward

Establishing an eastern Middle East sub-regional framework—among Turkey, Iran, Iraq, Syria, and the GCC—would be an important goal in itself; as well as a stepping stone toward the eventual establishment of a full Middle East regional cooperation framework including the Maghreb countries—Egypt, and Israel. In September 2008 Sheikh Khalid bin Ahmed al-Khalifa, foreign minister of Bahrain, called for the establishment of a regional organisation that would include the Arab states, Iran, Turkey, and Israel.[10] The former German foreign minister Joschka Fischer has also called for a new Middle East order for bringing the Arab countries and their neighbours together. And which could learn from Europe's evolution from the ECSC to the EU.[11] The Madrid Security Working Group also discussed the establishment of a new regional security platform, a CSCME (Conference on Security and Cooperation in the Middle East).

The current standoff over Iran's nuclear programme makes the building of a sub-regional framework very difficult. However, even in the context of unresolved threats, working toward such a framework is possible and necessary. The Helsinki Process and the OSCE were launched during the Cold War to build trust and cooperation in the context of mistrust and mutually perceived threats. Yes, there are escalating tensions in the eastern Middle East today, but this is precisely why the region's leaders should increase their communication and interaction. Such contacts could help find middle ground for a negotiated solution, limit conflict if the confrontation escalates further, or help pick up the pieces and rebuild regional trust and cooperation if a military confrontation occurs.

Obviously a negotiated settlement to the nuclear issue would provide impetus for regional trust-building and cooperation. However, regardless of whether Iran moves into military confrontation with Israel and the West or ends up with a nuclear weapon, the states of this sub-region are going to remain neighbours and must work out the future of their relations together.

> The first step in establishing a sub-regional framework would be to establish a regular contact and consultation platform such as a purpose-built, recurring sub-regional conference. This strategy has worked in Europe and with other emerging regional frameworks. The conference would provide the venue for communication and the exploration of common interests as well as offer a chance to share perspectives on areas of disagreement. If preliminary progress were made, participants could move toward and attempt to draft a declaration of principles.

Such an initiative could lead to the establishment of working groups. These could be on a number of issues including security, investment and trade, transport and labour, energy, human development, and cultural exchange.

Turkey has proposed such initiatives for the Caucasus and the Black Sea area, and a similar initiative by Israel and the Arab countries have led to the Madrid conference and its aftermath.

The first such a conference should involve, Iraq, Iran, Turkey, and Syria the GCC members. It would differ from the 'Neighbours of Iraq' meeting in terms of its objectives and agenda, although it would include a number of the same actors.

The meetings could be institutionalised on an annual basis and, if progress were made, participants could consider setting up a more permanent and structured cooperation organisation—a necessary layer of sub-regional cooperation among important players. Membership of such an organisation would not preclude membership in other organisations such as the Arab League or NATO; nor would it preclude the subsequent establishment of a full Middle Eastern framework including the Maghreb, Egypt, and Israel. One form of sub-regional cooperation does not preclude another.

Acknowledgements

I would like to thank research assistants Michael Huijer and Saria Francis, and interns Zina Azem and Sinan Bolak of the Carnegie Middle East Centre for their invaluable help in researching sources and information for this paper.

Notes

1. This is a version of a paper published by Carnegie in 2010.
2. In this paper I use "region" to refer to the Middle East, which I consider to include the Arab countries, Israel, Turkey and Iran. I use "sub-region" to refer to the Eastern Middle East:, in which I am including the key states of Turkey, Iran, Syria, Iraq, and the GCC countries.
3. For an excellent exposition of the global trends toward regionalism as well as a discussion of the recent past of regionalism in the Middle East see Louise Fawcett and Andrew Hurrell, eds. *Regionalism in World Politics: Regional Organization and International Order,* (New York: Oxford University Press, 1995).
4. Uri Dadush, "Regionalism and the Gulf Countries," presentation at Emirates Center for Strategic Studies and Research, 2009
5. Figures are for 2008, from the World Bank, "World Development Database indicators, April 2010".
6. Compiled from official statements and news reports. [There is no one global official source for this: Indeed, they were compiled by my ra's from multiple statements and news reports of officials and press in the region.]
7. Nimrod Raphaeli, "The Growing Economic Relations between Iran and Turkey," Inquiry & Analysis No. 414, (Washington, D.C.: Middle East Media Research Institute), 2008.
8. Habib Shaikh, "Plan to Link Mecca with Ankara Through Rail Line Gets a Fresh Boost," *Khaleej Times,* February 6, 2009.
9. For a discussion of Iranian threats to the gulf see Kristian Ulrichsen, "Gulf Security: Changing Internal and External Dynamics," working paper, (London: London School of Economics, The Centre for the Study of Global Governance, Kuwait Programme on Development, Governance and Globalization in the Gulf States,), 2009, p.13.
10. Ian Black, "Bahrain Calls for New Regional Organisation With Israel,"*The Guardian,* October 1, 2008.
11. Joschka Fischer, "The Emerging 'New Middle East,' " Project Syndicate, April 30, 2008, http://www.project-syndicate.org/commentary/fischer25/English

References

Mustafa Aydin and Damla Aras, "Political conditionality of economic relations between paternalist states: Turkey's interaction with Iran, Iraq, and Syria," *Arab Studies Quarterly,* vol. 27, nos. 1 and 2, (Winter-Spring 2005) .

Anoush Ehteshami, "Competing Powerbrokers of the Middle East: Iran and Saudi Arabia," *Emirates Occasional Papers Series no. 67* (Abu Dhabi: Emirates Center for Strategic Studies and Research), 2008.

Mehran Kamrava, "Iran's Regional Foreign and Security Policies in the Persian Gulf," in the Center for International and Regional Studies, ed., *The International Relations of the Gulf—Summary Report,* (Doha, Qatar: Georgetown University School of Foreign Service in Qatar, 2009).

Louise Fawcett and Andrew Hurrell, eds. *Regionalism in World Politics: Regional Organization and International Order,* (New York: Oxford University Press, 1995).

Fred H. Lawson, *Constructing International Relations in the Arab World,* (Palo Alto, California: Stanford University Press, 2006).

James Lindsay and Ray Takeyh, "After Iran Gets the Bomb," *Foreign Affairs,* vol. 89, no. 2, (March/April 2010), pp. 33-49.

Marina Ottaway, "Iran, the United States, and the Gulf: The Elusive Regional Policy," occasional paper no. 105 (Washington, D.C.: Carnegie Endowment for International Peace, Middle East Programme), 2009.

Nimrod Raphaeli, "The Growing Economic Relations between Iran and Turkey," *Inquiry and Analysis no. 414,* (Washington, D.C.: Middle East Media Research Institute), 2008.

Kristian Ulrichsen, "Gulf Security: Changing Internal and External Dynamics," working paper, (London: London School of Economics, The Centre for the Study of Global Governance, Kuwait Programme on Development, Governance and Globalization in the Gulf States,), 2009.

Fareed Zakaria, *The Post-American World (*New York: W.W. Norton & Company Inc., 2008).

11

The East Asia Summit and the Regional Security Architecture

Joseph Chinyong Liow

Introduction

On December 14, 2005 representatives from 16 countries gathered in Kuala Lumpur, Malaysia, for the inaugural session of the East Asia Summit (EAS). Participants included the 10 members of the Association of South East Asian Nations (ASEAN), China, Japan, South Korea, Australia, New Zealand and India. As in the case of China, Japan and South Korea, the latter three states were dialogue partners of ASEAN and had either joined or indicated their willingness to become part of ASEAN's Treaty of Amity and Cooperation (or TAC), the principal document that has governed multilateral institutions such as ASEAN and the ASEAN Regional Forum (ARF). The summit was exceptional not merely because of its composition: more important perhaps was the unique opportunity afforded its participants, through the EAS framework, to collectively shape the contemporary East Asian region in ways that would best maintain its economic dynamism, enhance regional security, and preserve peace and stability among themselves.

The purpose of this essay is to consider the emergence and evolution of the EAS against the backdrop of the trends in East Asian regionalism. It argues that while the EAS is proving to be a major pillar of the emergent regional architecture, it is still confronted by several major challenges arising from the three unresolved tensions in East Asian regionalism between: inclusive and exclusive regionalism; process and membership; and ASEAN centrality and major-power leadership.

The East Asia Summit: Origins, Objectives, Obstacles

Opinions differ, but many regional observers concur that the origins of this summit go back to the 1990 proposal for an East Asian Economic Grouping (EAEG) floated by the Malaysian prime minister Mahathir Mohamad. The rationale behind the EAEG concept was economic, ostensibly devised in response to the apparent post–Cold War formation of regional trade blocs in Europe and North America. China had strongly supported the Malaysian initiative for an EAEG that would have excluded non-Asian states.[1] The EAEG lapsed, however, in the wake of Japanese reticence and strong American objections as Washington and Tokyo favoured more inclusive models of East Asian economic cooperation. Reactions from several ASEAN members, such as Indonesia, were also lukewarm. In response, Mahathir modified it to an East Asian Economic Caucus (EAEC) in October 1991. The project was later revived through the ASEAN+3 (China, Japan and South Korea) summit of heads of state and government that first took place in Kuala Lumpur in December 1997, and eventually found further expression through the creation of the EAS in December 2005. Following its formation, the EAS was rapidly defined as a platform where regional leaders could advocate and promote various issues ranging from energy security to transnational threats.

The EAS grew out of the proposal President Kim Dae Jung of South Korea made at the second ASEAN+3 meeting in Vietnam in November 1998, for the formation of an East Asian Vision Group (EAVG) that would explore the possibilities for the formation of an East Asian community.[2] Comprising 26 civilian experts, the group was tasked to research and recommend concrete measures that ASEAN+3 could take to foster East Asian regional cooperation. It was obvious that the general consensus among all 13 participants of ASEAN+3 was a keenness to further develop East Asian regionalism.

In 2001 the EAVG released the findings of its study, titled *Towards an East Asian Community: Region of Peace, Prosperity, and Progress*.[3] Among its recommendations was the establishment of an East Asia Summit. Based on its observations of regional events and the developing shape of regionalism, it envisioned East Asian nations moving towards establishing some permutation of a regional community. It argued that such a community would benefit the states in the region, by building on the existing regional cooperatives and strengthening the processes of institutionalisation. The East Asia Summit, it surmised, would be a useful vehicle to build this community and pre-empt or resolve any regional challenges that might arise.

Initial reactions to the proposal were cautious but positive. While the states concerned broadly embraced the EAS idea as another step towards community building in the region, opinions differed on the channels through which this was to be actualised. Some states such as China remained inclined towards ASEAN+3, and believed that the existing ASEAN+3 framework would be the best means of bringing the EAS to fruition, whereby the ASEAN+3 would become the EAS itself and the chair would be rotated among the members. Malaysia saw initiatives such as ASEAN+3 and EAS as extensions of Mahathir's proposals, made 15 years earlier and endorsed by the Chinese, for the creation of an exclusive East Asian grouping. Japan, in response to initial indications that states like India, Australia and the United States might be excluded from the EAS, expressed its preference for a more inclusive approache to regionalism.[4] As regards ASEAN, concern over the erosion of its centrality in the building of regional architectures informed its response, and explained its reluctance to simply have ASEAN+3 transformed into EAS. Eventually, a consensus was reached that the EAS would take the form of a separate institution altogether, complete with its own summit meeting.

Following the proposals of the EAVG, the Malaysian prime minister Abdullah Badawi at the ASEAN+3 summit in 2004 put forward the proposal to act on the recommendations and set up the EAS and even offered to host the first meeting in Kuala Lumpur in the following year. In December 2005, the EAS comprising the ten members of ASEAN, China, Japan, South Korea, India, Australia and New Zealand met for the first time on the sidelines of the annual ASEAN ministerial meeting in Kuala Lumpur, with Russia also present as observer at the invitation of the Malaysian hosts. At the conclusion of the meeting, the chairman's statement and Kuala Lumpur Declaration clarified that the EAS was to be an "open" forum "for dialogue on broad strategic, political and economic issues of common interest and concern with the aim of promoting peace, stability and economic prosperity in East Asia."[5] The chairman's statement also reaffirmed the role of ASEAN as the primary driver of the EAS, and that the gathering would be a leader-led meeting "for strategic discussions on key issues affecting the region and the evolving regional architecture". To give the meeting some substance, an *East Asia Summit Declaration on Avian Influenza Prevention, Control, and Response* was also signed, that committed members to deepen cooperation on policies to combat avian flu.

Originally scheduled for Cebu in December 2006, the second EAS was postponed because of concerns of disruption following Typhoon Utor. The summit eventually convened in January 2007 where further progress in trade and regional integration was made with the announcement of studying the possibility for a CEPEA (Comprehensive Economic Partnership in East Asia) agreement at a Track Two level, with support from an Economic Research Institute for ASEAN and East Asia (ERIA), Its creation was later confirmed at the third EAS meeting later that year in Singapore.[6]

Together with the ASEAN Summit and ASEAN+3, the fourth instalment of the EAS to be hosted by Thailand was postponed and rescheduled on several occasions because of the political crisis in Thailand. The September 2006 military coup in Thailand had plunged the country into a political quagmire. The coup precipitated widespread political protests as supporters and opponents of deposed Prime Minister Thaksin Shinnawatra tussled for primacy in Thai politics. By September 2008, the animosity between the two factions turned decidedly hostile with the outbreak of violence, forcing the government to declare a controversial state of emergency. The 14th ASEAN Summit scheduled for December was subsequently postponed because of attendant security concerns. ASEAN leaders eventually reconvened in the central resort towns of Hua Hin and Cha-am between February 27and March 2, 2009. While the ASEAN summit proceeded without distraction, the Thai authorities took the decision to postpone the Association's meetings with the dialogue partners yet again, rescheduling them to be held in Pattaya between April 10–12. As ASEAN and regional leaders gathered in Pattaya, the deepening of Thailand's political malaise saw pro-Thaksin "red-shirt" supporters storm the summit venue on April 10, forcing the besieged government of Prime Minister Abhisit Vejjajiva to suffer the ignominy of having to cancel the meeting at the eleventh hour and evacuate the visiting leaders by helicopter to a nearby military airbase. ASEAN, ASEAN+3 and the EAS eventually met in Hua Hin and Cha-am in October 2009.

The sixteenth ASEAN summit in Hanoi set the stage for the fifth EAS by outlining the focus of the institution. Specifically, discussions centred on connectivity (as a feature of community building) and CEPEA. Of further note at the summit was the explicit interest expressed by ASEAN leaders to draw both the United States and Russia into the EAS. To that end, the chairman's statement noted:

> We recognised and supported the mutually reinforcing roles of the ASEAN+3 process, the East Asia Summit (EAS), and such regional forums as the ASEAN Regional Forum (ARF), to promote the East Asian cooperation and dialogue towards the building of a community in East Asia. In this connection, we encouraged Russia and the US to deepen their engagement in an evolving regional architecture, including the possibility of

> their involvement with the EAS through appropriate modalities, taking into account the Leaders-led, open and inclusive nature of the EAS.[7]

ASEAN thus saw the EAS as a "channel" for Washington to deepen its engagement with the region.[8] Aside from providing another venue for the United States to demonstrate its commitment to the development of regionalism in East Asia, overtures to Washington and Moscow to join the EAS were also driven by the strategic move to generate a countervailing balance against increasing Chinese assertiveness in regional affairs.[9] Following this, senior ASEAN officials also decided to impose a moratorium on EAS membership by introducing a period of "consolidation and reflection".[10]

When the EAS was established in 2005, two issues informed ASEAN's shaping of the new institution's role and its underlying premise. First, as in the case of the ARF, the ASEAN members were keen to further institutionalise great-power relations within a multilateral cooperative structure. A core driver of Asia-Pacific regionalism has historically been the institutionalisation of ties between these powers through and with the support of ASEAN-led arrangements. ASEAN members have traditionally hoped to secure the commitment of the great powers to the promotion of regional peace and security in South East Asia and beyond. In this regard, the EAS was no different.

Second, when establishing the EAS, ASEAN was keen to preserve its dominant role in East Asian regionalism. Advocates of multilateralism continued to claim the relevance of ASEAN for East Asia's stability and security and to see its cooperative initiatives as an essential contribution to regional community formation. ASEAN's assigned leading role in the EAS was more by default than anything else since the prospect of either China or Japan at the helm would in all likelihood have been unacceptable, at least for the foreseeable future. This readily put to rest, at least in the short to medium term, early concerns regarding which country or "coalition of the willing" other than ASEAN could conceivably have emerged to power the summit. With the leadership of the EAS resting primarily with ASEAN, a *fait accompli* of sorts had already been established. Indeed, the modalities of the EAS were understood to be similar to those of other ASEAN-led institutions and defined by the norms and principles stipulated in the TAC. Unsurprisingly, proponents of the Association argued that the ASEAN-led formula had proven to be feasible in Asia, and hence should be applied to the summit;[11] its critics understandably offered more circumspect and dismissive views.

In the scholarly community, the plan to materialise the East Asia

Summit met with considerable scepticism. After all, the existing ASEAN-led institutions have drawn considerable flak over the years for their arguable failure to achieve stated aims or progress beyond the confidence-building stage. The creation of the EAS ignited the common criticism of the proliferating "alphabet-soup" of Asian groupings—a growing number of acronyms for a myriad regional configurations with incoherent overlaps leading to a murky, amorphous, ambiguous state of regional affairs.

Nevertheless, it should be noted that the EAS was envisioned as a form of response by ASEAN—a demonstration of its commitment to the development and progress of regional architecture. Instead of adding to the "alphabet soup" of groupings, the EAS was justified as a means for streamlining the ASEAN-plus meetings. For instance, the ASEAN+3 meetings with China, South Korea and Japan as well as ASEAN+1 meetings with other dialogue partners such as India, have been taking place alongside ASEAN summit meetings leading to extremely packed schedules for government officials, ministers and heads of state with series upon series of back-to-back meetings. For greater institutional efficiency and to limit the potential overlaps, ASEAN members argued that incorporating the ASEAN+ meetings into the EAS would be a prudent measure. By getting all of ASEAN's dialogue partners in the Asian and closer Pacific region into the same forum for discussion and cooperation, regionalism in Asia seemed to be heading in a positive direction—that complemented the pre-existing structures. The Asian region, it seemed, finally had a formal forum that was region-centric and focused, with a mix of membership that better reflected the state of play in the region in terms of indigenous stakeholders.

The EAS and Ongoing Debates on East Asian Regionalism

The politics surrounding the creation of the EAS reveals much about the several ongoing debates on the nature and model of East Asian regionalism. While this essay does not intend to resolve any of these debates, a review of the key points of contention would allow us to locate the EAS within the broader trends of East Asian regionalism, and its future evolution and development. Three debates are of particular salience in this context:

ASEAN Centrality: Is it Sustainable?

Regardless of the heated debate in academic circles, the centrality of ASEAN in Asia's institutional architecture has hitherto not been seriously questioned by regional decision-makers. However this is no longer the case. Officially, ASEAN remains in the "driving seat" of several existing regional arrangements which it helped found, notably, the ARF, ASEAN+3 and the EAS. In May 2010, the ASEAN Defence Ministers' Meeting Plus Eight (ADMM+8) defence track comprising of the ten ASEAN countries, Australia, China, India, Japan, New Zealand, South Korea, Russia and the United States, was established—yet another "ASEAN+" arrangement in the burgeoning ASEAN institutional complex. Yet, as recent institutional developments at both the global and regional levels—in particular, arrangements such as the G20 and the China-Japan-Korea Trilateral Summit—suggest, non-ASEAN states have no inhibitions in undertaking diplomatic and economic initiatives, which can have potentially significant ramifications for regional architecture and cooperation in Asia, without the involvement of ASEAN.

Proposals relating to Asia's institutional architecture have been floated from time to time. Where the recent spate of proposals appear to differ fundamentally from their predecessors is in their view of the place and role of ASEAN in Asian regionalism. At least three have stemmed from heads of government:

- In May 2008, Kevin Rudd, prime minister of Australia, kicked off the process by proposing an "Asia-Pacific Community". For the most part, the APC idea, as it came to be known, was nebulous and remained so for a long period. Indeed, even Australian foreign ministry officials were "caught off guard" by the announcement of this initiative.[12] To the extent that it was elaborated, Rudd's vision consisted of an overarching institutional architecture, an umbrella organisation that would serve as a "one-stop shop" for all things Asia, as it were. Furthermore, the APC was to be based on a concert of regional great and middle powers—including Indonesia—with no visible role for ASEAN.[13] Other than the expected chorus of protest from ASEAN leaders, the Rudd proposal evoked ambivalent reactions from China and the United States.[14] China's assistant foreign minister, Hu Zhengyue, announced that the conditions were "not ripe" for Rudd's envisioned mechanism to be placed on the regional agenda.[15] And while Kurt Campbell, the US assistant secretary of state for Asia and the Pacific, did not—as misreported by an Australian daily[16]—oppose Rudd's vision, neither did he offer the outright endorsement Canberra had possibly hoped for.
- In March 2009, President Lee Myung-bak of South Korea presented his "New Asia Initiative", which also envisaged key

responsibilities for regional powers in Asian regionalism and a presumably global role for his own country.[17]

- Not to be outdone, Japanese premier Yukio Hatoyama, during a visit to Beijing in September 2009, called for a European Union-like community for East Asia, one centred on a Sino-Japanese core that presupposed a potential reconciliation??? PL. CHECK between the two Asian powers, not unlike the European regionalism's reliance on the Franco-German base.[18] Principally focused on North East Asia rather than ASEAN, the Hatoyama vision, which came to be known as the East Asian Community, included the United States but not Australia. Much like the Australian proposal, the Japanese proposal received a cool response from Beijing and ASEAN.[19]
- There is a fourth proposal, one considerably more substantial than the preceding three. A report drafted by a Pacific Economic Cooperation Council (PECC) regional task force on regional institutional architecture called for the establishment of an Asia-Pacific Summit comprising the nineteen APEC participant states and India for high-level discussions on key issues.[20] The report also urged the creation of an informal "G10", a caucus consisting of the Asia-Pacific members of the G20, which could ostensibly voice the collective concerns of all Asia-Pacific countries.[21] Noting that the effectiveness of both the ARF and ASEAN+3 have been hamstrung by their "Southeast Asia-centric leadership", the report conceded that ASEAN's position in the "driving seat" of Asian regionalism is by default, given that it is the only option politically acceptable to all regional stakeholders.[22] At the time of writing, the report remains unpublished, which implies that possible disagreements could have arisen among PECC member countries about its conclusions.

Without exception, the four visions sketched out above signal a growing disenchantment, within and outside the region, over East Asia's regional architecture—its extant structures, conventions, modalities, and leadership. They share at least four commonalities.

First, reflecting the academic debates alluded to earlier, they share an evident concern for the lack of an overarching coherence in East Asia's regional architecture. This concern is by no means new, not least for those who worry that Asia's cluttered institutional landscape—its so called "variable geometry",—reflects the absence of any broad strategic vision and rationale among the architects of East Asia's regionalism. Critics readily insist that the fault lies with the ASEAN, which, as per conventional wisdom, went on an institution-building spree that privileged form over function and content.[23] The consequence, as such, was a messy, incrementally enhanced, and heavily compartmentalised architecture comprising institutions formed on an ad hoc basis to meet specific aims—APEC for trade liberalisation, ARF for confidence-building, and ASEAN+3 for economic and financial cooperation. Yet non-ASEAN states must bear a measure of responsibility as well. For example, little is made today of China's initial enthusiasm for the EAS in the light of its current coolness towards the summit, or of Japan's inability to demonstrate regional leadership by pushing through its Asian Monetary Fund proposal in the face of protests from Washington as well as the World Bank and IMF.[24]

Second, while they accept that Asia's existing regional organisations—most of them, at least—are here to stay, they also call for the practical reform of those arrangements to ensure overall architectural coherence.[25] Proposed reforms include clarifying and streamlining institutional roles and responsibilities. Yet here too there is a fundamental disagreement. Hinted at in the Hatoyama proposal, but most forcefully asserted in the Rudd and PECC task force proposals, is the perceived need for a major architectural alteration to end the compartmentalisation of existing regional arrangements.[26] This is usually expressed in terms of an overarching structure at the heads of government level—be it Rudd's Asia-Pacific Community or the PECC task force's Asia-Pacific summit idea—along with a comprehensive, region-wide agenda. On the other hand, others insist that compartmentalisation in itself is not a problem, but concede that existing institutions have suffered from the rapid increases in membership and agenda, and as such have not been as effective as initially hoped. They thereby argue for reclaiming the APEC mandate as an economic forum focused primarily on trade liberalisation and investment—and, by implication, keep security issues within the ARF—and strengthening these institutions as problem-solving or "action-oriented" mechanisms rather than being just talk-shops. Significantly, it is the leaders of ASEAN countries, as much as anyone else, who are making such observations today.[27]

Thirdly, they believe that no regional architecture can be possible without first establishing a regional order. Crucially, the proposals share the view that order can only be underwritten by a concert or coalition of regional great and middle powers, although they differ slightly on which powers should be included. Japan's proposal on the East Asia Community stood out for its explicit acknowledgement for Sino-Japanese rapprochement and collaboration as the anchor for the regional union, not unlike the Franco-German anchor for European integration. But the Asian powers in question coming up with their own version of the Élysée Treaty, which effectively sealed reconciliation between the two European power houses and provided the framework for bilateral cooperation, remains unlikely so long as Japan resists making a comprehensive apology and related restitutions demanded by its Asian neighbours, most notably, China. As two analysts have noted elsewhere: "The future of the region depends on the rise of China and the revitalisation of Japan; one cannot happen without the other. In other words, the future now depends on China and Japan thinking together."[28] Understandably, the respective progenitors of the four architectural visions see key roles and responsibilities for their own countries, whether in those power coalitions or (in their view) as the logical representatives and mouthpieces of Asian or Asia-Pacific interests in the wider world.

Fourth and crucially for our purposes, they relax the extant assumption of ASEAN centrality in regionalism in East Asia. Importantly, none dispute the Association's significant contribution to Asia's regional architecture. Indeed, ASEAN's success lay in its ability, as a grouping of mostly small (other than perhaps Indonesia) and weak states, to institutionalise and regularise security relations among North East, South East and South Asia and the world's great powers—China, Japan, Russia, the United States—beginning with the ARF in 1994. But even as ASEAN's subsequent struggles with internal disunity and impotence raised doubts about its leadership and relevance to regional security, the consensus on ASEAN's centrality, held, for the most part, among its regional partners. While this obviously did not stop other regional states from starting their own multilateral initiatives, the courtesy of consulting with ASEAN was more or less observed. Of late, however, there have been worrying signs that the centrality consensus can no longer be taken for granted, such as in the case of recent key decisions taken by the "+3" countries—China, Japan and South Korea—in the context of the Trilateral Summit. And if the proposals considered above are any indication, the willingness to rethink, the hitherto seemingly non-negotiable—ASEAN as the core of Asia's institutional architecture and regional cooperation—has grown considerably.

ASEAN is clearly concerned about the apprehensions regarding its role and relevance to the regional architecture. Its report on the 43rd ASEAN Ministerial Meeting in Hanoi in July 2010, made much of ASEAN's dialogue partner countries' "unequivocal support for ASEAN centrality", as well as their declared hope that "ASEAN would continue to play a central role in the emerging regional architecture".[29] The need for such public expressions of support for ASEAN only serves to highlight the likely angst felt by the Association following the incessant questions over its ability to "drive" Asian regionalism. At the same time, it is unlikely that the criticism of ASEAN centrality will abate any time soon, so long Asia's regional architecture remains as it is.

It should be noted too that debates over ASEAN centrality are not only taking place outside of the Association. While all ASEAN states undoubtedly continue to press for ASEAN centrality, within the Association there have been rumblings of discontent as a result of differences between those who see ASEAN centrality as a means to attain greater regional stability and integration, and others who favour ASEAN centrality as an end in itself; and in so doing could frustrate the malleability that is critical for the building of a regional architecture on an immensely fluid geo-strategic terrain.

Inclusive versus Exclusive Regionalism

Another bugbear for regionalism in East Asia has been the issue of whether Asia's institutions ought to be region-centric. This concern goes as far back as Mahathir's proposal for the EAEG, later changed to EAEC, which, as noted earlier, some have regarded as the precursors to both ASEAN+3 and EAS.[30] Memorably, the Malaysian leader sought to keep Australia and the United States out of Asian regionalism.[31] On its part, China has sought indirectly to limit the role and influence of the United States in regionalism. From a reluctant participant in the ARF, which it initially perceived as a tool for a Washington-led encirclement, China today has evolved into a clever connoisseur of multilateralism in the service of its own interests. ASEAN+3 and the depth of ASEAN-China ties are good illustrations of Beijing's growing influence in Asia as facilitated by regionalism.

At the same time, Beijing's initial enthusiasm for the EAS fizzled out when it became clear that others wanted the summit to include Australia, India and New Zealand, presumably to widen the area of economic integration, but also as a countervailing attempt against the possible replication in the EAS of the perceived Chinese dominance of ASEAN+3.[32] Subsequent indications of a possible interest on the part of the United States to join the EAS, especially following Washington's accession to the ASEAN TAC in July 2009, likely cooled Beijing's commitment to the EAS even further. In this respect, while China has not openly sought to exclude the participation of the United States in East Asian regionalism, it continues to be very serious about what David Shambaugh calls the "balance of influence" in Asia,[33] and has worked assiduously to enhance its image and influence over Asian countries relative to that of America. For example, Beijing's hosting of the first ever summit of the Mekong River Basin countries (including Myanmar) in April 2010 has been seen by some as a public diplomacy effort to balance out the goodwill earned by Washington from its accession to the TAC.[34]

Regardless of several attempts to define regionalism in exclusive terms as an Asian affair, the fact remains that a key feature of Asia's institutional landscape continues to be its open and inclusive character. As Prime Minister Lee Hsien Loong of Singapore remarked in his keynote address at the 2006 edition of the Shangri-La Dialogue: "The most robust and stable configuration for regional cooperation is an open and inclusive one."[35] Analysts have also taken note of both the economic and security versions of "open regionalism" that ostensibly define Asia's brand of regionalism.[36] It bears reminding that ASEAN's early years following the Confrontation (Indonesia's undeclared war-1963 to 1966—opposing the formation of the Federation of Malaysia,) and the Vietnam War were marked by concerns about the potential interference of extra-regional powers in South East Asia. The concern found a rhetorical expression in the Zone of Peace, Freedom and Neutrality (ZOPFAN) idea in the 1970s, which was never realised. Indeed, the regional aspiration for neutrality did not prevent ASEAN states from remaining robust allies of the United States (Philippines and Thailand) and members of the Five Power Defence Arrangement (Malaysia and Singapore). But more than these, the developments that perhaps best embody ASEAN's proactive engagement with external interactions are its "ASEAN+1" dialogue partner and Post-Ministerial Conference (PMC) formulae, which laid the groundwork for wider regional arrangements such as APEC and ARF. If anything, the rocky road from ZOPFAN to ARF highlights the gradual acceptance by the ASEAN states, particularly Indonesia, of the need to proactively engage the big powers rather than eschew them.

Open regionalism implies that regionalism in East Asia is by no means a purely indigenous and hermetically contained enterprise. It also means that it was possible to accommodate different models of regionalism—admittedly, uneasily at times—within open regionalism's flexible parameters.[37] Regional ambivalence has allowed traditionally "non-Asian" states to participate in the various "Asian" regionalisms. As Julia Gibson has noted: "The changing shape of East Asian institutions and debates over the very meaning of the region provide ample space for their legitimate inclusion."[38] The seemingly off hand way in which the ASEAN sometimes responds to questions regarding institutional architecture and membership composition can be attributed to the absence of a collective ASEAN position and policy on these issues. At the same time, it also suggests that ASEAN's ultimate concern is to ensure its centrality in those institutions: "The core concern of the ten Southeast Asian members of ASEAN is not so much the scope of membership of regional institutions as in ensuring the centrality of Southeast Asia's position within them. The approaches of individual Southeast Asian countries differ but they all want to secure a role for ASEAN in any regional architecture."[39] At the very least, Asian regionalism has not been confined to geography for political as well as functional reasons. Indeed, the relative inconsequentiality of geography in some regional arrangements only underscores the issue-specific or functionalist rationale that underpins them.[40]

Process versus Outcomes

The foregoing discussion takes us to another important consideration referred to earlier: so far as ASEAN is concerned, should its "default centrality" in Asian regionalism be a means to an end or purely an end in itself? This question is at the heart of the debate over whether East Asian regionalism is oriented towards process or results.

In the immediate post-Cold War period, with its standing as *primus inter pares* in Asia-Pacific more or less assured, ASEAN was celebrated as a regional organisation *par excellence* owing to a most unusual development in the annals of international relations: a Third World regional organisation defining the terms of regional security in a region which was of acute interest for big powers.[41] With its leading role in the creation of the ARF, a multilateral security forum servicing the entire Asia-Pacific region, ASEAN became not only the hub of a regional initiative involving the world's major powers (China, Japan, Russia and the United States), but in this instance it in effect shaped the form and content of Asian regionalism.

Consequently, in attempting to explain this apparent structural anomaly, much has been made of the ASEAN Way, the model of regional security to which ASEAN bestowed the ARF. With confidence-building very much at the core of the ARF's activities, it was argued that the ASEAN Way—which emphasizes consultation, flexible consensus, informality, institutional minimalism, cooperative security, and non-interference—so crucial to ASEAN's own success story, was the appropriate model for Asia-Pacific security cooperation if not community building.[42] And while newer regional institutions such as ASEAN+3 and EAS do not claim (not publicly at least) the ASEAN Way as their diplomatic model, their status as ASEAN-centric arrangements effectively ensures that ASEAN's way of doing business—no different really from how the world of sovereign states conducts its affairs, according to a former secretary-general of ASEAN[43]—is equally the *modus operandi* of regionalism in Asia in general. Even so, analysts generally supportive of the ASEAN Way have also argued for the need for the model to evolve to satisfy the demands of an increasingly transnational Asia, where the logics of state sovereignty and non-interference are no longer universally applicable.[44] In fact, the formulation of the ASEAN Charter is indicative that the Association is itself aware of the need to recalibrate aspects of its traditional norms.

Others argue that neither the ASEAN Way nor ASEAN centrality deserve the valorisations accorded them by ASEAN's backers. Critics contend that no matter the extent of regional aspiration expressed in the seminal documents and evolutionary plans of action common in Asian regionalisms, concrete institutional progress is not what those regionalisms are designed to achieve because ASEAN-led regional initiatives are all about maintaining ASEAN's pride of place and little else—a conflation of means and end, as it were.[45] To that effect, the ASEAN way is seen by its detractors as a key reason for the yawning gap between upbeat institutional visions and harsh regional realities. Hence the contention that East Asian regionalism is all process, since—so the reasoning goes—few to no concrete results can come about so long as the tacit aim of the ASEAN Way is essentially to preserve, not revise, the regional status quo. Or, as the late Michael Leifer once put it, ASEAN—and, by extension, its complex of region-wide institutions—is all about conservation, not innovation.[46]

One criticism of the "East Asian regionalism is all process" claim is, that institutional efficacy is sacrificed at the altar of regional cooperation and integration. Regionalism in East Asia continues to be characterised by institutional minimalism. For example, the region's institutions operate under few explicit procedural rules and delegate few tasks to standing secretariats—indeed, most institutions have no secretariats; the ASEAN secretariat provides administrative support to the ARF, ASEAN+3 and EAS, not to mention its parent organisation, ASEAN—and specify few obligations for their members. Second, despite a growing web of overlapping institutions, regionalism in Asia is still predominantly intergovernmental in nature rather than supranational where sovereignties are pooled, or where authority is delegated to nongovernmental bodies. Third, as the early confusion over whether ASEAN+3 or EAS is the appropriate regional vehicle for building the East Asian Community clearly showed, Asia's overlapping institutions are potentially in competition rather than complementary to one another.[47] Fourth, reliance on a consensus-based decision-making system—flexible rather than unanimous, it should be said—and the emphasis on preserving the "comfort level" of all participants, when combined especially with sizeable memberships, complicate attempts to reach agreement on effective cooperation.[48] Much in the same vein, Richard Woolcott, the Australian special envoy tasked to promote Rudd's Asia-Pacific Community vision, speaking about the ARF said: "many believe it is too large and has made insufficient progress since its inception."[49] Finally, Asia's institutions are primarily confidence-building exercises that aim, at the most, to create an environment conducive to incremental integration.

Given these "limitations", critics argue that the region's arrangements are not designed for efficiency, concrete action, and progress commensurate with their self-professed ambitions.[50] For the most part, they are not persuaded by the ASEAN's charter of November 2007, and are concerned that this apparent shift to a rules-based regionalism would further entrench South East Asia's intergovernmental character by consequence codifying ASEAN's longstanding sovereignty and non-interference norms.[51] As noted earlier, the chorus of appeals to streamline the extant regional institutional architecture stems from the need for more effective institutions. "The Asia-Pacific region has too many organisations, yet they still cannot do all the things we require of them", as a prominent analyst has noted. "Instead of focusing on what we've got, we should look at what we need."[52] More recently, some have sought to reconcile the tension between process and outcome. Mostly unapologetic about regionalism in Asia, they note that a process-based regionalism is a central feature of regional security management by a grouping of small states with particular security calculations and domestic power configurations.[53]

Conclusion: What to Expect from the EAS?

What might we expect of the EAS, particularly given that it is still in its infancy? The EAS should at best be regarded as another confidence-building exercise, at least in the short to medium term, especially in view of the probable inclusion of the United States and Russia. Critics see little institutional change being driven by the EAS due to its inability to meet regularly and to agree on a roadmap and a set of collaborative issues at this stage of its development. They add that the only point of convergence among its 16 participants—18, once the Americans and Russians are formally included—might well be their willingness to let ASEAN assume the leadership of this latest institutional forum. It can be argued, however, that the relevance of the EAS to Asia's stability will depend on two crucial factors.

First, the EAS should neither be regarded as a replacement for the ARF or ASEAN+3 nor as an embryonic structure eventually leading to the creation of an alternative security architecture for East Asia. To be sure, there remains a low possibility that the EAS could inadvertently sideline the ARF, if only because the EAS is a leader-driven forum whereas the ARF is only a foreign ministers' conclave. But given that the ASEAN serves as the hub for both arrangements, it is unlikely that the Association would let the ARF lapse into irrelevance (although some might argue that has already taken place). On the contrary, the EAS ought to complement existing cooperative arrangements and contribute to the emerging security architecture defined by the trends and driving forces discussed above. However, much would be required not just of the ASEAN but of all regional participants to ensure that the agendas, interests and roles of the EAS and its institutional counterparts in East Asia are neither at odds nor a duplication of the others'.

Second, the EAS should be viewed not only as a confidence-building enterprise—a central feature of all forms of Asia-Pacific institutionalism—but also as a future platform for substantive cooperation. In other words, the EAS will need to graduate from a nascent institution for addressing broad concerns and generalised confidence-building to a regional mechanism armed with a thematic and problem-oriented agenda. This is a major challenge for not only the EAS, but for all East Asian regionalisms. In this regard, it is noteworthy that the ARF is seeking to move beyond dialogue to practical security cooperation in the area of transnational and/or non-traditional security concerns.[54] Likewise, the latest regionalism to be added to the mix, ADMM+8, is aimed at facilitating and enhancing regional defence cooperation between and among the militaries of its member countries in humanitarian assistance and disaster relief efforts, maritime security, and the like. While these examples hint at an emerging regional consensus on the need to move beyond talking and confidence-building, it remains to be seen whether these institutions will realise these new expectations, or (as their critics would insist) go the way of regionalism at large in emphasising process over results. Similar questions confront the EAS, particularly for the medium and the long term.

That said, it is imperative, at this early stage of the game, that members of the EAS establish a level of comfort amongst themselves. While the ASEAN countries have had four decades of collective experience in regional reconciliation, such opportunities have not been extended to the North East Asian members of the EAS, whose relations with each other have largely been confined to bilateral ties and the Six Party Talks. Likewise, countries such as Australia, India and New Zealand also require time to establish a level of confidence with their counterparts from East Asia. Needless to say, institutions with no other aim than confidence-building do not get far. It is therefore critical that the EAS move forward in due course to a substantive collaboration on the complex issues and challenges that affect the region.

All this is to say that while the creation of the EAS should be lauded, it will also at some stage have to redefine itself in functional and issue-specific terms in order to help in establishing a firmer foundation for the regional architecture. Issues of interest and great urgency would include terrorism, maritime security, energy challenges and climate change, as well as health security. Importantly, such initiatives should be undertaken in greater cooperation with the ARF to promote a more effective regional security cooperation.

Notes

1. See Joseph Liow Chinyong, "Malaysia-China Relations in the 1990s: The Maturing of a Partnership", *Asian Survey* 40(4) (July/August 2000): 672-91.
2. Kim Dae Jung, "Regionalism in the Age of Asia", *Global Asia* 1(1) (2006): 11.
3. The document is available at www.mofa.go.jp/region/asia-paci/report2001.pdf
4. See Toru Oga, "Open Regionalism and Regional Governance: A Revival of Open Regionalism and Japan's Perspectives on the East Asia Summit", *Interdisciplinary Information Sciences* 15(2) (2009): 181-2.
5. See "Kuala Lumpur Declaration on the East Asia Summit", Kuala Lumpur, 14 December 2005 http://www.aseansec.org/18098.htm
6. Item 12 of the Chairman's Statement read: "We welcomed ASEAN's efforts towards further integration and community building, and reaffirmed our resolve to work closely together in narrowing development gaps in our region. We reiterated our support for ASEAN's role as the driving force for economic integration in this region. To deepen integration, we agreed to launch a Track Two study on a Comprehensive Economic Partnership in East Asia (CEPEA) among EAS participants. We tasked the ASEAN Secretariat to prepare a time frame for the study and to invite all our countries to nominate their respective participants in it." The Chairman's Statement of the Second EAS is available at http://www.aseansec.org/19302.htm
7. Available at http://www.aseansec.org/24509.htm
8. Interview with a senior Thai Foreign Ministry official, Bangkok, September 30, 2010.
9. "US, Russia to join EAsia group", *Straits Times*, 20 July 2010.
10. Interview with a senior Foreign Ministry official, Singapore, September 13, 2010.
11. *Ibid.*

12. Interview with Australian Foreign Ministry official, Canberra, 7 April 2010.
13. See Seng Tan, "The 'Asia-Pacific Community' Idea: What Next?" *RSIS Commentaries* 49, 25 May 2009, http://www.rsis.edu.sg/publications/Perspective/RSIS0492009.pdf
14. See Joseph Chinyong Liow, "Southeast Asia in 2009: A Year Fraught with Challenges" in Daljit Singh (ed.), *Southeast Asian Affairs 2009,* Singapore: Institute of Southeast Asian Studies, 2010.
15. "What is Australia's Asia-Pacific idea all about?" *Reuters*, 24 October 2009, www.reuters.com/article/idUSSP389695
16. Brad Norington, "Barack Obama's man Kurt Campbell junks Kevin Rudd's Asia-Pacific plan", *The Australian*, June 12,2009.
17. Cheong Wa Dae, "President announces 'New Asia Initiative' ", *Korea.net*, March 8, 2009, http://www.korea.net/news/News/NewsView.asp?serial_no=20090308001&part=101&SearchDay=
18. "Japan's new premier pitches East Asia union", *China Daily*, September 29, 2009, http://www.chinadaily.com.cn/world/2009-09/23/content_8724372. htm
19. Aurelia George-Mulgan, "Hatoyama's East Asia Community and regional leadership rivalries", *East Asia Forum*, October 13, 2009, http://www.eastasiaforum.org/2009/10/13/hatoyamas-east-asia-community/
20. The PECC was formed in 1980 as an epistemic community consisting of government officials, private sector representatives, and academics from 26 Asia-Pacific countries. Today, the PECC provides information and research support for APEC ministerial meetings and working groups.
21. Hadi Soesastro, Allan Gyngell, Charles E. Morrison and Mr. Jusuf Wanandi, "Report of Regional Task Force on Regional Institutional Architecture", prepared for the PECC Standing Committee, June 2009, unpublished.
22. *Ibid.*, p. 16.
23. See Seng Tan (ed.), *Do Institutions Matter? Regional Institutions and Regionalism in East Asia*, RSIS Monograph No. 13, Singapore: S. Rajaratnam School of International Studies, 2008.
24. Alan D. Romberg, "The East Asia Summit: Much Ado About Nothing—So Far", *Freeman Report* by the Center for Strategic and International Studies (December 2005), http://www.stimson.org/pub.cfm?id=264
25. This concern also recalls the longstanding debate over the relevance and efficacy of existing institutions, whether their respective mandates and agendas are congruent or competitive, and so forth. See, See Seng Tan, "Introduction", in See Seng Tan, *Regionalism in Asia, Vol. III: Regional Order and Architecture in Asia*, London: Routledge, 2010, p. 2-4.
26. Carlyle A. Thayer, "Kevin Rudd's multilayered Asia Pacific Community initiative", *East Asia Forum*, June 22, 2009, http://www.eastasia

forum.org/2009/06/22/kevin-rudds-multi-layered-asia-pacific-community-initiative/

27. See Seng Tan, "ARF: Ad hoc Regional Forum?" *The Straits Times*, July 30, 2010.
28. Shiping Tang and Haruko Satoh, "Can China and Japan Think Together?" *PACNET* No. 52, December 29, 2007.
29. "ASEAN Centrality on Centre-Stage, Hanoi, July 23, 2010", in *ASEAN Bulletin* (July 2010), http://www.aseansec.org/24842.htm
30. Masahiro Kawai, "Evolving Economic Architecture in East Asia", *ADB Institute Discussion Paper*, No. 84 Tokyo: Asian Development Bank Institute, December 2007 p. 18.
31. Gregore Lopez, "Mahathir's Regional Legacy", *East Asia Forum*, June 17, 2010, http://www.eastasiaforum.org/2010/06/17/mahathirs-regional-legacy/
32. Masahiro Kawai and Ganeshan Wignaraja, *ASEAN+3 or ASEAN+6: Which Way Forward? ADB Institute Discussion Paper*, No. 77, Tokyo: Asian Development Bank Institute, 2007; Frank Frost and Ann Rann, "The East Asia Summit, Cebu, 2007: issues and prospects", E-Brief, 1 December 2006, updated 20 December 2006, Canberra: Parliamentary Library, Parliament of Australia, 2006. www.aph.gov.au/library/intguide/FAD/eastasia_summit 2007.htm.
33. David Shambaugh, "China Engages Asia: Reshaping the Regional Order", *International Security* 29(3) (Winter 2004/05): 64-99.
34. This observation was made by a prominent Thai journalist, Kavi Chongkittavorn. Cited in Marwaan Macan-Markar, "China, US angle for Mekong influence", *Asia Times Online*, June 15, 2010, http://www.atimes. com/atimes/China/LF15Ad01.html
35. Keynote Address by Prime Minister Lee Hsien Loong at the fifth International Institute for Strategic Studies (IISS) Asia Security Summit (Shangri-La Dialogue), June 2, 2006, Shangri-La Hotel, Singapore, http://www.iiss.org/conferences/the-shangri-la-dialogue/shangri-la-dialogue-archive/shangri-la-dialogue-2006/2006-plenary-session-speeches/keynote-remarks-he-lee-hsien-loong/
36. On open economic regionalism, see Ross Garnaut, *Open Regionalism and Trade Liberalization: An Asia-Pacific Contribution to the World Trade System*, Singapore: Institute of Southeast Asian Studies, 1996. On open security regionalism, see Acharya, "Ideas, Identity, and Institution building: From the 'ASEAN Way' to the 'Asia-Pacific Way'", The Pacific Review 10(3)(2005): 319-346.
37. Joseph Camilleri, "East Asia's Emerging Regionalism: Tensions and Potential in Design and Architecture", *Global Change, Peace and Security* 17(3) (2005): 253-261; James R. Kurth, "The Pacific Basin versus the Atlantic Alliance: Two Paradigms of International Relations",

Annals of the American Academy of Political and Social Science 505 (September 1989): 34-45.

38. Julia Gibson, "Regionalism and security in East Asia", in Anthony Burke and Matt McDonald (eds.), *Critical Security in the Asia-Pacific,* Manchester: Manchester University Press, 2007, p. 39.
39. Soesastro *et al*, "Report of Regional Task Force on Regional Institutional Architecture", n. 21, p. 13.
40. Nonetheless, functionally oriented institutions could impede regionalism because they do not hold strictly to a geographically bounded conception of region and specifications. Andrew Hurrell, "Regionalism in Theoretical Perspective", in Louise Fawcett and Andrew Hurrell (eds.), *Regionalism in World Politics: Regional Organization and Regional Order,* Oxford: Oxford University Press, 1995, pp. 37-73.
41. This "structural anomaly" is discussed in Leifer, *The ASEAN Regional Forum: Extending ASEAN's Model of Regional Security*, and Yuen Foong Khong, "Review Article: Making Bricks without Straw in the Asia-Pacific?", *The Pacific Review* 10(2) (1997): 289-300.
42. Acharya, "Ideas, Identity, and Institution building ...", n. 36. In fact, the ASEAN Way was already being feted as early as 1989; see Michael Haas, *The Asian Way to Peace: A Story of Regional Cooperation* New York: Praeger, 1989.
43. Against the notion that sovereignty, non-interference and the like are specific to the ASEAN Way, Severino writes, "It must be noted at the outset that the so-called 'ASEAN Way' is actually the way of the world, the world being, like ASEAN, made up of sovereign states that are not subject to any extra-national authority." Rodolfo Severino, "Regional Institutions in Southeast Asia: The First Movers and Their Challenges", Background Paper 24, Asian Development Bank (ADB) Study Finalization Workshop on Institutions for Regionalism in Asia and the Pacific, Shanghai, 2-3 December 2009.
44. Mely Caballero-Anthony, *Regional Security in Southeast Asia: Beyond the ASEAN Way* Singapore: ISEAS, 2005.
45. David Martin Jones and Michael L.R. Smith, "Making Process, Not Progress: ASEAN and the Evolving East Asian Regional Order", *International Security* 32(1) (Summer 2007): 148-84.
46. Michael Leifer, *ASEAN's Search for Regional Order,* Singapore: Faculty of Arts and Social Sciences, National University of Singapore, 1987 p. 21.
47. Tan, *Do Institutions Matter?,* n. 23.
48. According to Olson's seminal work on group size and collective action, the larger the group *ceteris paribus* the less likely its members will choose to cooperate for some joint gain. Mansur Olson, *The Logic of Collective Action: Public Goods and the Theory of Groups,* Cambridge, MA: Harvard University Press, 1965.

49. Richard Woolcott, "Towards an Asia Pacific Community", *The Asialink Essays* 9 (November 2009): 3.
50. On institutional design and efficacy, see Acharya and Johnston, "Comparing Regional Institutions: An Introduction", in Amitav Acharya and Alastair Iain Johnston (eds.), *Crafting Cooperation: Regional International Institutions in Comparative Perspective,* Cambridge: Cambridge University Press, 2007 p. 1-31; Etel Solingen, "The Genesis, Design and Effects of Regional Institutions: Lessons from East Asia and the Middle East", *International Studies Quarterly* 52(2) (June 2008): 261-94.
51. Eugene K.B. Tan, "The ASEAN Charter as 'Legs to Go Places': Ideational Norms and Pragmatic Legalism in Community Building in Southeast Asia", *SYBIL* (Singapore Year Book of International Law) 12 (2008): 171-98.
52. Alan Gyngell, "Design Faults: The Asia Pacific's Regional Architecture", *Lowy Institute Policy Brief* (July 2007).
53. See Alice Ba, *(Re) Negotiating East and Southeast Asia: Region, Regionalism, and the Association of Southeast Asian Nations*, Stanford: Stanford University Press, 2009; Deepak Nair, "Do Stated Goals Matter? Regional Institutions in East Asia and the Dynamic of Unstated Goals", RSIS Working Paper No. 199, Singapore: S. Rajaratnam School of International Studies, June 2010.
54. See Jürgen Haacke, "The ASEAN Regional Forum: From Dialogue to Practical Security Cooperation?", *Cambridge Review of International Affairs* 22(3) (2009): 427-49, and Rodolfo Severino, *The ASEAN Regional Forum,* Singapore: ISEAS, 2009.

Section Four

WIDENING THE SECURITY AMBIT

12

Technology, Security and State Development in East Asia and India

Tai Ming Cheung

From South Asia to North East Asia, states across the Asia-Pacific region regard technological innovation as vital to their economic competitiveness and national security. Technological drivers have played a central role in the transformation of many regional countries into advanced economic powers over the past sixty years. Japan, Taiwan and South Korea (Republic of Korea) have successfully turned from industrial latecomers into technological frontrunners, while China is rapidly catching up. India is also beginning to realise its considerable technological potential, especially in areas such as information technology (IT) and software development. Technological considerations loom large in some of the region's most acute security problems, from nuclear proliferation and rivalry on the Korean Peninsula and South Asia to military tensions in the Taiwan Strait.

This paper examines the role that technological development has played in shaping the relationship between economic development and security among states in East Asia over the past sixty years. Particular attention will be paid to the evolution of grand strategic thinking on technology and innovation and the impact this has had on economic development and national security, especially in Japan, South Korea, China and India. What are the core strategic and ideological values and interests that shape national approaches? What role have nationalist and globalist factors played? What are the key long-term technological trends and how will they affect East Asia's economic and security dynamics?

Defining Techno-nationalism, Techno-globalism and Techno-hybridism and Their Place in Grand Strategic Thinking

A useful starting point for examining the role that technological considerations play in East Asia's international relations is to understand how the grand strategies that states pursue are forged. A state's grand strategy represents the overarching vision of how a ruling coalition coordinates and utilises economic, military, diplomatic, technological and other capabilities to achieve national goals.[1] Achieving such a consensus is often contentious and requires extensive deliberation among contending domestic political coalitions with diverging views as to what constitutes the state's core interests, the nature of external threats, and what are the most suitable means and ends to pursue.

In her examination of how competing groups in industrializing states approach external cooperation and conflict, Etel Solingen argues that these domestic coalitions may be categorised into two ideal types. The first group may be labelled as "internationalists", as they are in favour of economic liberalisation, promotion of regional peace, and access to foreign capital and technology. On the opposing side are "statist-nationalists" who resist internationalisation, liberalisation and peace and who support the maintenance of strong military and defence-industrial establishments by raising tensions.[2]

This coalitional framework thesis may also be applied to the technological realm where similarly aligned factions of techno-globalists compete against techno-nationalists. Within this model, a third, more centralist, coalition exists in which moderate elements of the two other groups come together to advocate a techno-hybrid approach. These three schools of thought may be described as follows:[3]

- Techno-nationalists believe that only a state-controlled and closed-door approach to technological innovation can safeguard national security, economic competitiveness and international status. Emphasis is placed on nurturing indigenous capabilities through adoption of highly regulated protectionist regimes that prevent foreign direct investment (FDI) but encourage one-way importation of advanced technology and knowledge.
- Techno-globalists advocate a bottom-up market-defined approach that stresses borderless collaboration and unrestricted flow of knowledge, technology, talent, trade and foreign investment. Firms are the dominant drivers of the innovation process, while governments play a secondary role in supporting fundamental research and development (R&D).

- Techno-hybridists are pragmatists who seek to integrate the duelling techno-nationalist and techno-globalist ideologies into a balanced operational model. They believe that the state and firms/markets play complementary roles either working together through public-private partnerships or occupying different components of the innovation system. State involvement should be focused towards nurturing basic R&D capabilities such as national laboratories and universities while firms can participate in the global marketplace, especially multinational production and innovation networks.

Table 1: Defining Characteristics of the Three Schools of Technological Grand Strategy

	Techno-nationalism	*Techno-globalism*	*Techno-hybridism*
Dominant Players	The State	Multinational firms and global markets	Balanced relation-ship between the state, firms and mar-kets
Comparative International Relations Theory	Realism	Liberal institutionalism	None
Innovation Model	National innovation systems	Global production and innovation networks	Multiple drivers of innovation at region-nal, national and global levels
Variants	Commercial or military	Commercial	Primarily commercial, secondary military

Economics, Security and Technology in East Asia from the 1950s to Early 1990s: The Era of Strong Techno-nationalism

For states in East Asia, the competing coalitional setup during the second half of the twentieth century was not between techno-nationalists and techno-globalists but among different sub-sets of techno-nationalist thinking. As the region emerged from war and foreign occupation in the late 1940s and '50s, the focus of national policymakers was on state building and economic development. Their thinking and approaches to these priorities were framed by two common and powerful influences: nationalism and security. As Meredith Woo-Cummings has persuasively argued, the rise of the developmental state in Japan, South Korea and Taiwan, along with China and North Korea (Democratic People's Republic of Korea) was decisively shaped by war, colonialism and the fear of war.[4] This could also be applied to India. These experiences accounted for a distinctive overarching regional pattern of industrialisation and technological development that was geared towards a "perpetual mobilisation for war" during the Cold War.[5]

In the technological arena, these shared security concerns—although viewed from opposing sides—led the East Asian states to become ardent proponents of the techno-nationalist school of grand strategic thinking in their initial stages of industrialisation, especially in placing national security considerations at the very top of their development priorities. This accounted for the region-wide emphasis on building heavy, strategic and technology-intensive industries.

Two distinct types of techno-nationalist grand strategies and models were adopted by East Asian states during this period: commercial and military techno-nationalism. The commercial variant emphasises the importance of technological autonomy for economic security, especially to allow home-grown vertically integrated firms to be competitive in the international marketplace. Military techno-nationalism stresses the overriding importance of military-security priorities in technological development. How states made their choices depended on the nature of their political and economic systems, whether those in power supported or opposed international engagement, and their strategic alignments within the Cold War system (Table 2).

States allied with the US and with leaderships that supported international economic liberalisation, such as Japan, South Korea and Taiwan, pursued the commercial techno-nationalist model.[6] Communist regimes in China and North Korea adhered to military techno-nationalist models of development because of the dominance of statist-nationalist groups in charge that favoured the building of command economies and feared attack from the US and its regional allies.

These techno-nationalist coalitions may also be differentiated by how assertive or accommodating they have been in the pursuit of their goals. Those who advocated aggressive, uncompromising approaches in their technology development policies such as emphasising complete self-reliance and focusing overtly on military priorities may be defined as strong techno-nationalists. Those who favoured a more accommodative and open posture may be labelled as pragmatic techno-nationalists. This differentiation may be likened to the offensive-defensive realist divide in international relations theory.

Table 2: Key Features of the Military and Commercial Versions of the Techno-Nationalist Development Model

	Military Techno-Nationalist Development Model	*Commercial Techno-Nationalist Development Model*
Key Regimes	Pre-1978 China, North Korea, pre-World War II Japan	Post-1945 Japan, South Korea, Taiwan, post-1978 China
Development Pri-orities	Defence industrial base and heavy industries	Heavy, export-oriented and high-technology industries (iron and steel, chemicals, electronics)
Threat Environ-ment and Security Posture	Acute external threats, militarization of entire state	Severe external threats, signify-cant defence outlays, heavy reliance on security protection from the US
Orientation and Nature of Economy	Closed autarkic system; command-style economy	Semi-open economy with exten-sive trading links with Western countries; market-oriented economic structure
Science and Technology Drivers	Overwhelmingly state driven	Important roles played by the state, firms, and importation of foreign knowledge and techno-logy transfers
Technology Outputs	Military equipment, heavy industrial equipment	Consumer electronics and retail goods, heavy industrial equip-ment

The Strong Commercial Techno-nationalist Model of Japan and South Korea

With their successful economic takeoff in the 1970s and '80s, Japan and South Korea, though the latter lagged behind Japan, were able to bridge the technological gap and join the ranks of the world's leading technological powers through the pursuit of aggressive commercial techno-nationalist strategies, became role models for industrial latecomers. Both countries targeted the development of capital-intensive, technologically advanced mass manufacturing industries in which innovation was focused on incremental process refinement, foreign interaction was confined predominantly to the importation of technologies and know-how, and the state and major business conglomerates played a central role in guiding this development.

Both countries had an insatiable appetite for technological imports and transfers. They initially concentrated on the manufacturing and heavy industrial sectors that offered economies of scale. For Japan, the focus was on manufacturing technologies in sectors such as iron and steel, automobiles, electronics and communications equipment.[7] Through intensive efforts in reverse engineering, diffusion, assimilation and heavy investment in the development of indigenous R&D capabilities, Japanese and South Korean firms were able to master these technologies and begin to make gradual improvements. Japanese firms in particular excelled at incremental process innovation because of the high degree of integration of their R&D, production and management systems. But a sizeable number of Japanese industries also failed to become internationally competitive despite extensive government and institutional support: they include the chemical, civil aviation, software and mainframe computer industries. Some of the key reasons for the poor performance of these sectors included a shortage of trained talent, inadequate R&D investment and misalignment between the domestic and global markets.[8]

South Korea's efforts in catching up were still more uneven. A major reason was that it was far behind Japan technologically and organisationally.[9] From this lower baseline, South Korean firms had to make a steep and concerted ascent up the technological ladder that was extremely costly and time-consuming. When South Korean companies began to make noticeable technological advances in the 1980s and early 90s, however, they only had limited resources and had to selectively target a narrow range of areas in which to make additional progress in catching up. Moreover, South Korea's technological output was primarily for export as the local market was too small to absorb rising levels of production. This severely restricted the ability of different industries to coordinate and link their development with each other. For example, the semiconductor industry grew autonomously and had little interaction with the computer or telecommunications sectors. In these circumstances, a handful of industries that enjoyed sustained support from the country's large family-owned business conglomerates (known as *chaebols*) and the state, such as the DRAM, CDMA, automobile and shipbuilding sectors, emerged to become world leaders in their technological areas, while the personal computer and consumer electronics industries struggled and ultimately failed in their catch-up efforts. Some analysts credit the achievement of South Korean firms and industries that were able to narrow—and in some instances eliminate—the technology gap, to their ability to effectively integrate domestic and external technological resources under a strict management regime. Outsourcing of R&D and production activities both domestically and overseas was also an important route for catching up.[10]

Japan and South Korea, while eager to acquire foreign technology, adopted stringent protectionist policies that shut out FDI and restricted the ability of foreign companies to operate in their domestic markets or even to establish joint ventures. Up until the Asian financial crisis in the late 1990s, foreign ownership and control of major Japanese and South Korean firms was virtually impossible because of complex cross shareholdings among domestic institutions. Moreover, the Japanese government set challenging domestic industrial and technological standards regimes that essentially excluded the participation of non-Japanese firms.

Institutionally, the Japanese and South Korean governments worked closely with banks and major business groups to shepherd the catch-up process among firms because they were able to effectively coordinate private activities with public policies.[11] In Japan, the Ministry of International Trade and Industry (MITI) and *Keiretsu*—business groups—cooperated to ensure that there was adequate investment in R&D to promote long-term product and process innovation. In South Korea, the *Chaebols* emerged in the 1980s to play the same role as the *Keiretsu* and they were the force behind the country's technological take off.

While Japan and South Korea were allowed to pursue techno-nationalist policies in the civilian sphere, this did not extend into the defence arena. Both countries were required to enter into collaborative co-production and co-development projects, as this was the only way that the US would allow transfer of sensitive defence-related technologies, especially from the 1990s as products became increasingly more complex. The Japanese and South Korean defence industries and military establishments strongly opposed this dependence. Attempts were made by both countries to develop their own weapon systems, with mixed results. Both countries were most successful in defence electronics, armoured vehicles and warships as they were able to leverage their commercial strengths in these technologies, but both had to rely on the US for critical equipment such as fighter aircraft, air defence systems and missiles.[12]

Defence technological and industrial cooperation between the US and Japan and South Korea often resembled shotgun marriages between unhappy and distrustful partners, given that techno-nationalist sentiments ran strong and deep in the defence industries and military establishments of all these countries. This was most vividly demonstrated in the US-Japan FS-X fighter aircraft programme, a poorly conceived and badly executed co-development project in the 1990s set up to produce a new combat aircraft for the Japanese Air Self-Defence force. The US wanted Japan to acquire an existing fighter plane, the F-16C, and undertake some limited modifications; but Japan sought to develop essentially a brand new fighter using the F-16 as a baseline reference.[13] The result was a long-delayed and enormously expensive programme that produced the Mitsubishi F-2 fighter, which was only a modest incremental improvement over the original US version although it was refitted with Japanese avionics and technologies. But the cost of the F-2 was almost three to four times higher than the F-16, that Japan has been able to buy only a limited number.

Although an aggressive commercial techno-nationalist grand strategy was instrumental in effecting the transformation of Japan and South Korea from imitative laggards to innovation frontrunners in the late industrial era, this rigid and closed approach started becoming a serious obstacle to their continuing progress and competitiveness in the 1990s. Globalisation and the emerging IT revolution began to decisively shift the drivers responsible for economic, technological and military development from states and national economies to multinational corporations and global production and innovation chains. Analysts argued that this meant that Japan and South Korea's techno-nationalist development model was no longer relevant. Liu Xielin pointed out that "the features of Japan and Korea's catch-up process seem outdated and in some critical ways inappropriate given recent economic, technological and social changes in the global environment."[14] The era of strong commercial techno-nationalism appeared to be over at the same time as the closing of the twentieth century.

The Strong Military Techno-nationalist Model: China and North Korea

In contrast to the commercial focus of its North East Asian neighbours during the Cold War, China and North Korea went down a development path that stressed military priorities in the relationship between economics, security and technology in their early industrialisation drives. This was not surprising as both states were preoccupied with regime survival in the face of acute security threats between the 1950s and the end of the 1970s. In addition, the militaries in China and North Korea wielded extensive political and policymaking power and influence that ensured that they had priority access in resource allocations and technology and economic development plans.

The key features of the military techno-nationalist grand strategy that was adopted by China—some elements of which were also followed by North Korea—were:[15]

- Technological development is strategic and has implications for the relative position of the state in the global military and economic balance, or what Chinese strategists term as comprehensive national strength.

- The state must invest in critical technological sectors because of the high risks and long time cycles involved in high-technology R&D.
- The state should pursue import-substituting indigenisation.
- The state must nurture an indigenous capacity to innovate.
- Technology diffusion, whether through spin-offs or spin-ons, should be a central long-term goal.

From the 1950s to the end of the 1970s, China applied this grand strategy to building conventional and strategic defence industrial sectors that eventually overshadowed the rest of the national economy. The Chinese techno-nationalist model was known as the "Two Bombs, One Satellite" or *Liangdan Yixing* ideology.[16] This referred to the defence economy's crowning success in the development of nuclear weapons and space capabilities in the 1960s and 70s.

Another important characteristic of *Liangdan Yixing* was its overwhelming emphasis on big science, especially the undertaking of large-scale and highly complex projects. The nuclear weapons, missile, space and nuclear submarine projects required a massive and sustained mobilisation effort by the state to provide the necessary technological, financial and engineering resources.

When Deng Xiaoping assumed power in 1978, one of his first tasks was to sideline the Maoist military techno-nationalist grand strategy and focus on civilian economic and technological development.[17] At the same time as he rolled back the pervasive militarisation of the economy and society, adopted an economic "open door" policy and launched economic liberalisation initiatives that ended the country's autarkic isolation and heralded the beginning of the building of a market economy.

During his rule between 1978 and the early 1990s, Deng did not explicitly replace the defunct military techno-nationalist model with a new strategic framework. A flexible and ad hoc approach was implemented instead, which Barry Naughton and Adam Segal have described as "purely adaptative, opportunistic policies of 'muddling through'."[18] This could be termed techno-pragmatism. Chinese decision-makers tend to be instrumentalist in how they view and make use of nationalism, which is an imported concept. [19] Despite the heavy emphasis on science and technology (S&T) development before the 1980s, the excessive attention devoted to strategic nuclear and missile priorities meant that technologically the country lagged well behind virtually every other civilian and military sphere. Consequently, the Chinese government promoted a highly opportunistic strategy focused on seeking technology imports and FDI to fill gaps that the domestic base could not meet.

A strong military techno-nationalist ideology has been the cornerstone of North Korea's national development priorities since the founding of the regime in 1948. However, between the 1950s and 70s, North Korea was heavily dependent on the Soviet Union, and to a much lesser extent on China, for its military and technological needs, especially in its reconstruction following the Korea War. From the early 1970s, a concerted effort was made to establish a self-reliant military technological and industrial base. This initially focused on building conventional military S&T and manufacturing facilities to produce tanks, warships and aircraft, but was expanded in the 1980s and 90s for the developing of ballistic missiles, nuclear weapons and chemical and biological warfare capabilities.[20] Also in the 1980s, North Korea began to focus attention on the development of civilian S&T capabilities, including computers and electronics.[21]

After Kim Jong Il assumed power in 1994, the military techno-nationalist ideology was further strengthened and elevated to become the overarching state doctrine through the adoption of the S*ongun Chongchi* (Military-First) and *Kangsong Taeguk* (Building a Rich and Strong State) ideologies. A 2003 commentary by the North Korean newspaper *Nodong Sinmun* explained the pivotal importance of military priorities in the country's economic and technological development: "In a powerful state, the defence industry takes a leading and key position in the economy. As long as the imperialists continue to exist, the defence industry is a lifeline for the country and the nation." The obvious conclusion, the commentary noted, is that "developing the defence industry with priority while developing the light industry and agriculture simultaneously is the best way to satisfactorily ensure the people's independent and creative life while actively engaging in the fierce anti-imperialism struggle."[22]

Limited Techno-nationalism: India's Poor Track Record of Technological Development

Like its East Asian counterparts, India also enthusiastically embraced a state-led, techno-nationalist approach to economic development after independence in 1947. Under Jawaharlal Nehru and P.C. Mahalanobis in the 1950s and 60s and Indira Gandhi in the 1970s and 80s, closed import-substituting industrialisation was the cornerstone of economic policy, with an emphasis on heavy and strategic industries such as steel, chemicals, space, nuclear technology and defence. For numerous political, economic, historical and organisational reasons, however, India's ability to pursue and implement this strategy was far less effective and successful than Japan and South Korea, and was more comparable to pre-reform China.[23] The underpinnings of India's techno-nationalist thinking also leaned towards the military techno-nationalist approach of China than with the commercial impulses of the Japanese or South Korean model, although the Indian approach was far less militarised because it faced a less hostile external security environment than China.

The strong economic nationalist ideology that prevailed during this period was deeply suspicious of foreign trade and investment and this was also apparent in its attitude towards S&T development. Among the first priorities for the country's fledgling post-independence government was the building of an indigenous state-controlled S&T system and this saw the rapid establishment of key institutions such as the Atomic Energy Commission, Defence Research and Development Organisation and the first Indian Institute of Technology. But a chronic lack of well-qualified scientists and engineers, the poor level of overall national technological development, excessive bureaucratic compartmentalisation and limited funding meant that India's technological capabilities were severely constrained. Urgent national security needs during this period meant that the government in New Delhi had little choice but to seek and become dependent on Soviet military equipment and defence technology while its domestic defence industry struggled to develop indigenous products.[24]

This huge gulf between political rhetoric emphasising indigenous techno-nationalism and the actual reality of a floundering, under-resourced S&T system that was heavily dependent in many critical areas on foreign assistance persisted throughout the second half of the twentieth century. In the country's first technology policy statement issued in 1983 by Indira Gandhi's government, for example, the leading objective was "to attain technological competence and self-reliance to reduce vulnerability, particularly in strategic and critical areas, making the maximum use of indigenous resources".[25]

The Advent of Techno-globalism in the 1990s and Techno-hybridism in the Twenty-first Century

A series of transformative events beginning in the late 1980s led to a far-reaching reshaping of the international order in the 1990s and called into question the continuing relevance of the techno-nationalist approach. This included the end of the Cold War, the accelerating spread of economic globalisation, the advent of the information age and the arrival of the revolution in military affairs. These developments steadily eroded the importance and reach of the state and national systems as powerful new actors and processes unbounded by national borders emerged, especially global corporations and transnational networks of production and innovation. Techno-globalism gradually found its way into the thinking of scholars, policy planners and decision-makers across East Asia to either complement or replace techno-nationalist agendas.

Techno-globalism may be compared to liberal institutionalism in which competition among states and firms is viewed as a non-zero-sum game. This ideology has a number of attributes:

- Allowing FDI in manufacturing and services and international mergers and acquisitions with limited restrictions.
- The rise of global firms with horizontally integrated R&D and manufacturing facilities.
- The growth of transnational networks of production and innovation through outsourcing and modularisation.
- The erosion of economic sovereignty of states, especially in the making and implementation of macroeconomic policies.
- Adopting open, free trade principles that encourage capital and labour mobility and supported by low tariffs and lightly regulated regimes.
- Rolling back the role of the state in S&T development to primarily supporting basic research and encouraging firms to assume responsibility for applied and commercial R&D.

While techno-globalism is primarily focused on civilian activities, the process has also spilled over into the military arena to a limited extent. Following the end of the Cold War, allied states in Western Europe and North America opened their defence industries to cross-border mergers and acquisitions with few restrictions; and this led to the establishment of a few huge global defence and aerospace firms such as EADS and BAE in Europe and Boeing and Lockheed Martin in the US.[26] However, states in East Asia and other parts of the world did not join this process and their defence and strategic sectors remain closed to foreign investment.

Singapore and Hong Kong are archetypical examples of techno-globalist regimes in East Asia. Both have market-friendly regulatory and trade systems with low tariffs, unfettered capital mobility and transparent and non-discriminatory legal systems. This has allowed them to become hubs for the regional headquarters of multinational corporations. Their innovation systems are open and have extensive collaborations with foreign counterparts in the academic, corporate and governmental spheres.

Singapore has pursued a techno-hybrid approach in the development of its defence economy. The government has sought to nurture an indigenous defence R&D base, but has sought to confine its activities to basic research while pushing domestic companies to engage in applied development and commercialisation. In addition, Singapore defence firms have been actively engaged in joint ventures with foreign counterparts, especially as the country is heavily reliant on imports for its military equipment.[27]

The Wariness of Japan and South Korea Towards Techno-globalism

Japan and South Korea did make some efforts in the 1990s to relax their techno-nationalist policies and promote techno-globalist activity, but this was a carefully managed and highly selective process. Japan's MITI, for example, opened up several of its large-scale research programmes to foreign participation at the end of the 1980s in areas such as intelligent manufacturing, biological research and computing, but the overall external involvement in R&D was modest and more of an exception.[28]

These two countries remained committed to the techno-nationalist model until the onset of the Asian financial crisis in the late 1990s. When the crisis triggered severe economic turmoil in South Korea, the government under intense international pressure undertook a major overhaul of its techno-nationalist regime by allowing foreign multinationals to acquire major stakes in local firms, including in the technology and industrial manufacturing sectors, and establish R&D facilities within the country. However, the authorities in South Korea have continued to "use every means to retain its own system of technological innovation and industrial production".[29] South Korea has shifted from the strong techno-nationalist model to a limited or more pragmatic variant.

Japan was less severely impacted by the Asian financial crisis and so did not need to take any of the market-opening steps required of its neighbour. A small number of sectors were opened up to foreign investment and participation, such as in the financial and automobile industries, but foreign firms were only allowed to acquire minority stakes. Most of the Japanese economy and its S&T sector remained protected from the outside world. Indeed, some analysts pointed out, Japanese firms went on the offensive during the Asian financial crisis by increasing their FDI in South Korea and South East Asia to thereby "strengthen and extend Japanese techno-nationalism and industrial policy throughout the region".[30]

Japan Inc. was unable to respond in the same manner during the 2008 global economic meltdown and so could not take advantage of the foreign investment opportunities available to further strengthen its overseas networks. Weakened by the prolonged anaemic state of the country's economy and intensifying foreign competition over the past decade, Japanese firms, especially those in the manufacturing and industrial assembly sectors, could ill afford to expand their foreign investment.[31]

In contrast to the reluctance of Japan and South Korea to respond to the changing external strategic, economic and technological landscapes in the 1990s, China was more adaptable. It had already shifted from a strong techno-nationalist ideology under Mao Zedong to a more commercial and pragmatic techno-nationalist orientation with Deng Xiaoping. Chinese economic policymakers became attracted to key elements of the Japanese and South Korean technology and industrial development models. This initially focused on reverse engineering and large-scale importation of technological and industrial capabilities during the 1980s, but broadened in the 1990s to the emulation of the Japanese *Keiretsu* and Korean *Chaebol* corporate models of building a small core of sprawling, highly diversified, vertically integrated state-owned firms able to compete internationally.[32]

China's Manoeuvring between Techno-globalism and Techno-nationalism since the 1990s

China embraced key elements of techno-globalist thinking in the early 1990s by undertaking a far-reaching opening up of the national economy to FDI and rolling back state dominance of the S&T system by allowing the participation of private and other non-state firms.[33] This led to massive inflows of foreign investment into the medium- and high-technology sectors by multinationals and soaring imports of technology goods by foreign and Chinese firms throughout the 1990s. In addition, large numbers of entrepreneurial domestic and foreign joint-venture new technology enterprises were established or spun off from the state sector and quickly emerged to become major players in the domestic technology marketplace.

By the turn of the twenty-first century, the Chinese technology development model had become a patchwork quilt of techno-nationalist and techno-globalist thinking stitched loosely together. These dual sentiments came to the fore in the drafting of China's new medium- and long-term S&T development plan (MLP). Thousands of Chinese scientists, engineers, corporate executives and a small number of Chinese-born foreign scholars were consulted to examine issue areas deemed to be vital to China's S&T competitiveness, including basic science, energy and resource S&T, national security and the national innovation system.[34] Heated debates took place over the fundamental orientation of the MLP.

One revolved around the relationship between indigenous innovation and technology imports. Leading economists argued that China should retain its existing technological development model of attracting foreign multinationals and encouraging technology transfers to boost its S&T competence. Many scientists, though, opposed this view and insisted that China could not depend on getting core technology from other countries. The scientific community eventually prevailed and the MLP reflects the changed focus in the Chinese technology development model towards indigenous development.

The MLP was issued in 2006 and is avowedly techno-nationalistic in nature. In assessing the fundamental relationship between national security and technological development, the MLP defined this linkage in stridently realist, almost zero-term terms. The plan pointed out that the only way that China could advance against international competition was by improving "its independent innovative capabilities and master a number of core technologies, own a number of proprietary intellectual property rights and groom internationally competitive enterprises in important fields."

A central concept in the MLP is the notion of indigenous innovation (*zizhu chuangxin*). This term is viewed "by some as a regression to the self-defeating techno-nationalist notions of self-reliance from the Maoist era", although the MLP seeks to define this concept not only from an ideological viewpoint but also from a functional perspective.[35] Specifically, indigenous innovation is seen to be a way of promoting original innovation, re-assembling existing technologies in different ways to produce new breakthroughs, and the absorption and upgrading of imported technologies.

Although the MLP was a triumph for the techno-nationalist lobby, techno-globalist advocates remain vocal and influential. They continue to argue that China should take advantage of the opportunities opened up by the IT revolution and position itself for a radically different techno-globalist development model. Liu Xielin, a technology policy scholar at the Chinese Academy of Sciences and adviser to the government, has written that China's development path for catching up in the twenty-first century has been fundamentally altered from the 1990s and is also significantly different from the traditional approaches taken by Japan and Korea in the late industrial age. He points to several developments that have caused this shift.[36] First, the IT revolution has dramatically changed the rules of the game for catching-up, which in turn has given rise to new business models, products and services. Second, modular production, or outsourcing, has become a key feature of the global economy. Modularity allows the outsourcing of design and production of components and subsystems. Third, global technological outsourcing provides a new channel for Chinese companies to become involved in product innovations at an earlier stage than in the previous techno-economic paradigm. Whereas innovation in the industrial era relied on absorption and adaptation, the strategy in the information age is based more on market-oriented innovation and technology outsourcing. Many leading Chinese technology firms, such as Haier and Huawei, have adopted this approach.[37]

Although the enactment of the MLP may have suggested that Chinese policymakers were retreating from their techno-globalist leanings, the evidence on the ground shows otherwise. FDI continued to pour into the country and foreign firms were assuming growing prominence in R&D activities in China. The number of foreign-invested firms in the electronics and communications industry jumped from 350 in 1998 to more than 1150 in 2004, while there were more than 930 foreign R&D centres in China by the end of 2006 compared to 600 in 2004, with around 200 new centres opening annually.[38]

China's technology development strategy has evolved into a pragmatic and sometimes messy techno-hybrid model that incorporates both competing and complementary strands of techno-nationalist and techno-globalist thinking. With these two camps firmly entrenched and wielding extensive influence among policymakers and key state agencies, there will be occasional zigzags in technology-related policies to accommodate their different interests.[39] A prime example of this occurred in 2009 when China issued indigenous innovation procurement guidelines requesting that government agencies should buy only technology products that were based on domestic intellectual property. The move drew strong condemnation from foreign companies and governments who argued that the policy went against established international norms.[40]

The procurement guidelines controversy reflects a broader effort by China to promote its own home-grown technical standards within the global technology system. The setting of international technical standards has traditionally been in the hands of multinational corporations, but China has begun to raise its profile since the beginning of the twenty-first century, especially in areas such as telecommunications, digital audio and video, computer microprocessors, wireless local area network security and Internet protocols. The MLP explicitly calls for China to "actively take part in the formulation of international standards and drive the transferring of domestic technological standards to international standards." Richard Suttmeier and Yao Xiangkui argue that China's technical standards strategy should be understood as a modified form of techno-nationalism, in which "technological development in support of national economic and security interests is pursued through leveraging the opportunities presented by globalisation for national advantage."[41] How China flexes its growing technological power in this area will be a good indicator of where the balance between techno-nationalist and techno-globalist impulses lies in its evolving techno-hybrid framework.

Chinese technology strategy in the national security arena has also witnessed major changes since the 1990s with the end of the Cold War and the advent of the revolution in military affairs. The beginning of the information age in the early 1990s forced the Chinese military and defence industrial establishments to critically re-examine their development strategies. Mechanisation was the central goal of modernisation efforts in the industrial era through the development and acquisition of conventional weapon systems such as tanks, warships and aircraft. But with the emergence of a new information-based techno-military paradigm, the debate shifted to how to adapt to these far-reaching changes.[42]

Military chiefs and defence industrialists have been urged to come up with new approaches to meet this paradigm shift. At the 16th Communist Party Congress in 2002, the country's senior leadership proclaimed that the first twenty years of the twenty-first century would be a period of "important strategic opportunities" that would allow China to make rapid advances in its goal of building a prosperous society.[43] The PLA and the defence economy were required to "firmly seize this period of strategic opportunities" and leapfrog ahead in their military modernisation efforts.[44] President and Commander-in-Chief Jiang Zemin called for the fulfilment of the "dual task of building mechanisation and informationisation" of the country's armed forces.[45] This meant that the building of information systems and systems integration technology would be accorded the same priority as the development of conventional weapon systems.

Finding ways to catch up and leapfrog ahead have become the central tenets in this search for a new military development model. This is because a key assumption of this debate is that the new information paradigm offers a unique chance to leap ahead and significantly narrow the military technological gap with the world's advanced military powers.[46] If this window of opportunity is not decisively grasped, some military strategists warn, "the gap between China and military powers may widen … and we may find ourselves in a passive position in future military struggle."[47]

While Chinese leaders urge the PLA and the defence economy to catch up with the world's technological leaders as quickly as possible, defence planners are more cautious and do not envisage developing the mix of capabilities required to be an advanced military information power until at least the middle of the twenty-first century at the earliest.[48] These assets include command, control, communications, computers, intelligence, surveillance and reconnaissance (C4ISR) systems, sophisticated electronic warfare systems and integrated command automation networks. As an interim goal though, the PLA will seek to equip and connect its mechanised platforms with sensors, navigational positioning systems, infrared detectors, computers and other devices that would allow them to conduct network-enabled operations within the 2020 timeframe.[49]

These military requirements represent an unprecedented technological challenge for the defence scientific and industrial communities. Their ability to adequately fulfil these needs will depend crucially on the nature of the developmental catch-up models they adopt. For the defence industry, the evidence from the structural and organisational reforms that have taken place since the late 1990s indicates that it will follow a path similar to that taken by Japan and Korea with the building of a closed developmental model spearheaded by an elite of vertically integrated companies. These ten defence corporations are huge vertically integrated behemoths that have a near-monopoly control of the domestic defence market.

However, reliance on these conglomerates has risks. First, their track record of technological innovation is mixed. Second, their institutional cultures are deeply rooted in the legacy of the central planning system. Third, their overwhelming focus is on industrial-era practices and processes, and much of their industrial equipment and plants are not suitable for the development or manufacture of military products of the information age, such as electronic components, software and precision-guided munitions. The authorities have sought to introduce a modicum of competition and outsourcing into this closed apparatus, especially through the establishment of a dual-use system. However, this only applies to non-critical products and technologies and will not include technological projects intended for frontline combat use.

Under this closed vertically integrated developmental model, the prospects for achieving a technological catching up by the defence economy may be limited. The general pattern of innovation under this model, as shown by the Japanese and Korean experiences, is incremental, which suggests that any technological advance would be gradual and deliberate. The more open and vertically disintegrated developmental catch-up model that is favoured by the fledgling dual-use economy may offer a better recipe for success in narrowing the technology gap. The information and communications technology (ICT) sector, which is a central player in building military information technology capabilities, has embraced this alternative approach that allows for greater market-oriented innovation and the ability of firms to learn and absorb from global technology outsourcing, not only in manufacturing but also increasingly in R&D work. Chinese firms such as Huawei, ZTE and Datang have emerged to become market leaders in the ICT sector in China while also making inroads into global markets.

While these two alternative catch-up models operate along separate trajectories, growing interaction and convergence between the defence economy and dual-use economy is gradually leading to the emergence of yet another development path. This is a semi-open hybrid model as per which the defence economy selectively opens itself up to the domestic civilian economy for investment and joint partnerships, and makes use of the dual-use economy as an intermediate conduit for cooperation with foreign firms. Although this will be primarily with Chinese firms, it could also extend to local subsidiaries of foreign companies.

This hybrid approach began to take root at the end of the 1990s. The defence economy has been eager to tap into the domestic and foreign capital markets to raise funds and form strategic partnerships with leading foreign defence companies and it has sought to pursue this strategy through the listing of its converted civilian portions of its major defence conglomerates. How to balance these opposing techno-national and techno-globalist tendencies will be a major long-term challenge for the Chinese leadership. While the Chinese leadership in the post-1979 reform era has made the opening up its economy to the outside world a core pillar of its policies, there has been an inherent caution over excessive foreign dependence, especially in strategic areas related to national security and technology.

Taiwan offers yet another example of a state that has embraced a techno-hybrid approach in its technology development. Keller and Samuels refer to Taiwan as the "paradigmatic example of the techno-hybrid regime", in which the formula is "not quite liberal" yet "not quite mercantile".[50] Taiwan has encouraged the forging of close linkages and mutual dependence between its leading technology companies and foreign multinationals through FDI and the seamless integration of Taiwanese firms into global production and innovation systems through outsourcing. At the same time, there has been a strong emphasis by the Taiwanese government on building robust domestic innovation capabilities such as state-backed research institutes that act as hubs for the diffusion of R&D resources and knowledge and the nurturing of university-industry partnerships.[51] This hybrid model has been critical to the global success of Taiwan's electronics and semiconductor industries.

India's Embrace of Techno-globalism and Techno-hybridism since the 1990s

India also began to open up and liberalise its closed state-dominated economy in the beginning of the 1990s in response to domestic fiscal crises and international pressures. By the mid-1990s, far-reaching market reforms and the enactment of policies promoting foreign investment and technological collaboration with foreign firms had "effectively altered the balance between globalisation and nationalism" within the Indian economy in favour of the former.[52] In a new S&T policy statement in 2003, the government toned down its longstanding insistence on indigenous innovation and pointed to the need to master "national needs in the new era of globalisation", in view of the importance of promoting international S&T cooperation.[53]

Under this changed development paradigm, new sources of technological innovation began to emerge, such as the private sector-led IT services sector. The rise of the Indian IT and software industry is directly due to techno-globalist drivers that have allowed Indian firms to integrate with the global technology and knowledge economy. Foreign firms have also flocked to India to take advantage of a large and well-educated pool of human talent. More than three hundred international companies had established R&D centres in India by the late 2000s, with specialisation in areas such as pharmaceuticals, biotechnology and microelectronics.[54] While strategic sectors such as defence, space and nuclear technology remain firmly under state control, there are signs of growing technological collaboration with the outside world. The defence industry, for example, is forging expanding technological linkages with US defence companies in response to improving US-India strategic relations.

Prospects for Technological Change in East Asia and its Impact on the Security-economic Nexus

The gales of technological progress that have already had such a profound impact on East Asia will intensify in the coming decades. How states respond and adapt to these changes will determine whether they keep up or get left behind in the ongoing race for technological leadership. Regimes such as Japan, and to a lesser extent South Korea, that continue to adhere to outdated techno-nationalist notions of self-sufficiency and closed doors are likely to struggle compared to those with more flexible techno-globalist and techno-hybrid approaches, of which China and Taiwan are prime examples.

Over the long-term horizon, a couple of technological trends and their implications for East Asia's economic and security dynamics should be examined. The first issue is East Asia's changing place and role in the global technology hierarchy. The transition from the late industrial era to the information age since the 1990s has heralded a transformation in the nature of technological globalisation. One of the principal developments during the 1990s and the early years of the twenty-first century, especially in East Asia, was the rise of global production networks, in which multinational firms relocated or outsourced their manufacturing requirements to lower-cost areas elsewhere in the world.

In the past decade, an emerging trend has been the formation of global innovation networks (GIN) that "integrate dispersed engineering, product development, and research activities across geographic borders". There has been debate whether multinationals in the US, Europe and Japan would allow their innovation activities to be expanded abroad or kept in-house within their national borders because of their strategic importance. Recent research on the global electronics industry, which is often a harbinger of change for the rest of the high-technology economy, indicates that GINs are rising rapidly and will lead to far-reaching structural changes in the geography of innovation and production in the high-technology sector within the next decade.[55] Dieter Ernst offers some distinct features of the new global technology order shaped by the arrival of GINs:

- While GINs originated in the US, Europe and Japan, they have now expanded into less developed regions, especially across Asia. China, Korea, Taiwan and India have become leading hubs of this reconfigured geography of offshore innovation.[56]
- This new global technology order cannot simply be left to market forces. Governments, especially from catch-up economies, have a critical role to play to ensure that their countries reap the benefits of this globalisation and are not left behind.

Although competition for global technology leadership is intense and requires access to extensive resources and a robust R&D base, many East Asian states have invested heavily in these areas and are well placed to continue to move up the innovation ladder. Some of the less developed countries such as China have to overcome critical bottlenecks such as improving intellectual property protection, so they are likely to require more time to make progress. But there are strong prospects that the rest of East Asia and India will join Japan in the next one or two decades and become a new pole of advanced innovation in the global technology order.

The second major long-term trend is China's effort to become a world-class technological leader. The stated goal of China's leadership is to catch up technologically by 2020 and this has led to a concerted mobilisation of China's burgeoning economic, political, organisational, military and scientific resources to meet this challenge. While the 2020 target may be overly ambitious to bridge the still wide gulf in innovation infrastructure and capabilities between China and the US, Europe and Japan, but China is making major strides and could catch up in a few select areas.

China's eventual emergence as a technological powerhouse, especially in defence-related sectors, has profound implications for regional and global security. While China's military technological capabilities still lag behind the US by at least one or two generations, this gap will gradually narrow, and China may be in a position to challenge US military technological dominance in several strategic areas in the next ten to fifteen years. This include outer space, aviation, the maritime commons and the digital domain. Technology-related tensions have already been growing between the US and China in the past few years, especially in cyber security and China's development of asymmetric military capabilities such as anti-satellite weapons and ballistic missiles.

Of even more near-term concern is its impact on the military situation in the Taiwan Strait. The sustained build-up of the PLA's military and technological capabilities against Taiwan over the past fifteen years has seriously affected the United States' ability to militarily intervene in the event that an armed conflict takes place between China and Taiwan. Japan and South Korea may also find themselves scrambling to keep pace with China's military technological progress. This raises the question whether they will move closer to the US to counter a stronger China or forge better security relations with Beijing to hedge their bets.

Notes

1. See Barry Posen, *The Sources of Military Doctrine* (Ithaca, NY: Cornell University Press, 1984).
2. Etel Solingen, *Regional Orders at Century's Dawn: Global and Domestic Influences on Grand Strategy* (Princeton, NJ: Princeton University Press, 1998). See also Kevin Narizny, *The Political Economy of Grand Strategy* (Ithaca, NY: Cornell University Press, 2007), who puts forward a similar argument but uses the cases of the US and Great Britain in the nineteenth and twentieth centuries.
3. William W. Keller and Richard J. Samuels, "Innovation and the Asian Economies", in William W. Keller and Richard J. Samuels (eds.), *Crisis and Innovation in Asian Technology* (Cambridge: Cambridge University Press, 2005).

4. Meredith Woo-Cummings, "Back to Basics: Ideology, Nationalism, and Asian Values in East Asia", in Eric Helleiner and Andreas Pickel (eds.), *Economic Nationalism in a Globalizing World* (Ithaca, NY: Cornell University Press, 2005), pp. 91-117.
5. *Ibid.*, p. 98.
6. See Morris Low, "Displaying the future: techno-nationalism and the rise of the consumer in Postwar Japan", *History and Technology*, 19(3) (2006): 197-209.
7. Hiroyuki Odagiri and Akira Goto, "The Japanese System of Innovation: Past, Present and Future", in Richard Nelson (ed.), *National Innovation Systems* (Oxford: Oxford University Press, 1993), pp. 76-114.
8. Michael E. Porter, Hirotaka Takeuchi and Mariko Sakakibara, *Can Japan Compete?* (London: Macmillan Press, 2000), pp. 31-5; Adam S. Posen, "Japan", in Benn Steil, David G. Victor and Richard R. Nelson (eds.), *Technological Innovation and Economic Performance* (Princeton, NJ: Princeton University Press, 2002), pp. 101-2.
9. Jang-Sup Shin, *The Economics of the Latecomers* (London: Routledge, 1996), pp. 122-37.
10. See Yong-Ho Bae, Sungsoo Song, Mi-Jung Um, Dae-Hee Lee and Michael Hobday, *Case Study of Technological Innovation of Korean Firms* (Seoul: Science and Technology Policy Institute, November 2002).
11. Shin, *The Economics of the Latecomers*, n. 9, pp. 53-7.
12. Michael W. Chinworth, *Inside Japan's Defense* (Washington, DC: Brassey's, 1992), pp. 168-80.
13. Mark Lorell, *Troubled Partnership: A History of U.S.-Japan Collaboration on the FS-X Fighter* (Santa Monica, CA: RAND Corp., 1995).
14. Xielin Liu, *China's Development Model: An Alternative Strategy for Technological Catch-up* (Hitotsubashi University, Working Paper, 22 March 2005), p. 8.
15. Evan Feigenbaum, "Soldiers, Weapons and Chinese Development Strategy: The Mao Era Military in China's Economic and Institutional Debate", *The China Quarterly*, No. 158, June 1999, pp. 285-313. Feigenbaum further points out that closely related to this integrative economic-security developmental approach is the importance of organizational flexibility.
16. See Liu Yanqiong, *"Liangdan Yixing Gongcheng De Chenggong Jingyan Yu Qishi"* [The Experience and Enlightenment from the Success of the Two Bombs and One Satellite Project], Masters Thesis, Guofang Keji Daxue [National Defence Science and Technology University], 2002.
17. See Tai Ming Cheung, *Fortifying China* (Ithaca, NY: Cornell University Press, 2009), Chapter 3.

18. Barry Naughton and Adam Segal, "China in Search of a Workable Model", in Keller and Samuels, *Crisis and Innovation in Asian Technology*, n. 3, p. 186.
19. See Suisheng Zhao, "Chinese Nationalism and its International Orientations", *Political Science Quarterly,* 115(1) (2000): 1-33.
20. Hong So'ng-p'yo, "North Korea's Military Science and Technology", *Kunsa Nontan,* 29 April 2005; Hong Ung-ch'o, "The Immortal Accomplishment of Great Leader Comrade Kim Il Sung Who Founded the Brilliant Tradition of Juche National Defense Industry Construction", *Kyo-nje Yo'ngu*, Vol. 3, 2008, pp. 5-7, translated by *Open Source Center*.
21. Rian Jensen, *State over Society: Science and Technology Policy in North Korea*, Johns Hopkins University SAIS, U.S.-Korea Working Paper Series, August 2009.
22. "Military-First Ideology Is an Ever-Victorious, Invincible Banner for Our Era's Cause of Independence," *Nodong Sinmun*, 21 March 2003.
23. See Atul Kohli, *State-Directed Development: Political Power and Industrialization in the Global Periphery* (Cambridge: Cambridge University Press, 2004), Chapter 7.
24. Anil Athale, "Defence Research: India's Achilles Heel", *Indian Defence Review*, 25(3) (September 2010): 32-7.
25. Quoted in Roddam Narasimha, "Science, Technology and the Economy: An Indian Perspective", *Technology in Society*, No. 30, 2008, p. 336.
26. See Richard Bitzinger, *Towards a Brave New Arms Industry?*, Adelphi Paper (International Institute for Strategic Studies), No. 356, 2003; Stephen G. Brooks, *Producing Security: Multinational Corporations, Globalization, and the Changing Calculus of Conflict* (Princeton, NJ: Princeton University Press, 2005).
27. Robert Karniol, "Singapore: Defence Eco-System", *Jane's Defence Weekly*, 15 February 2010.
28. Gregory Corning, *Japan and the Politics of Techno-Globalism* (Armonk, NY: M.E. Sharpe, 2004).
29. Linsu Kim, "Crisis, Reform, and National Innovation in South Korea", in Keller and Samuels, n. 3, p. 87.
30. Keller and Samuels, p. 230.
31. Motoko Iwami, "Japanese FDI: Recent Development and Outlook", *Japan Institute for Overseas Investment Bulletin,* No. 15, 2009.
32. Edward S. Steinfeld, "Chinese Enterprise Development and the Challenge of Global Integration", *MIT Special Working Paper Series*, January 2002, pp. 23-5; Wei Xie and Steven White, *From Imitation to Creation? The Critical Yet Uncertain Paradigm Shift for Chinese Firms* (INSEAD, 2004), http://ged.insead.edu/fichiersti/inseadwp2004/2004-07.pdf
33. Naughton and Segal, "China in Search of a Workable Model", n. 18, pp. 170-6.

34. For background to the drafting of the MLP, see Cong Cao, Richard P. Suttmeier and Denis Fred Simon, "China's 15-Year Science and Technology Plan", *Physics Today*, December 2006, pp. 38-43; Hao Xin and Gong Yidong, "China Bets Big on Big Science", *Science,* Vol. 311, 16 March 2006, pp. 1548-9. See also Hao Xin and Jia Hepeng, "China Supersizes its Science", *Science,* Vol. 315, 9 March 2007, pp. 1354-6; Sylvia Schwaag Serger and Magnus Breidne, "China's Fifteen-Year Plan for Science and Technology", *Asia Policy* (National Bureau of Asian Research), No. 4, July 2007, pp. 135-64.
35. Cong Cao *et al.*, *ibid.*, p. 40.
36. Xielin Liu, *China's Development Model*, n. 11.
37. Xie and White, *From Imitation to Creation?*, n. 32, pp. 13-14.
38. *OECD Reviews of Innovation Policy: China* (Paris: Organization for Economic Cooperation and Development, 2008), pp. 268-73; and "Foreign R&D Centers in China on the Rise", *People's Daily Online*, 20 August 2004, http://english.peopledaily.com.cn/200408/20/eng20040820_153974.html
39. State entities that are most supportive of techno-nationalist approaches include the National Development and Reform Commission, Ministry of Industry and Information Technology, and People's Liberation Army. Organizations that are more techno-globalist in orientation include the Ministries of Commerce and Foreign Affairs.
40. Loretta Chao, "China's Curbs on Tech Purchases Draw Ire; U.S. Government, Dozens of Global Industry Groups Speak Against Push for State Agencies to Buy 'Indigenous Innovation' ", *Wall Street Journal*, 11 December 2009.
41. Richard P. Suttmeier and Yao Xiangkui, *China's Post-WTO Technology Policy: Standards, Software, and the Changing Nature of Techno-Nationalism,* National Bureau of Asian Research Special Report, No. 7, May 2004, p. 3. See also Christopher S. Gibson, "Technology Standards—New Technical Barriers to Trade?", in Sherrie Bolin (ed.), *The Standards Edge: The Golden Means* (Ann Arbor, MI: Bolin Group, 2007).
42. See You Ji, "Learning and Catching Up: China's Revolution in Military Affairs Initiative", in Emily Goldman and Thomas Mahnken (eds.), *The Information Revolution in Military Affairs in Asia* (New York: Palgrave Macmillan, 2004), pp. 97-123.
43. See http://news.xinhuanet.com/english/2002-11/18/content_633685.htm. See also Robert L. Suettinger, "China's Foreign Policy Leadership: Testing Time", *China Leadership Monitor*, No. 9, Winter 2004.
44. Zhang Zhaoyin, "Firmly Seize the Period of Important Strategic Opportunities to Promote Leap-Type Development", *Jiefangjun Bao*, 25 February 2003.

45. See Xu Xiaoyan and Zhai Tongzheng, "Hasten Informationization Building within Our Forces", *Jiefangjun Bao*, 10 December 2002, p. 6.
46. You Ji, "Learning and Catching up", n. 42, pp. 104-6.
47. "Chinese Generals on Opportunities and Challenges", *Liaowang*, 14 July 2003, pp. 21-4.
48. *Hong Kong Wen Wei Po*, 14 March 2005.
49. Wu Yujin, "Explore Effective Ways of Securing Leap Forward in Armaments Development", *Jiefangjun Bao*, 30 September 2003, p. 6.
50. Keller and Samuels, "Innovation and the Asian Economies", n. 3, p. 12.
51. Pao-Long Chang and Hsin-Yu Shih, "The Innovation Systems of Taiwan and China: A Comparative Analysis", *Technovation*, Vol. 24, 2004, pp. 529-39.
52. Baldev Raj Nayar, *Globalization and Nationalism: The Changing Balance in India's Economic Policy, 1950-2000* (New Delhi: Sage, 2001), p. 130.
53. Cornelius Herstatt *et al.*, *India's National Innovation System: Key Elements and Corporate Perspectives* (Working Paper No. 51, Hamburg University of Technology, April 2008), p. 27.
54. R.A. Mashelkar, "Indian Science, Technology, and Society: The Changing Landscape", *Technology in Society*, No. 30, 2008, p. 305.
55. Dieter Ernst, *A New Geography of Knowledge in the Electronics Industry? Asia's Role in Global Innovation Networks*, East-West Center, Policy Studies No. 54, 2009.
56. Ernst defines the global hierarchy of innovation as follows: top tier comprising centres of excellence located in the US, Japan, and the European Community; second tier of advanced locations in Israel, Taiwan, Ireland and South Korea; a third tier of catch-up regions such as Beijing, the Yangtze and Pearl river deltas in China, and Bangalore, Chennai, Hyderabad and Delhi in India; and a fourth "new frontiers" tier in lower-tier cities in China and India as well a Vietnam and Eastern Europe.

13

Energy Security and Asian Regional Architecture

Xia Liping

In the contemporary world, major challenges to security come more from non-traditional security problems. Non-traditional security threats are also called global problems, transnational problems or low-politics problems. They include terrorism, piracy, environmental pollution, global warming, shortage of energy, population explosion, drug trafficking, international crimes, HIV/AIDS, and so on. Non-traditional security threats exist in the whole world and concern all human beings; and they threaten the existence and development of human beings.

Security problems can be transformed into non-traditional security problems, and the latter can lead to armed conflicts. For example, weapons of mass destruction are traditional security problems. However, if terrorists have such weapons, it becomes a non-traditional security threat and a trans-national problem. Similarly, cross-boundary attacks by terrorists may lead to armed conflict. Because of the large-scale armed race between the erstwhile super powers during the Cold War, long-term exploitation and the sidelining of developing nations by developed nations, and the over-exploitation of natural resources by human beings, non-traditional security threats have been growing. Given the growing impact of non-traditional security threats on international relations and rising common interests between various countries, the states of the world, especially the major powers, have to resolve disputes through consultation and coordination.

Major Features of Energy Insecurity in Asia

Energy security as a non-traditional security issue has grown in importance over the years. In recent years, Asian countries have made some progress in bilateral and multilateral dialogue, cooperation and coordination in the energy field, but they are still far away from a regional architecture of energy security. Rapid economic development and population growth, in Asian countries is creating energy insecurity.

- Over the next two decades, energy demand in Asia will keep growing.
- Energy efficiency of developing countries in Asia is lower than that of developed countries.
- Oil production has become more difficult than before. Countries and companies have to go to the deep sea, land-locked areas and politically and socially difficult areas to get oil.
- Sea lanes for transport oil are plagued by piracy.
- "Energy nationalism" in Asian countries may be increasing.

Unless Asian countries cooperate, energy insecurity will have important long-term geo-political implications for the region. Asian powers are responding more with "energy nationalism" than with energy cooperation as they reach out to secure future supplies. The risks of terrorist attacks on vital energy infrastructure are rising, fuelling fears of supply disruption and political instability.

Energy Security Relations between China and India

China has only five billion tons of oil resources, less than seven per cent of the global reserves. Since China became a net importer of oil in 1993, its purchases of foreign oil have increased annually. In 2008, China's crude oil imports reached 178 million tons, while refined oil imports were at 39 million tons, a rise of 9.6 per cent. China's dependency on imports for its energy needs was 48.5 per cent in 2008. In 2009, China imported nearly 200 million tons of crude oil, increasing its dependency on imports to more than 50 per cent.

In 2007, the Indian petroleum minister Mani Shankar Aiyar announced that the nation had 16.6 billion tons of oil reserves. Even so, with its 1.1 billion population, the per capita reserves in India are only 1.5 tons, less than 1/42 of the world average. India's current dependency on foreign oil exceeds 75 per cent. Increase in imports, with the surge in oil prices, has cost India a great deal in foreign exchange assets. It has been estimated by India that for every rise $5 rise in oil price, India's economic growth rate will drop 0.5 per cent, and its inflation rate would increase by 1.4 per cent.

The International Energy Agency has estimated that China's and India's foreign oil dependency was 61 per cent and 85 per cent respectively in 2010. At present, China is the world's second-largest consumer of oil. India is the world's fifth-largest consumer but will soon become the fourth-largest. By 2030, India will likely overtake Japan and Russia to become the world's third-largest energy consumer. The governments of China and India have begun a dialogue and cooperation on energy. In April 2005, during Premier Wen Jiabao's visit to India, the two governments issued a joint declaration. Article 9 of the declaration states that the two countries agree to cooperate in energy security and energy savings. It also encourages collaboration to explore and exploit oil and natural gas resources in third countries.

In December 2005, oil companies from the two countries teamed up for the first time to purchase 37 per cent of oil assets in Syria that belonged to the Canadian Oil Co. at a cost of $573 million. India's Oil and Natural Gas Corporation (ONGC) is also working with China Petroleum and Chemical Corporation in Iran to exploit the Aveda Varan oil field, in which China holds 50 per cent and India holds 20 per cent in shares.

On February 22, 2005, India Gas Co signed an agreement to invest $243 million in China Gas Co. This was the first-ever cooperation between listed companies of the two countries. In August 2006, China and India worked together again to buy a 50 per cent share of an oilfield in Colombia. In January 2006, petroleum minister Mani Shankar Aiyar led a natural gas delegation to China, which signed five agreements.

Major Features of Multilateral Dialogue on Energy Cooperation in Asia

Dialogue among Oil-importing Countries

In early 2005, India hosted a round table meeting of Asian Ministers on oil co-operation in which major oil-consuming countries in Asia, including China, Japan, South Korea and India, participated. They agreed to adopt a unified position to deal with the "Asian premium" and oil security issues together. On December 16, 2006, energy ministers of China, the US, Japan, South Korea and India met in Beijing to discuss energy security issues, how to maintain the stability of the international energy market, and form a relationship of cooperation instead of competition. The summit and dialogue of foreign ministers between China, India and Russia also called for strengthened energy cooperation between their countries.

APEC Dialogue on Energy Security

In 2001, leaders of the Asia-Pacific Economic Community (APEC) endorsed the Energy Security Initiative (ESI). In 2005, APEC leaders agreed to reduce the ratio of energy use to economic output by at least 25 per cent from 2005 levels by 2030. The APEC Leaders' Declaration of 2009 committed APEC countries to rationalise and phase cut over the medium term fossil fuel subsidies that encourage wasteful consumption, while recognising the importance of providing those in need with essential energy services.

On June 19, 2010, the Ninth Meeting of APEC Energy Ministers, held in Fukui, Japan, issued the "Fukui Declaration on Low Carbon Paths to Energy Security: Cooperative Energy Solutions for a Sustainable APEC". The declaration committed the APEC economies to enhancing regional energy security in the midst of emerging concerns about the global environment and world economy. It underlined that more efficient use of energy and a cleaner energy supply would simultaneously boost energy security, help economic growth and lower emissions. The declaration instructed the Energy Working Group (EWG) to implement the studies and initiatives in its Work Plan, with support from the Asia Pacific Energy Research Centre (APERC), Expert Group on Clean Fossil Energy (EGCFE), Expert Group on Energy Data and Analysis (EGEDA), Expert Group on Energy Efficiency and Conservation (EGEEC), Expert Group on New and Renewable Energy (EGNRET), Biofuels Task Force (BTF), and Energy Trade and Investment Task Force (ETITF), as well as through cooperation with relevant multilateral forums. The EWG and its subsidiary bodies were instructed to continue their important contributions to formulating and implementing the APEC Growth Strategy and its sustainable growth pillar. The EWG also was instructed to establish a Task Force to implement an APEC Low-Carbon Model Town Project. The project will present successful models for coordinated use of advanced low-carbon technologies.

In order to enhance energy security and accelerate deployment of clean energy technologies throughout the APEC region, APEC energy ministers encourage enhanced cooperation with other multilateral forums and with the private sector in the region. The first-ever joint meeting of high-level APEC energy and transport officials is to be held in the United States in 2011. In 2012 an APEC energy ministers' meeting will be held in Russia. The Asia-Pacific Partnership on Clean Development and Climate comprises the United States, China, Australia, Canada, India, Japan, and the Republic of Korea. The APP works with private sector partners to address national air pollution reduction, energy security and climate change concerns. The United States and India serve as chair and co-chair, respectively, of the Coal Mining Task Force.

Sub-regional Multilateral Dialogue

Until now, in South Asia, the multilateral dialogue on energy has mainly been held within the framework of the South Asian Association for Regional Cooperation (SAARC). The second meeting of SAARC energy ministers was held in New Delhi on March 7, 2007. If relations between India and Pakistan can be improved, an Iran-Pakistan-India gas pipeline could deliver natural gas from Iran to Pakistan and India.

Multilateral Dialogue on and Cooperation of Energy in Central Asia

In Central Asia, multilateral dialogue on energy has been conducted by Shanghai Cooperation Organization (SCO). The SCO's energy cooperation has great strategic significance for energy security in Asia. After several years under construction, the oil pipeline from Kazakhstan to China became operational in 2006. This pipeline will develop into the SCO's multilateral energy cooperation project involving Russia and other Central Asian countries. The designed handling capacity of the pipeline is 20 million tonnes per year, which will be a big jump over the 500,000 tonnes being currently handled by railways each year. Gas pipelines from Central Asia and Siberia to China will also be constructed. If connected with the Xinjiang-Shanghai Gas Pipeline, the Central Asia-China lines will also help in the implementation of the Energy Eastward Transportation Programme. Japan and South Korea can also take part in this project. Energy supplies from Central Asia and Siberia, as distinct from the Middle East, Africa and Latin America, do not require the protection of a blue-water navy. For the first time in history, Central Asian countries have also got an eastward energy pipeline which does not go through Russia or the Caucasus or the Middle East, but goes across China, to finally reach the Pacific Ocean.

Multilateral Dialogue on Energy in North-East Asia

Multilateral organisations that can be used to institutionalise energy security cooperation in North East Asia include the China-RoK-Japan trilateral cooperation and the Six-Party Talks. Spurred by the first China-RoK-Japan Summit in 1999, the trilateral mechanism has promoted cooperation in the economic, financial and environmental areas for prosperity in the region. At their summit held in Cambodia in November 2002, the three countries agreed to further deepen "cooperation for prosperity" based on trust and promote trilateral cooperation in a wide range of areas such as economics and trade, information and telecommunications, environmental protection, human resources development and cultural cooperation.

During the summit meeting, some economic research institutions of the three countries submitted a Report on the Proposal of Joint Policies and a feasibility study for a China-RoK-Japan free-trade zone and its possible economic impact in the future. The summit endorsed this report in principle and supported the proposed feasibility study. At the summit held in October 2003, at the initiative of Chinese Premier Wen Jiabao, the summit issued a joint declaration on promoting trilateral cooperation. In December 2003, they agreed to establish a trilateral committee, to be led by foreign ministers of the three countries, to research, plan and observe cooperation in different fields. The advancement of trilateral cooperation in a wide range of areas such as "cooperation for prosperity" and "cooperation for stability" is the driving force for promoting North-East Asian cooperation.

The Six-Party Talks meanwhile have provided a suitable mechanism to resolve the North Korean nuclear issue through dialogue and negotiations. The arrangement has the potential to develop gradually into an important mechanism for sub-regional security dialogue and cooperation in North East Asia, if the North Korean issue can first be resolved. However, the North Korean nuclear issue is too complicated to be resolved in the short run.

Cooperation Scenarios

In North East Asia, cooperation for energy saving and clean-coal technology can include:

(i) Hydrogen energy technology; wind technology; solar energy; tidal power; garbage power; clean-coal technology.
(ii) Joint exploration of oil by China and Japan in the East China Sea.
(iii) Mechanism of coordination of oil price between oil consumers.
(iv) Pipeline from Russia through North Korea to South Korea.
(v) Electricity grids from South Korea to North Korea.
(vi) Mechanism of coordination between oil producers and consumers.

Major obstacles to achieving multilateral cooperation in energy are the North Korean nuclear issue and irrational nationalism. Cooperation in energy saving and clean-coal technology, coordination of oil price between oil consumers, coordination between oil producers and consumers are more feasible than the other three in the current situation. If the North

Korean nuclear issue can be peacefully resolved through the Six-Party Talks, the pipeline from Russia through North Korea to South Korea and electricity grids from South Korea to North Korea can be implemented as one of the incentives to lure Pyongyang to abandon its nuclear weapon programme.

Main Fields of Multilateral Energy Cooperation in Asia

Natural Gas

Natural gas can ease the transition to a low-carbon economy since gas has far lower carbon footprint than other fossil fuels. Unconventional gas resources can also boost the region's energy security by making it more self-sufficient. The EWG has been instructed by the APEC Meeting of Energy Ministers to conduct an unconventional gas census to evaluate the potential of unconventional resources and to recommend cooperative actions which could increase natural gas output, boost natural gas trade and use, and moderate natural gas prices to the extent appropriate both for producers and consumers in the APEC region, with assistance from APERC, EGCFE and EGEDA.

Improving Energy Efficiency

This is one of the quickest, greenest and most cost-effective ways to address energy security, economic growth and climate change challenges at the same time. More efficient transport, industry, buildings and equipment can help limit direct fossil fuel needs as well as the demand for electricity, which continues to be generated in large part from natural gas and coal. The EWG and EGEDA have been instructed to continue collecting complete, accurate and timely oil and gas data for contribution to the Joint Oil Data Initiative (JODI) in cooperation with other relevant bodies where necessary, to extend their capacity building in energy statistics to interested economies, and to support multilateral initiatives for more transparent and less volatile energy commodity markets. The EWG and APERC have also been instructed to keep promoting energy efficiency through Peer Review on Energy Efficiency (PREE) and the Cooperative Energy Efficiency Design for Sustainability (CEEDS), and to consider follow-up efforts including capacity-building activities, policy research support and processes to gauge the success of member economies' efforts to implement the recommendations of these programmes.

Energy-efficient Buildings and Appliances

The building sector accounts for two-fifths of energy use in the region. Net zero energy buildings are being developed using energy-efficient appliances, components and systems. Trade and investment in energy-efficient appliances can be encouraged through more harmonised standards and testing methods.

Fuel-efficient Vehicles

Vehicles using lightweight materials and other advanced technologies can greatly reduce both oil consumption and carbon emissions. Electricity-driven and other alternative fuel vehicles also provide a major opportunity to shift transport fuels from oil to other energy sources.

Cleaner Energy Supply

It boosts both sustainable development and energy security. Low emission power sources—renewable, nuclear and fossil fuels with carbon capture and storage—can allow electricity generation to be increased in a sustainable fashion without the risk of needing to be curtailed to cope with climate change; their deployment should be promoted. Bio fuels from sustainable biomass sources to some extent can replace oil use and crude oil imports for transportation, and they have a far smaller carbon footprint. The EWG have been instructed to elaborate an Initiative of Capacity Building for Promoting Market Development of Green Energy Products in cooperation with the Committee on Trade and Investment (CTI).

Renewable Energy Technologies

These include solar, wind, geothermal and bio-energy for electricity and bio-fuels for transport. Continued technology development efforts are needed to further reduce their cost, standardise products, develop supply sources, and share best practices to accelerate their use in electricity generation, buildings and transport sectors. The EWG and BTF have been instructed to continue assessing the resource potential for bio-fuels to displace petroleum-based fuels, the relative costs of bio-fuels, sustainable development practices for bio-fuels and strategies for expanding bio-fuels infrastructure, in cooperation with the Automotive Dialogue and the Transportation Working Group (TWG). The EWG has also been instructed to continue its assessment of renewable energy options for reducing carbon emissions, spurring investment and creating new jobs, in cooperation with EGNRET and the Small and Medium Enterprises (SME) Working Group.

Nuclear Power

Nuclear power can diversify the energy mix and limit carbon emissions. More Asian countries are reaffirming their international commitment to safety, security and non-proliferation as the fundamental elements for the peaceful use of nuclear energy. Solid financial frameworks, as well as cooperation among Asian countries and with relevant multilateral organisations, can help support new nuclear power plant construction consistent with this commitment. The EWG has been instructed to undertake a Nuclear Power Emissions Reduction Potential Study (NUPERPS) on the potential for existing and planned nuclear power plants in interested APEC economies to reduce carbon emissions, and to consider possible cooperation with other relevant organisations such as the International Atomic Energy Agency (IAEA), including its Asian Nuclear Safety Network (ANSN).

Cost-effective Technologies for Carbon Capture and Storage (CCS)

They are essential for reducing carbon emissions from power generation within the many Asian countries that still rely on coal and other fossil fuels for a significant portion of their electricity generation.

Clean-coal Technologies

They can make the use of coal more efficient and low emission. Asian countries should redouble their efforts to develop and deploy such technologies and share information on them through multilateral forums. The EWG has been instructed to extend and reinforce its analysis of technology options for CCS and its dissemination of best practices for applying these technologies to new and existing power plants, working with the EGCFE and other multilateral bodies. The EWG and EGCFE have also been instructed to develop an initiative for deploying advanced clean coal technologies such as Ultra Super Critical (USC) and Integrated Gasification Combined Cycle (IGCC) to make coal-fired power plants more efficient.

Smart Grid Technologies

They include advanced battery technologies for highly efficient and cost-effective energy storage, can help to integrate intermittent renewable power sources and building control systems that let businesses and consumers use energy more efficiently, and they can also help to enhance the reliability of electricity supply, extend the useful life of power system components, and reduce system operating costs. The EWG has been instructed to start an APEC Smart Grid Initiative (ASGI) to evaluate the potential of smart grids to support the integration of intermittent renewable energies and energy management approaches in buildings and industry.

Energy Security and Anti-Terrorism

Terrorists may disrupt energy supply by blowing up oil or gas pipelines, or bombing power stations including nuclear power stations. As one of the measures of energy security, the Meeting of Energy Ministers of the APEC has urged APEC economies to continually strengthen their ability to respond to oil supply disruptions through further development of emergency response mechanisms, improved real-time information sharing and establishment and effective management of strategic oil stocks as appropriate.

Asian countries should also regard anti-terrorism as one of the means to improve energy security. A comprehensive approach adopted to address both the symptoms and root causes of terrorism, enhanced capacity building of developing countries against terrorism, and rejection of double standards of linking terrorism with specific countries, ethnicities and religions. In order to do so, Asian countries should take an active part in regional or sub-regional cooperation against terrorism.

The SCO has developed into an important mechanism of promoting regional security, stability and development. In 2003, China participated in joint military exercises under the framework of the SCO. In September 2003, the Foreign Ministers of the six member countries of the SCO signed the Shanghai Convention on Combating Terrorism, Separatism and Extremism and the Agreement between the SCO Member Countries Concerning the Regional Counter-Terrorism Agency. In 2003, the Chinese also participated in the meeting of the attorney general and in the first meeting of the Council of the Regional Counter-Terrorism Agency of the SCO, playing an important role in guiding and strengthening cooperation within the organisation. In January 2004, the SCO Secretariat and the SCO Regional Counter-Terrorism Agency were formally inaugurated in Beijing and Tashkent respectively. In June 2004, during the summit meeting of the SCO in Tashkent, leaders of the member countries of the SCO signed the Agreement on Cooperation of Anti-illegal Transportation of Narcotic Drugs, Mental Drugs and their Materials. The SCO also established the mechanism of regular meetings of Secretaries of National Security Councils of member countries, to enable them to strengthen their security cooperation.

In September 2004, the Chinese Ambassador to Portugal, Ma Enhan, attended the second meeting in Lisbon, as the representative of the SCO, to the OSCE (Organisation of Security and Cooperation in Europe) on preventing and combating terrorism. He briefed the meeting on the principled positions of the SCO on preventing and combating terrorism and the concrete actions it had taken, and exchanged views with the OSCE and other international organisations on anti-terrorism cooperation.

Following the signing of the Shanghai Convention of the SCO, China began to negotiate and conclude bilateral agreements for combating terrorism, separatism and extremism with other member states of SCO. China has signed agreements in this regard with Tajikistan and Uzbekistan, Kazakhstan and Kyrgyzstan. Cooperation between countries on non-traditional security threats has been increasing, so that they will seek to resolve traditional security problems more through dialogue and negotiations. The process of establishing and developing international and regional mechanisms of security cooperation and dialogue will continue. International, regional and sub-regional mechanisms of security cooperation will become important parts of the future system of international relations.

14

Towards an Architecture of a Global Disasters Management Regime: A Road Map

Yaacov Y.I. Vertzberger

The Rationale*

The December 2004 Indian Ocean earthquake and tsunami, as well as a set of other devastating natural disasters that have since struck other regions, including the recent catastrophic earthquake and tsunami in Japan, vividly reminded us of the fragility of human existence.[1] Furthermore, the post-disaster worldwide out pouring of compassion and offers of assistance to the affected countries and people, also reminded us of the need to effectively harness and channel international cooperation, generosity, resources, and capabilities in a manner that will result in the most efficient practices for pre-disaster preparedness, for the containment and management of the impacts of disaster on people and their social and physical environments. In fact, although disasters are fortunately not every day events, both man-caused and nature-caused disasters are not as rare as we would wish them to be, and require systematic attention and planning to put in place coping mechanisms be they unilateral, mini-lateral or

* An earlier version of this paper was conceived while the author was a visiting scholar at the East-West Centre, Honolulu. It was triggered by the Centre's involvement in the post-tsunami relief efforts and the wide ranging discussions that accompanied these activities. The author thanks Dr. Charles Morrison, East-West Centre President and Dr. Muthiah Alagappa, distinguished senior scholar, for suggestions and comments on an earlier draft. Sole responsibility for opinions and ideas expressed and/or errors lies with the author.

multilateral, or a combination thereof. More specifically, in the long run the collective objectives of the international community should focus on proactive measures to anticipate, mitigate and prevent the worst outcomes of disasters through early warning and by providing quick, efficient and comprehensive relief assistance. These goals will be best served by reliance on the combined resources of the international community as a whole including states, multilateral institutions, and non-governmental organisations (NGOs) rather than relying on unilateral effort or even mini-lateral efforts. It can only be achieved through the carefully designed and constructed architecture of an international regime that will ensure a systematic and broadly integrated response to the threats and consequences of disasters, from early warning through relief efforts to recovery that will be institutionalised, rather than relying on repetitive ad-hoc national and international improvisations, irrespective of how compassionately and generously motivated.[2]

Whatever the intentions of aid providers, the predominantly bilateral context of current aid efforts in post-disaster situations, no matter how well disguised and framed, gives rise to suspicions concerning the "real" motives of aid providers. The fact that foreign military forces often play a major role in relief operations, for obvious reason, tends to evoke a nationalistic backlash from individuals and groups in countries and societies that are aid recipients. These negative responses could be substantially moderated if relief aid was offered through an institutionalised multilateral regime mechanisms that will both engage and empower recipient countries.

The discussion that follows sets out the broad outlines of the fundamentals of the architecture of disasters management regime (DMR) the centrepieces of which are resilience capacities, and the parameters of the main issues that flow from it, and that need to be addressed and agreed upon by participants working towards building an effective and sustainable global early warning and disaster management regime. It is also essential, taking into account the nature of such regime, that states, multilateral organisations and non-state actors will all be assigned an active role in the regime's operations, and that "free riding" would be reduced to a minimum. The regime's grand design should adopt a top down approach, i.e. from the global level, through the regional and national levels to the local operations' level. Yet, the actual construction of the regime should be a step by step from the bottom up, i.e., from locally based capacities to national, regional and global capacities.

The operational focus of DMR is to improve all-around-resilience, in particular of those societies that have neither the material and financial resources nor the knowhow to construct and maintain resilience in the face of disasters. The DMR aims to make the resilience of all societies and states not only the responsibility of their governments, but also the collective responsibility of the international community. The regime will ensure that disaster mitigation capacities are more evenly and broadly distributed across states and will also reduce the costs of recovery and shorten the time required for going back to "normal".

Resilience: The Foundation of Disaster Mitigation and Recovery

Resilience is defined as: "…the intrinsic ability of an organisation [system] to maintain or regain a dynamically stable state, which allows it to continue operations after a major mishap and/or in the presence of continuous stress".[3] It also reduces the impact of surprise and response time urgency of decision-making. It is, thus, the key pivotal factor for containing and mitigating the primary (direct) and secondary (indirect) dangers and consequences from disasters. By implication resilience will positively affect the recovery pace and costs.

Resilience is thus a capacity that is not acquired by default, but has to be constructed by design, as a central pillar of preparations for contingent disasters. Such design would take into account a broad range of likely, even if low probability, disaster contingencies rather than merely focusing on a single most likely set of events. Because of the range of outcomes and costs that such capacity building will have to address it will have to depend on cooperation among civil society organisations, state institutions and the resources of the international community. The extent of self-reliance versus the dependence or reliance of others will vary across states and disaster contingencies. The specificities of this allocation will be regulated by the DMR, that will also set standards of what is the level of acceptable resilience and will monitor its sustainability. Finally, building resilience may result in a false sense of confidence that overlooks the remote chance and very low probabilities of extreme contingencies that would result in cascading resilience failures, that could turn disasters into catastrophes. Resilience planning and construction should consider such mega-emergencies in spite of their very low probabilities.

The effectiveness of resilience in achieving disaster-mitigating outcomes is closely linked to the defining basic features of the disaster that the community faces. These features include three reference coordinates: (1) the nature of the disaster (e.g., tsunami, terror attack, financial meltdown etc.); (2) its geographic scope and reach (e.g., regional-local, national, trans-boundary); and (3) within this geographic scope, the nature and extent of vulnerabilities that are affected by the impact of the disaster: physical (e.g., infrastructure), political (e.g., survival of governance authority), social (e.g., the cohesion of the community), psychological (e.g., individual and collective trauma effects), or any combination there of. These components frame and define any specific disaster and the shared understanding process of sense making, and the resulting comprehension of what is at stake, across the organisations (within and between states) that are tasked with coping and disaster management. In the absence of such a shared understanding, the coordination among these organisations is likely to be flawed and ineffective.

Individual perceptions and shared understanding of the severity and consequences of disasters are tightly coupled to perceptions of resilience. The valuation and perceptions of resilience shape pre-disaster predictions and evaluations of the severity of the threats posed by potential disasters. In other words, the threats from similar disasters (in terms of the above three defining components—essence, geographic range and affected vulnerabilities) will be perceived, by both decision makers and the public, quite differently, to the extent that these are associated with qualitatively different perceptions of resilience capacities. That would also imply that resilience shapes not only the construction and perception of the nature and gravity of the threats posed by the disaster, but also the estimates of the demands made by these threats on the decision makers' time, stress and skills, and on how the public will respond to a known impending threat of a catastrophe (panic vs. calm and confidence). Perceptions of resilience are, thus, central to risk assessments and perceptions of potential disasters. Over estimation of resilience could result in the under estimation of risk assessment, that could then negatively affect the sense of urgency and the extent of the preparations for an anticipated crisis. The complexity of correctly analysing and precisely understanding resilience effectiveness could thus result in unrealistic assessments of risks posed by potential crises, that would result in inadequate preparations, and consequently higher than necessary resulting damages, flowing from a biased evaluation of resilience.[4] This would mean that the two main sources of errors in preparing for disaster events are: either under estimation of the extent of the risk represented by the disaster in question; or over estimation of the resilience—physical, political, social and psychological. The synergy between these types of biases can only make matters worse.[5]

These biases and errors could occur at each of the phases that make up the evolution of the interaction between the impending disaster and those affected by it such as: anticipation, resistance, absorption and restoration. Each requires a built-in institutional on going commitment for preparatory investment in resilience.[6] In the case of potential major disasters, the commitment to resilience would indicate the need for nothing less than high-reliance organisations to devise, implement and monitor the effectiveness of all-around-resilience. The term connotes resilience across issue-areas, phases and vulnerabilities-type, which is a gargantuan task for any single organisation or set of organisations, that involves both division of labour and at the same time a tight coordination that revolves around shared perceptions and sense-making of the situation at hand, as well as a common risk assessment, and a resulting shared view of appropriate policy and operational responses.

Furthermore, resilience as-a-goal should be approached as a process rather than as a static end-state. It should be constantly monitored and requires on going maintenance, upgrading and adjustment to improve the capacity to respond to new environmental or man-made challenges. Some of them are either not fully recognised (due to imperfect information), or not fully understood due to the novelty of the situation (as, for example, the case with the 2010 oil spill in the Gulf of Mexico, and even more so, in the 2011 unprecedented disaster that struck Japan's Fukushima prefecture). Thus constructing resilience is also, metaphorically speaking, an ongoing Sisyphean task.

The implications from this discussion are that the construction of effective all-around sustainable systematic resilience is a goal that is beyond the capabilities and resources of most countries. It is, therefore, a task that should be pursued collectively, that is, viewed as a global problem that should be coped with and managed globally, through the institution and mechanisms of a disaster management regime. The following sections will briefly outline and map the issues and problems of the building and maintenance of such a regime. Obviously, such a regime, like many others, will require an asymmetrical division of labour, responsibilities and costs sharing. The conflict over burden-sharing is bound to add to the inherent technical problems of regime-building, which in its turn will be embedded in conflicting views of what is fair in this context, and how far can the moral obligations of the richer and better resources-endowed participants, be stretched.

The Essentials for Regime Building

1. Underlying Premises

1.1. Mankind will continue to face the threats of both nature-caused and human-caused disasters. The former are inevitable although the effects can be moderated, and the number and magnitude of the latter can be controlled but cannot be completely prevented.[7]

1.2. Disasters of both kinds will vary in magnitude from extreme disasters (such as unprecedented hard-to-anticipate earthquake and tsunami) to ordinary disasters (e.g., seasonal typhoons and hurricanes) and "normal" human-caused accidents (e.g., fires in oil and gas fields, nuclear plant accidents, etc.).

1.3. The import of these disasters may be local or regional with occasional global second order consequences. Even local events could have long-term regional impacts (e.g., natural disasters in Japan, China or Taiwan that would disrupt economic growth and have extensive consequences regionally for the growth of other economies).

1.4. Vulnerabilities and resilience to disasters are unevenly distributed among countries between and within regions. At the same time the current state of coping capabilities with respect to disasters (early warning systems) or with their outcomes are also variable across these countries depending on technological and financial resources, governments' priorities in allocating national resources for building resilience, and the quality of governance mechanisms to exploit most efficiently all available capabilities.

1.5. With very few exceptions (the United States and Japan) it is unlikely that any single country is adequately prepared to deal, on its own, efficiently and comprehensively with the dual tasks of pre-disaster early warning and other preventive measures, as well as post-disaster relief and reconstruction. For many countries other perceived needs have priority over preparations for what is often perceived to be the low probability threat of rare disasters, which is regarded as prohibitively expensive and at the same time seems to provide in the short-run a low return on the required investment.

1.6. In the absence of their own national capabilities for dealing with both extraordinary disaster, and even ordinary disaster situations, governments in effect rely on policy gambles. This assumes that either no serious disaster will befall their country or that, if it did, improvised responses and spontaneous external aid would contain the worst consequences. This approach leaves their populations exposed to grave risks and eventually places an added burden on the international community.

1.7. Despite the absence of a clearly articulated alternative to this essentially fatalistic policy gamble, governments are rarely held fully accountable for their inadequate preparedness. In part this is because there are no agreed upon international standards for adequate national disaster preparations and governments' accountability.

1.8. This paper represents a first attempt to articulate a superior alternative approach to the policy gamble approach. It calls attention to the urgent need for a collective, integrated, systematic, comprehensive, resource-rich, pre-planned, and institutionalised set of mechanisms for disaster management. It is premised on a pre-designed combination of building up national resilience capabilities that will be coordinated with multilateral cooperation, stemming from the recognition that human security is at the same time a shared universal responsibility and governments' prime priority. This premise is at the heart of the proposal for a management regime that covers the whole gamut from early warning to disaster relief and recovery. It is not claimed that an international regime for dealing with all the phases of disasters is the ultimate solution to pre-disaster prevention measures and post-disaster coping problems, and it may not be free from some embedded weaknesses. But it has the potential to by far outperform unilateral and mini-lateral current strategies for managing disasters. The broad design of the new institution as specified below aims to provide a starting point for a policy debate among scholars, decision-makers and practitioners in governments, multilateral institutions and non-governmental organisations. The agenda that is discussed is comprehensive but by no means intended to be exhaustive.

2. The Participants

2.1. The prime actors and movers of the regime will be nation states. They will hold decision-making powers for the regime according to set rules.

2.2. The second circle of members, with advisory and coordination roles, will be multilateral international organisations, such as the United Nations, World Bank, Asian Development Bank, and other similar organisations.

2.3. The third tier of members will consist of NGOs who will have observer-participant status, and will be empowered by the regime institutions to carry out specific tasks within specific geographic boundaries.

It follows that these different participant groups will have dissimilar representation, authority, and roles in the regime's decision-making process. States will have voting rights on all major policy issues.

International organisations will have the authority to put items on agenda for discussion and vote, but shall not have voting rights. NGOs will have the right to attend and observe at all meetings and will have access to all available information, but shall not have the right to put items on the agenda or vote.

3. The Norms

The normative foundations of the regime are the pillars on which international cooperation is constituted. These norms include:

3.1. *The Human Security Norm.* This norm recognises the duty and responsibility of each government to do the utmost to safeguard the lives and welfare of its citizens, and to devote the resources required to achieve these objectives. The duty is not conditional on the rarity of the life- and welfare-threatening causes but on their gravity. Disasters caused by both nature and humans are among the major sources of individual and societal misery and poses acute threats to lives and property. They contribute to under development, poverty, and the degradation of access to basic needs.

3.2. *The Non-Discrimination Norm.* Regime participants must recognise the right of every member of their societies—irrespective of race, religion, ethnic origin, political views or persuasions—to equal access to all resources and capabilities concerning warning and disaster relief. There is one caveat. This norm will not apply to any groups or individuals that pose a threat to those assisting and providing humanitarian aid.

3.3. *The Accountability Norm.* Those at the helm of national and local authority-structures who neglect to carry out appropriate measures for disaster preparedness, and the building of resilience capacity should be held personally and politically accountable to their people. As such they could be subject to legal prosecution by local courts for the consequences of negligence and failures.

3.4. *The Permissive Sovereignty Norm.* This norm represents a commitment on the part of states to allow full and free access through their land territory, air space, and maritime territorial waters to all vehicles, ships, planes, their crews, and any other personnel on board, whether military or civilian, while on missions relating to either early warning or disaster relief.

3.5. *The Temporary Free Shelter Norm.* Countries bordering on other countries that are devastated by disaster wil! provide, when necessary, temporary shelter and other basic-needs assistance on their own or in tandem with the international community, to both early-warning evacuees and disaster refugees, as long as it is needed.

3.6. *The Common Solidarity Norm.* Recognising the uneven distribution of resources and wealth among the regime state members, the participants will accept that burden sharing for regime building and maintenance will be uneven. Wealthier countries have an obligation to help those in need in proportion to their larger resources. But none of the regime members is exempt from contributing its relative share of the cost burden. Free riding should be actively discouraged.

4. Regime Design and Architecture

Design of a global robust DMR poses an architectural challenge. We argue here for a multi-layered architecture. But before getting into the details of the offered design I shall define what precisely is meant by the commonly and loosely used term "architecture" as it applies to international politics. The definition of the term as used in this paper is conceptually and operationally defined as follows.

According to the conceptual definition, architecture is the design and construction of geographically-defined and issue-based international orders that bind agents, structures and processes with the purpose of increasing the probability of predetermined desired outcomes, or decreasing the probability of undesired ones.

According to the operational definition architecture is the end-result of the design and construction of a goal-driven social-political order within prescribed geographic and issue domains. It could grow from voluntary agreement among interested affected parties, involuntary coercion by powerful or hegemonic agents, or could emerge piecemeal over time. The emergent order charts domain boundaries, defines the targeted *agents* (states, alliances of states, ethnic groups, governmental and non-governmental organisations etc.), and their shared interests and preferences. The design of this order revolves around its *structure* and *process*. The former includes setting a desired distribution of power and authority, distribution of roles and the consequent division of labour among regime participants. The latter requires the construction of social institutions and setting their norms and rules. It also involves the building of steering mechanisms to regulate, constrain and control the interactions among the participating agents with predetermined shared expectations about desired outcomes. These are broadly characterised by a common desire for an understanding of the measures for reducing risk and uncertainty, and/or increasing cooperation, economic, scientific and technological advancement to minimize the consequence of potential disaster, both foreseen and unforeseen.

Taking into account the nature of disaster management a decentralised multilateral international regime should be structured to weave together and coordinate domestic (self-help) with external resources to produce high resilience as shall be discussed later in the paper. That implies a nested four-layered network of architectures:

(1) *The grand-architecture level* (global) will be centred on the structure and process of oversight and intervention in all pre-disaster and post-disaster planning and implementation of resilience-building efforts, through the allocation of external resources among affected countries in each disaster area, and in case of simultaneous occurrence of cascading disasters, to control and allocate the flow of resources across disasters and affected actors. The grand-architecture will be structured to posses material (e.g., satellite systems) or non-material (e.g., scientific knowledge) capabilities and resources that most individual countries (or even regions) are unlikely to be able to acquire or posses on their own, and will have the authority to make necessary operational and distributional decisions that will directly affect the disaster area.

(2) *The macro-architecture level* (regional) will be built around regional organisations (e.g., Association of Southeast Asian Nations [ASEAN], European Union [EU], Organisation of American States [OAS], and Organization of African Unity [OAU]) which will coordinate activities and allocate resources among the countries in that particular region that are affected by crises and disasters. Obviously, regions will differ in both material and non-material capabilities they are likely to have and bring to bear. The resource deficit of any region could be compensated by transfer of resources from institutions of the grand-architecture level. In the same manner surplus resources of the better equipped and resources-rich regional organisations could be lent or transferred to a resource-deficit region. The case of ASEAN is instructive. In 2005 ASEAN members signed an Agreement on Disaster Management and Emergency Response that covers all aspects of disaster management. An ASEAN coordination body the Centre for Humanitarian Assistance has been provisioned for, that will coordinate all national institutional arrangements of the ten member states.

These goals are also incorporated into the road map for an ASEAN community that was adopted in 2009.[8]

(3) *The micro-architecture level* (national) will be centred around individual national capabilities for dealing with disasters. National capabilities could be enhanced and complemented, both pre-disaster and post-disaster, by regional and/or global resources as circumstances would require.

(4) *The nano-architecture level* (local) will be focused on in-country locally mobilised resources that are available in the immediate area that would be affected by disaster. Such resources are available on location at the time of the event for an immediate response by local authorities. As the extent of the crises becomes better known it is most probable that in almost all cases of extensive disasters, local resources will have to be reinforced by national, regional and global resources and capabilities. The extent of the availability and accessibility of additional capabilities and resources and the extent of their adequacy will depend on the nature of the rules and procedures that govern the interaction between grand, macro, micro and nano-architectures as designed into the disaster-management regime.

In sum, it is argued that a top down pre-determined, nested architecture, from the grand to the local levels, that will be governed by an effectively coordinated command and control system and hierarchy is most likely to produce an efficient, cost-effective (low transaction costs) and timely disaster management operations, and will be reflected in the quality of resilience to disasters that will emerge over time. Such a multiple layered regime, in which the overall architecture would be designed so that lower level architectures are nested in higher level architectures, will be best positioned to universally serve mankind in its dire hours of need, by making the most of universal human, financial, scientific and technological resources available. In addition to this vertical *within-regimes* coordination, a disaster management regime could gain from horizontal *across-regime* coordination and cooperation, by taking advantage of the know-how and other resources of related regimes as the case will require (such as the monetary regime, the food regime etc.).

5. Regime Rules and Procedures

The implementation of regime norms and objectives requires a set of agreed rules, decision-making procedures and implementation instruments. These are linked to the anticipation of issues and problem-areas that the regime will have to address, and are necessary for effective preparation for coping with emergencies. Building shared anticipations through strategic communication is essential for the convergence of regime participants' expectations, which will reduce the space for unfounded suspicions, friction and disagreements over early warning operations and response to disasters, and serve to build the social capital, especially trust necessary for cooperation among disaster management organisations within single state and across boundaries.[9]

The following sections set out the anticipated issues and problem-areas that an early warning and disaster relief regime will have to address. The analysis takes a strategic management approach.[10] But at this stage we do not specify the actual rules and procedures, rather we only list the broad domains on which the countries participating in a DMR will have to reach agreement concerning rules and procedures. The list of issues and problem-areas is suggestive and open to the inclusion of additional items as part of future discussions of this proposal. The list is divided into issues concerning early warning, disaster relief, training of professionals for early warning tasks and disaster relief management, and finally the command and control structure of the regime. All of these are tightly linked to the scope and quality of material, social, psychological and political resilience that will result.

6. Early Warning: Issues and Problem-areas

- 6.1. *Hardware:* Access to satellites is a prime condition for timely early warning. In essence there are three options for the regime.
  - 6.1.1. Access to the commercial satellites that are either state owned or privately owned.
  - 6.1.2. Access to military (spy) satellites already in orbit by agreement with the governments owning these satellites.
  - 6.1.3. Investment by regime members in an independent satellite system that will be specifically built and owned by the regime participants and be technologically tailored to early warning and post-disaster tasks. In the short and medium terms only options 6.1.1 and 6.1.2 are viable.
- 6.2. *Satellite informational take:* Options 6.1.1 and 6.1.2 will require negotiated terms as to the type, quantity and cost of satellite information that will be transferred to regime members.
- 6.3. *Additional information sources:* Other important sources of relevant information, such as seismological data, need to be specified and identified.
- 6.4. *Information dissemination measures:* Means and procedures for real-time information dissemination among members in both

emergency and non-emergency times must be developed. These measures suggest at least two options of information processing and "packaging".

6.4.1. Establishing a single regional centre where all raw data will be received, processed, and disseminated to each country as an early warning or post-disaster information final product.

6.4.2. Distributing the raw data directly to local and state-level early warning centres, where it will be processed, interpreted, and disseminated all the way downstream.

7. Disaster Relief: Issues and Problem-areas

7.1. *National disaster management plans:* Participating countries need to prepare comprehensive national disaster management plans, including risk assessments and vulnerability studies. These plans have to share a common format and professional standards among all participating states. They will be monitored and periodically examined by the regime's technical agencies in order to upgrade and update them. To the extent possible, individual national plans should converge, and allow for inter-operability of cross-border use of national capabilities to assist neighbouring countries when necessary. An international team of experts could be assembled to assist countries that do not have the necessary expertise to prepare national disaster management plans on their own.

7.2. *Disaster relief task forces:* Participating countries will establish specialised units for disaster relief. The criteria for the size of that force in each country will have to be agreed upon among participating countries, and will obviously vary with each country's resources and risk exposure. National task forces would also be used to assist other member countries in need.

7.3. *Evacuation pre-planning:* As part of their disaster management plans participants need to review and plan evacuation priorities between and within potentially threatened areas.

7.4. *Law and order breakdown:* Disaster situations may lead to the collapse of law and order, which may then spread across national borders. Law enforcement agencies should prepare for quick restoration of law and order, including plans for obtaining assistance from other countries' law and order forces.[11]

7.5. *Legal jurisdiction:* Relief efforts could involve the presence of large numbers of relief workers, civilians, and military personnel from foreign countries. In some cases, their presence could be long-term. This requires a legal code agreed upon among regime participants that addresses jurisdiction in cases of crimes committed by foreigners or against them.

7.6. *Interoperability:* Effective relief work requires interoperability among the agencies operating. Two types of interoperability should be planned for, vertical and horizontal.

7.6.1. Vertical interoperability allows for agencies from the same country to communicate and coordinate their efforts. Particular attention should be given to interoperability between military and civilian agencies, especially in light of the fact that militaries play a key role in relief operations because they usually are the agencies best equipped with relief-relevant hardware, are best trained and in most cases are more experienced than other agencies.

7.6.2. Horizontal interoperability allows for agencies from *different* countries to communicate and coordinate their efforts. For the reason mentioned above, interoperability between military relief units from different countries and organizational cultures is of key importance.

7.7 *Addressing basic human needs:* Disasters challenge a society's ability to supply basic survival needs to affected populations, including food, water, sanitation, and medical treatment.

7.7.1. Water scarcity is the immediate most pressing problem and can be caused by the destruction of wells and pipeline systems or their contamination. The latter could have severe medical implications. This requires the establishment of multiple emergency water reservoirs in locations that are less vulnerable to disasters, and an emergency pre-planned system for transfer and distribution of the daily essential quantities of drinking water.

7.7.2. Food scarcity could be addressed nationally or internationally and has two dimensions: the availability of food supply and its transportation to needy areas. Experience shows that the more pressing problem is the transportation of food in a timely fashion. Underlying both dimensions is the need for accurate real-time information about the location of and the accessibility to areas where there are food scarcities. Another matter that requires attention is the orderly and fair distribution of food, which may also have law and order implications.

7.7.3. The scarcity of medical care has a number of aspects.[12] One is the dramatic increase in the number of people that need medical attention relative to the number of personnel and facilities available to provide treatment. Related to that is the scarcity of medical supplies and the capabilities to transport supplies to disaster areas and patients to hospitals (or bring hospitals to the patients). Such problems require creative thinking: for example, creating diffused storage systems for emergency supplies. Or perhaps adapting the Chinese model of training a large number of local "barefoot doctors" to handle immediate medical treatment needs until better trained professionals can arrive. It may also be desirable to provide local medics and doctors with satellite-linked laptops that will allow them to seek online medical guidance in the absence of access to major well-equipped medical facilities.

7.7.4. The sanitation issue posed by large displaced populations, besides being a problem in and of itself, could also exacerbate water contamination problems and medical needs arising from increased vulnerability and exposure to disease.

The synergy among the direct devastating effect of disasters, inadequate access to basic human needs by millions, and the limited capabilities of local and national governments to cope with these tasks make the basic needs problem-area one of the most important priorities for regime design.

7.8. *Post-disaster reconstruction:* The most important remedy for disaster, both materially and psychologically, is to shorten the time lag between loss and reconstruction. Therefore, the regime should incorporate measures that will make it possible to rebuild a devastated community's livelihood and lifestyle in as short a time as possible. Among other things, this requires a pre-disaster build-up of resilience, through a standing substantial emergency fund that can disburse reconstruction resources at short notice. It will also require a set of blueprints for reconstruction of different types of communities (e.g., fishing-dependent, agriculture-dependent, services-dependent, etc.). Psychologically, one of the major keys to building resilience to the stress imposed by disasters is the hope for a quick bounce back short time lag between disaster and economic recovery.[13]

8. Command, Control, and Training

This section deals with decision-making and other organisational issues that are likely to be especially contentious and difficult to resolve because of concerns about the implications that the mode of power sharing within the regime might have for national sovereignty.

8.1. *Command Structure:* One of the most important early decisions on regime design will be the choice between a hierarchical concentration of power (top-down command structure) vs. a diffused bottom-up command structure. Related to this is the question of whether to have a single or multiple command centres. A single centralised global centre of command would have power over non-state based resources, strategic decisions, and control most external intervention operations. The alternative possibility is to create regional command centres with the major powers, US, Russia, EU, China and Japan and key multilateral organisations represented at each. The diffused command structure may still require the appointment of a general secretary/ director general for the DMR to coordinate among these autonomous command centres.

8.2. *Control Measures:* The complexity of the problems and their breadth require the establishment of a system that will be capable to interface both with the command centre or centres, and the field agencies. The system will provide information for central decision-making and feedback from the field to "headquarters". It will also have to interface with local disaster managers in the field and provide them with information on issues ranging from medical information, through weather forecasts, to the status of relief and reconstruction projects etc. Achieving the goal of an effective real-time, two-way control system will require setting up a number of information clearing houses by issue-areas (e.g., satellite and seismographic data, medical data, financial data, etc.). These clearing houses will provide for a two-way flow of information between the command centre or centres and the field agencies, and from the field to the command centres that will complement other sources for real-time updates (e.g., satellite damage assessment) available to the decision-making centres.

8.3. *Training System:* An integral part of the regime should be an institutionalised training system. Training will encompass all aspects and stages of disaster control and management, from operating early warning hardware and interpreting early warning information, through food assistance disbursement, to reconstruction needs assessment. A uniform training system will ensure interoperability, and will take advantage of the most current information on best practices based on cumulative learning from experience. One technical issue that will have to be addressed is whether to have a single regional central training facility or have a few, smaller, sub-regional training facilities.

It should be noted that the agenda that is discussed above is a necessary, but not sufficient condition for building and maintaining resilience.

In Lieu of Conclusions

The above think-piece sets out the framework and components for what repetitive recent events have proven to be critically essential and urgent but is as yet missing, a disaster management regime. The paper does not claim to be all-inclusive and is open to suggestions, comments, and other relevant ideas. As experience with previous international regime building has proven, the process of regime building is time, cost and effort consuming, and never moves forward in a linear fashion. I do not expect that the regime in its entirety will emerge overnight. What is suggested, is a road map leading to a division of labour among groups of experts, where each can work on separate aspects, but with a single purpose and with the grand scheme of things in mind. This would allow for simultaneous modularity in regime design, where progress in one regime domain will be independent of the pace of progress or lack of progress in other domains. The paper also draws attention to the tight coupling between achieving resilience and implementing the agenda proposed by this paper.

The first necessary immediate step towards the construction of a disaster management regime is stock taking. It requires first a mapping all countries for their disaster vulnerabilities, that is, the likely nature or man-made types of disasters to which they might be exposed (volcanic eruptions, earthquakes, floods, forest fires plagues etc.). This step should be followed by establishing a repository of national capabilities for dealing with disasters for each participating state, to the extent that such capabilities exist. Such capabilities may be of a pre-disaster pre-emptive nature (population evacuation plans), or post-disaster responsive measures or both. In general these capabilities can consist of any combination of the three: disaster-relevant know-how, specialised equipment and trained personnel that are skilled in using the know-how, and operating the equipment. This information will be reported by each country to the regional organisation with which it is associated for disaster management and relief purposes. Plans for upgrading should start from this base line. It would also become the basic database for state-specific planning on how to increase resilience through national or international efforts and cooperation. The stored information should be made available online, once disaster strikes, to the organisers and coordinators of relief efforts.

Recent events and the globalisation of disasters have proven once more that the community of nations cannot afford procrastination or catching a "slow boat to China" as far as regime construction is concerned.[14] Too much is at stake. It is obviously in the interest of those who have been affected in the past and are likely to be affected in the future, as well as those countries and societies that were lucky to be out of harm's way, to establish a robust institutional mechanism, rather than continue to depend on a gambling-improvisation "strategy", that is, gambling on not being in harm's way and improvising if the gamble does not pan out. Enlightened governments should be therefore ready to accept the delegation of substantial national authority to an international institution that will be by far better equipped and suited to accomplish improved all around resilience and thus be more effective in coping with future disasters. As former Presidents George Bush and Bill Clinton reminded us in a joint television campaign for 2004 tsunami relief efforts, "No one can change what happened, but we can all change what happens next". This is exactly what this concept paper is about.

To summarise, this paper calls for a global regime-type approach for coping with the globalisation of disasters. A disaster management regime should be focused on containing and mitigating disasters by building and maintaining globally all-around resilience, that should address physical, political, social and psychological vulnerabilities. Achieving these goals requires, more than in any other existing regimes, close monitoring of the decision-making process and awareness of biases and errors that could critically undermine resilience, such as President Bush's management of the hurricane Katrina crisis.[15]

Attending and responding correctly to early warnings is not an obvious outcome, taking into account uncertainty and ambiguity involving such warning and the high costs of activating counter measures. This is particularly so if the warning is proven wrong. This is even more problematic when these measures involve collective action by actors from different cultures and building institutional cultures on mutual trust. The possibilities for biases and errors in the decision-making process during a major disaster are high and need round-the-clock monitoring. Faulty decision-making could turn a disaster into a catastrophe if resilience fails.[16] Moreover, taking into account how closely technological, financial and infrastructures are coupled in this era, an even more vicious problem could be posed by *cascading disasters* (earthquake followed by floods and collapse of major infrastructure complexes, such as the supply of electricity, water, food), followed by *cascading resilience* failures across domains, which could spin the situation completely out of control. Research on these eventualities has just scratched the surface.

Notes

1. For studies that map the potential for a broad and diverse range of potential disasters, and for a discussion of the difficulties in predicting these events and coping with them, see: Arjen Boin, "The New World of Crises and Crises Management: Implications for Policymaking and Research," *Review of Policy Research*, 26, 2009, pp. 367-377; Arjen Boin, Paul 't Hart, Eric Stern and Bengt Sundelius, *The Politics of Crisis Management: Public Leadership under Pressure*, New York, NY: Cambridge University Press, 2005; Kurt M. Campbell (ed.), *Climatic Cataclysm: The Foreign Policy and National Security Implications of Climate Change*, Washington D.C.: Brookings Institution Press, 2008; Patrick Lagadec, "A New Cosmology of Risks and Crises: Time for a Radical Shift in Paradigm and Practice," *Review of Policy Research*, 26 (4), 2009, pp. 473-486; Vaclav Smil, *Global Catastrophes and Trends: The Next Fifty Years*. Cambridge, MA: The MIT Press, 2008.
2. The literature on international regimes as global governance institution is extensive. For some of the more important writings see: John L. Campbell, *Institutional Change and Globalization*, Princeton, NJ: Princeton University Press, 2004; Andreas Hansenclever, Peter Mayer and Volker Rittberger, *Theories of International Regimes*, Cambridge: Cambridge University Press, 1997; Stephen D. Krasner (ed.), *International Regimes*, Ithaca, NY: Cornell University Press, 1983; Elinor Ostrom, *Governing the Commons: The Evolution of Institutions for Collective Action*, Cambridge: Cambridge University Press, 1990; Volker Rittberger with the assistance of Peter Mayer (eds.), *Regime Theory and International Relations*, Oxford: Clarendon Press, 1995; Oran D. Young, *Governance in World Affairs*, Ithaca, NY: Cornell University Press, 1999.
3. Eric Hollnagel, "Resilience: The Challenge of the Unstable," in Erik

Hollnagel, David D. Woods and Nancy Leveson (eds.), *Resilience Engineering: Concepts and Precepts*, Aldershot, Eng.: Ashgate, 2006, p. 16; see also Marc de Bruijne, Arjen Boin and Michael van Eeten, "Resilience: Exploring the concept and Its Meanings," in Louise K. Comfort, Arjen Boin and Chris C. Demchak (eds.), *Designing Resilience: Preparing for Extreme Events*, Pittsburgh, PA: University of Pittsburgh Press, 2010, pp. 13-32.

4. For discussions of definitions and complexities of risk and risks assessment see, for example: Baruch Fischhoff, Sarah Lichtenstain, Paul Slovic, Stephen L. Derby and Ralph L. Keeny, *Acceptable Risk*, Cambridge: Cambridge University Press, 1981; David Hillson and Ruth Murray-Webster, *Understanding and Managing Risk Attitude* (2nd ed.). Aldershot, Eng.: Gower, 2007; Sheldon Krimsky and Dominic Golding (eds.), *Social Theories of Risk*, Westport, CT: Praeger, 1992; Yaacov Y.I. Vertzberger, *Risk Taking and Decision-making: Foreign Military Intervention Decisions,* Stanford, CA: Stanford University Press, 1998.
5. The literature on misperceptions, biases and other errors is extensive. For some examples see: Max H. Bazerman, *Judgment in Managerial Decision Making* (3rd ed.), New York, NY: John Wiley and Sons, 1994; Mark Bovens and Paul 't Hart, *Understanding Policy Fiascoes*. New Brunswick, NJ: Transaction Publishers, 1996; Charles F. Parker and Eric K. Stern "Bolt from the Blue or Avoidable Failure? Revisiting September 11 and the Origins of Strategic Surprise," *Foreign Policy Analysis*, 1 (3), 2005, pp. 301-331; Yaacov Y.I. Vertzberger, *The World is Their Minds: Information Processing, Cognition, and Perception in Foreign Policy Decision-making*. Stanford, CA: Stanford University Press, 1990; Lawrence Woocher, "The Effects of Cognitive Biases on Early Warning," Paper presented at the International Studies Association Annual Convention, March 29, 2008.
6. Jerome M. Kahan, Andrew C. Allen and Justin K. George "An Operational Framework for Resilience," *Journal of Homeland Security and Emergency Management*, 6 (1), 2009, article 83. DOI: 10.2202/1547-7355.1675; Robert McCreight, "Resilience as a Goal and Standard in Emergency Management," *Journal of Homeland Security and Emergency Management*, 7 (1), 2010, article 15. DOI: 10.2202/1547-7355.1700. Available at: http://www.bepress.com/jhsem/vol.7/iss1/15; Karl E. Weick and Kathleen M. Sutcliffe, *Managing the Unexpected: Resilient Performance in an Age of Uncertainty* (2nd ed.), San Francisco, CA: Jossey Bass, 2007, p. 14.
7. Charles Perrow, *Normal Accidents: Living with High Risk Technologies*, Princeton, NJ: Princeton University Press, 1999.
8. Yang Razali Kassim, "From Ache to Chile: Is ASEAN Prepared for Another Disaster," *RSIS Commentaries*, 24 (1), March 2010.

9. E.g. Cecilia Cabañero-Vesoza and Helen R. Garcia, *Building Commitment to Reform Through Strategic Communication: The Five Key Decisions*, Washington D.C.: World Bank, 2009.
10. For the attributes of this approach see: Sang Ok. Choi, "Emergency Management: Implications from a Strategic Management Perspective," *Journal of Homeland Security and Emergency Management*, 5 (1), 2008, pp. 1-21.
11. On the criticality law enforcement see: Jeff Rojek and Michael R. Smith, "Law Enforcement Lessons from Hurricane Katrina," *Review of Policy Research*, 24 (6), 2007, pp. 589-608.
12. For a wide ranging coverage of public health issues following disasters see: Bruce W. Clements, *Disasters and Public Health: Planning and Response*. Burlington, MA: Butterworth-Heinemann, 2009.
13. On possible ways to address these issues see, for example: Nathan E. Hultman and Alexander S. Bozmoski, "The Changing Face of Normal Disaster: Risk, Resilience and Natural Security in a Changing Climate," *Journal of International Affairs*, 59 (2), 2006, pp. 25-41; Joanna Syroka and Richard Wilcox, "Rethinking International Disaster Aid Finance," *Journal of International Affairs*, 59 (2), 2006, pp. 197-214; Walden Bello, "The Rise of the Relief-and-Reconstruction Complex," *Journal of International Affairs*, 59 (2), 2006, pp. 281-296.
14. See David Alexander, "Globalization of Disaster: Trends, Problems and Dilemmas," *Journal of International Affairs*, 59 (2), 2006, pp. 1-22; Arjen Boin, "The New World of Crises and Crises Management: Implications for Policymaking and Research," *Review of Policy Research*, 26, 2009, pp. 367-377.
15. Charles F. Parker, Eric K. Stern, Eric Paglia and Christer Brown, "Preventable Catastrophe? The Hurricane Katrina Disaster Revisited," *Journal of Contingencies and Crisis Management*, 17 (4), 2009, pp. 206-220.
16. On the dichotomy between disasters and catastrophes see: Birkland (2009). Thomas A. Birkland, "Disaster, Lessons Learned, and Fantasy Documents," *Journal of Contingencies and Crisis Management*, 17 (3), 2009, pp. 146-156.

15

The Geopolitics of Himalayan-Tibetan Glacier Melt

Syed Iqbal Hasnain

Introduction

The "greater Himalayan region", sometimes called the 'Roof of the world' or the 'Asian water Tower'[1] holds the key to South Asia's ecological and social security by virtue of its being a centre of biological and cultural diversity, the final destination of its life-sustaining monsoons, and a storehouse for water and other natural resources. The *Fourth Assessment Report of the Intergovernmental Panel on Climate Change* (IPCC)[2] claim, that the Himalayan glaciers would disappear by 2035 has catapulted the region into the centre of global attention, as the glacier melt waters have enormous regional geopolitical significance. In January 2010, however, the IPCC officially withdrew the paragraph on Himalayan glaciers as the lead authors of the chapter had picked up" the quote from a newspaper and not from a peer-reviewed scientific journal. Notwithstanding, that the IPCC 'goofed up the scenario' glaciers in the region are, in reality, rapidly receding.[3] Approximately one third of the global population (1.5 billion) depend on the major rivers systems whose headwaters lie in the Hindu Kush-Himalaya-Tibetan mountains. At these headwaters are the thousand of glaciers and the seasonal snow-pack that provide a long-term storage of fresh water that feeds the major Asian river basins.

The region is home to thousands of glaciers in the tropical/sub-tropical region that exert a direct influence on social and economic development in the surrounding countries of as China, India, Nepal, Tajikistan, Pakistan, Afghanistan, Bhutan, Bangladesh, and Myanmar. It is subjected to the influences of multiple climatic systems, complicated geomorphologies and various internal and external geological impacts.

Glaciers melt also influences the downstream hydrological systems. The alpine ecosystems are affected by population growth, new economic activity and industrial emissions.[4]

All glaciers are showing negative mass balance which is reflected in increased discharges in melt water streams. More and more scientists now recognise the impact of black carbon on glacial melt.[5] Black carbon exacerbates the melting by contributing to atmospheric warming and increases the amount of heat absorption within the snow by turning it black. The sources of black carbon in the region include cement factories, diesel trucks and wood burning stoves. Although the Himalayan rivers have varying percentages of ice melt in annual flows; what makes glacial melt so critical, though it is a relatively small percentage of a river's annual flow, is the timing at which it occurs. If these flows come during the rainy monsoon season, they may lead to floods. But if they come during the hot, dry spring and fall months, the so-called "shoulder months" just before and after the monsoon, they keep the volumes of river water more constant and are highly beneficial to agriculture.

Melting Glaciers

Glacial melt is critical before and after the rainy season, when it supplies a critical proportion of flow in every river—from the Yangtze (irrigates more than half of China's paddy fields) to the Ganges and Indus (important for agricultural heartland of India and Pakistan). Chinese scientists[6] have monitored more than 680 glaciers on the Tibetan plateau and all of them are shrinking rapidly, with heaviest losses on its southern and eastern edges. The ice cover in the southern and eastern plateau has shrunk more than 6 per cent since the 1970s and the damage is still greater in Tajikistan and Northern India, with 35 per cent and 20 per cent declines respectively over the past five decades. If the current trend holds, it is believed that 40 per cent of plateaus glaciers could disappear by 2050. Full scale glacier shrinkage is inevitable and it will lead to ecological catastrophe.

The effects on the Indus and Brahmaputra river basins are likely to be severe owing to large populations and their high dependence on agriculture.[7] Continued, widespread melting of glaciers during the coming century will lead to floods/water shortages, declining crop yields and sea level rise threatening local communities and habitats. Thus far conclusions are being arrived at on the basis of limited data from satellite imagery and field research on selected glaciers, but the extreme conditions found in the region make field research expensive, time intensive, dangerous and physically challenging as work is to be done at a height of 4000 metres and above.

There is very little infrastructure at such heights to support research efforts and absolutely no historical data to use as a baseline. Nevertheless, glaciers are thinning at the rate of up to one metre per year and retreating at rates ranging from 10 to 60 metres per year. Many small glaciers (less than 0.2 km^2) have already vanished. During the last 30 years there has been a shrinkage of 5.5 per cent in the volume of glaciers on Tibetan plateau, and at simiiar rates in Nepal, India and Bhutan. Reports[8] indicate that by 2030 about 30 per cent of Himalayan glaciers would have disappeared; by 2050, 40 per cent; and by the end of century 70 per cent and that full scale glacier shrinkage in the region will eventually lead to an ecological catastrophe. These anticipated changes will have considerable effects on water and food security scenarios.

The Copenhagen and Cancun Accord's arrived at in December 2009 and 2010 respectively is on framework and does not include verifiable binding targets for reductions in carbon dioxide emissions. The 2°C barrier agreed at COP 15 translates into a 2.5 Wm^{-2} barrier for energy additions. However, the barrier poses a huge dilemma for policy makers because the blanket of man made GHGs that surround the planet as of 2005 has already trapped 3 Wm^{-2} and the barrier of 2°C has already been exceeded by 20 per cent.[9] The most crucial climate tipping point is the rise of temperature of up to 0.3°C per decade on the plateau—approximately three times the global warming rate.[10] Black carbon or soot deposition on snow will reduce reflectivity by 2.0 to 5.2 per cent—amounting to about 70-204 mm of water equivalent runoff from a typical Tibetan glacier. A study by NASA scientists[11] indicates that as a response to the imbalance in radiation energy by dust and black carbon in the Indo-Gangetic plain and Himalayan foothills, the atmosphere over the plateau is anomalously heated and moistened in an elevated heat pump (EHP) effect. The warm and moist atmosphere overlying the Tibetan plateau land surface, causes a reduction in surface sensible and latent heat fluxes from land to atmosphere, i.e. a net heat gain by the land surface. The net heat gained is melts more snow and ice on the Tibetan plateau.

Geopolitical Significance of Tibetan Plateau

Tibet between 1911 and 1950 was an independent state possessing all the attributes of statehood.[12] Mao knew that whoever holds Tibet dominates the Himalaya piedmont; and he who dominates the Himalayan piedmont, controls the water in the Asia.[13] The government of India, upon inheriting the past treaties signed by British, should have worn the British mantle, to its advantage. Unfortunately it failed to do this without giving serious thought to the consequences that would follow. The Indians were not only shocked at China's unprovoked military action; they were also much offended in view of the explicit assurances the Chinese government had given to them. In a sharp note to the Chinese government, the Indian foreign ministry expressed deep regrets and surprise at its decision to send troops into Tibet just after the Tibetans had initiated negotiations with China's ambassador in New-Delhi.[14] The United States and British governments supported the Indian position, and the US state department has informed New Delhi of its desire to help Tibet by whatever means possible. The Americans recognised that, in view of geographic and historic factors, the main burden of the problem rested on India and that its cooperation was necessary in any attempt to help Tibet effectively. The Indian foreign ministry, however, dissuaded the United States against supplying military aid to Tibet. Prime Minister Nehru even requested that Washington refrain from publicly condemning China for its action in Tibet. Whatever the considerations, India's attitude of washing its hand off Tibet deeply hurt and humiliated the Tibetan people and put India at a disadvantage.[15]

The strategic position of Tibet has became more obvious in recent years as climate change has the potential to reduce the snow pack and glacier mass which cascade down and alter hydrological system.[16] Now, China is has sovereign rights over the world's largest fresh water resources outside of the polar regions. These water resources vulnerable to global and regional warming are very critical for sustaining the food and water security of South Asia. Should China be the lone stakeholder over the fate of waters in Tibet? What happens to the lower riparian nations that depend heavily on these rivers? China has a robust glaciological programme and knows fairly well how long snow/ice resources will last. Therefore, it has embarked on integrated water resource management (IWRM) of all the rivers emanating from Tibetan plateau. China has already constructed the dam and water diversion projects on Rivers Salween and Mekong in spite of regional and global criticism.[17]

Brahmaputra (Yarlung Tsangpo) River

China aims to build 59 reservoirs on the Tibetan plateau to prevent glacier runoff. Construction is in full swing at Zangmu for a 540 MW power project and feasibility studies have been completed to construct five more similar projects further upstream on Yarlung Tsangpo. Tapping the power of the Tsangpo (Brahmaputra for Indians) river as it bends and plunges from the Himalayan roof of the world down towards Indian and Bangladeshi flood plains has long been a dream of Chinese politicians and hydro-engineers. Metog will be site of a mega project at huge bend inside canyon which is about 3.1 miles deep and 198 miles in length. This will involve construction of a series of tunnels, pipes, reservoirs and turbines to generate 40,000 MW power and exploit the spectacular 2000 metre fall of the river as it curls down towards India. The water diversion project was a part of its tenth five year plan.[18]

The Brahmaputra river water will be diverted through three channels in the eastern, central and western regions respectively. The entire labour from the Lhasa–Beijing railway line has been assigned to this mega project to execute it fast. The project will have ominous consequences for millions of Indians and Bangladeshis. Chinese conservationists have admitted that the canyon is home to more than 60 per cent of the biological resources of the Tibetan plateau and many indigenous communities reside in the canyon[13]. Notwithstanding, the reality China is officially denying that it is constructing any reservoir for water diversion or dam on Brahmaputra river. Surprisingly, the government of India is again accepting the official denials as Prime minister Nehru did it in 1950 prior to Tibet's annexation. Now the world has changed and in the 21st century China has emerged as the number two economic power in the world with enormous military and nuclear capabilities. In the integrated global economy China holds more than three trillion US dollars in strategic reserve. The great powers of the world treat China with respect and caution. Control over the water resources of Tibet in a warming world has given China a strategic edge over its neighbours. Whenever, the Indian media raises the issue China starts hardening its position on Arunachal Pradesh, a state in Northeast of India and the government of India hides itself behind the official statement issued by the Chinese government. The eastern flowing river Yarlung Tsangpo/Brahmaputra is more critical for China's strategic interest than west flowing rivers like Indus.

The Brahmaputra has always been considered the very soul of the Indian state of Assam as poets and ordinary people alike consider the river to be a part of their folklore and culture. China, in its own interests, could withhold water for power generation and irrigation during the dry season, with catastrophic consequences for the lower-riparian countries. Further, this whole region would be starved of nutrient-rich sediments that enrich the soil, but which would be held up in the reservoir instead of reaching the downstream.

The Chinese hold that dams would alleviate the problem of floods and erosion in the Brahmaputra. Ironically, this makes little sense, since flooding could actually get worse due to the relentless silting which, will be accelerated by the slowing down of the river flow. It may be noted that flooding normally happens not so much because of snow/glacier melt waters in the Tsango section, but more because of the monsoon rains from on the southern side of the Himalayas. Many tributaries join it in Arunachal Pradesh and make it into a huge water resource. At about this place the pre-rainy season flow averages well above 120,000 ft^3 per second, rising to 1 million ft^3 per second during the monsoon season[18]. All rainfall in the Assam hills is discharged into this river, making it ten kilometres wide at some places. Recently, China and India agreed on a joint mechanism to share hydrological data on rivers Brahmaputra and Sutlej for the rainy season only. From water resources point of view data sets during low flow periods are very critical and interestingly this is not the part of joint agreement in spite of India's insistence.

In India and Bangladesh, water resources are already overstretched by increasing demands from growing populations, economic growth and intensive agriculture. The proposed dam on the Tsangpo canyon is likely to reduce flows in the Brahmaputra by 20 to 30 per cent during the low flow periods.

There is no political will in India to take up issues related to reservoir construction and diversion of water on Tibetan plateau. The reality is that China has already commenced the project by using its stimulus package fund of $580 billion for infrastructure development to create more jobs during the economic downturn and India's muted protest is not going to change their intent.

Indus River

Tensions between India and Pakistan began soon after the two states gained independence in 1947. To date, the states have engaged in three armed conflicts with one another. The November 2008 terrorist attack on Mumbai considerably set back the progress that had been achieved by bilateral discussions. The Indus River, originates in western China, runs through Indian Kashmir and the Gilgit-Baltistan region of Pakistan. The Indus water treaty (IWT) signed by India and Pakistan in 1960, was brokered by the World Bank[19] to equitably allocate the rivers of Indus system to India and Pakistan. The treaty assigned three eastern rivers (Ravi, Beas, and Sutlej) to India for its exclusive use, which is about 20 per cent of the total flow of the six rivers. The treaty gave the flow of the Indus proper and two of its tributaries (Jhelum and Chenab) to Pakistan for its use, which is about 80 per cent of the total flow (with a provision that India could use western rivers for the development of energy through 'run of the river' projects without storing water).

All six rivers that belong to the Indus System originate in the Hindu Kush-Himalaya-Tibetan region and are fed by snow and glacier melt waters, which account for about 44.8 per cent of the total river flow

(Table 1). The proportion of glacial melt in the Indus, Jhelum, and Chenab is predicted to increase, as glaciers in Karakorum-Himalaya and Tibetan plateau recede. Broadly, the major controls on the climate are latitude, altitude, and position relative to Indian monsoon and westerly.

Table 1

All in %	*Indus*
Snow and Glacier Melt in total Q	44.8
Upstream Area > 2000 m in	40
Glaciated Area in	2.2
Upstream ppt. > 2000 m	36
Downstream ppt.	64
NMI: Upstream Q/Downstream Q (NMI=index)	151

Source: *Science*, V.328, June 2010[20].

According to the report above receding glaciers could trim the water supply in the upper Indus by more than 8 per cent by 2065, implying considerable shifts in stream flows that are expected to maintain ecosystems, feed irrigation schemes, and turn hydroelectric turbines. The study has suggested that 915 km^2 of Himalayan glaciers in the headwaters of the Chenab River (30-33° N, 76-79° N) thinned by an annual average of 0.85 m between 1990 and 2004. A USAID report of 2010 succinctly raises the dangers of glacier lake outbursts floods (GLOF) caused by melting "Himalayan/Karakorum glaciers involving large impoundments by short-lived, unstable debris/ice dams that block tributaries of Indus river system....causing outburst floods."[21] The rapidly tipping mass of glaciers and the unpredictability in the amount of runoff is causing great concern in water-dependent sectors, such as agriculture and power generation. Most Himalayan glaciers are in retreat for three reasons. First the overall warming tied to carbon emissions. Second, rain and snow patterns are changing, so that less new snow is added to replace what melts. Third, air pollution by cook stoves and trucks falling as black carbon or soot on surfaces became darker and less reflective—causing them to melt more quickly.

The World Bank's facilitation of negotiations contributed hugely to the success of the IWT and the bank continues to be an important player in resolving disputes over the construction of power projects by India on western rivers. Over the years, the World Bank has remained fully involved by appointing neutral experts and arbitration panels to resolve disputes. The IWT essentially does not have any provisions for joint management of the river basin. It only covers about the division of water. However, it has provisions to furnish hydrological, hydraulic, and engineering data to Pakistan, the lower riparian. India claims that it has scrupulously fulfilled its obligation to share data on all the projects undertaken on western rivers. Nonetheless, Pakistan denies that India shares such data. Both countries engage in endless debates about the equity of water allocation, both inside the commission meetings and in public forums. While India tries to use permissive provisions, Pakistan applies restrictive ones. This upper and lower riparian saga is going on and on.

The Indus and Sutlej rivers originate in the ice fields of the western Tibetan plateau, an area experiencing exceptionally high warming—0.3°C per decade warming above 4000 meters, twice the global average. Glacial retreat in the Tibetan Plateau is driven by warming due to the thickening of the global greenhouse gas blanket, but the rapidity of glacier retreat during the past 25 years suggests additional warming may be playing a critical role. Black soot in aerosol pollution can warm the troposphere, enhancing surface melt. According to a report by the United Nation Environment Programme (UNEP)[22], *Report on Integrated Assessment of Black carbon and Tropospheric Ozone,* the Himalayan and Tibetan plateau are regions where black carbon is likely to have profound impacts on the ice fields and glacier melting. Reducing emissions in the Indian sub-continent and China may help in reducing air pollution and adverse impact on glaciers.

Conclusions

Traditionally, trans-boundary water conflicts center on infrastructure built by upper riparian country/countries to alter hydrological flows, affecting the quantity and timing of flows, especially during low flow periods. Currently, India and Pakistan are gridlocked over the interpretation of various provisions under the IWT. Pakistan is currently objecting to India's energy project on the Kishenganga, a lower tributary of the Jhelum in Indian Kashmir, and declared this issue to be a 'dispute' to be taken to the court of arbitration. Nevertheless, Pakistan has not recognized the emerging realities of climate change's role in rapidly shrinking glaciers in the Tibetan Plateau, Indian Kashmir and Gilgit-Baltistan region. Tipping mass of all glaciers and variability in monsoon precipitation are driving huge flows in the Indus tributaries.

Uncertain information and a deficit of knowledge on glacier melt processes have fuelled a wider debate about the nature of climate science. However, a positive outcome of the debate has been the immense awareness generated among politicians, media and the general public on the role glaciers play in sustaining flows in the Indus river system during the dry months. Pakistan, however, feels vulnerable since India has the 'commanding heights' of the Indus and thus has the ability (or perceived ability) to threaten the water security of Pakistan. Reshaping the political narrative in both countries is a key factor in managing water resources—such reshaping can only take place by engaging policy communities and opinion leaders through (Track ll) hydro diplomacy in south Asia. The climate change platform provides an opportunity to India and Pakistan and India and China to measure, monitor and model glacier system across the region and ensure trans-boundary water security.

Notes

1. Ives, J.D.; Messerli, B.; *The Himalayan Dilemma: Reconciling Development and Conservation*", Wiley and Sons, London, 1989, p. 320.
2. IPCC," Impacts, adaptation and vulnerability. Contribution of working Group 11 to the *Fourth Assessment Report of the Intergovernmental Panel on Climate Change* (Parry, M.L.; Canzioni, O.F.; Palutikof, J.P.; Vonder, P; Hansen, C.E. (eds.), Cambridge; Cambridge University Press. 2007, p. 35.
3. Orville, S., "The Message from the Glaciers" *The New York Review of Books*, May 27, New York, USA. 2010, p. 9.
4. Tandong, Y; Thompson, L.G.; Mosbrugger, V., *"Report of the 1st Third Pole Environment Workshop. Institute of Tibetan Plateau Research* (ITP)" China Academy of Sciences, Beijing, 2009.
5. Yasunari, T.J.; P. Bonasoni; P. Laj; K. Fujita; E. Vuillermoz; A. Marinoni; P. Cristofanelli; R. Duchi, G. Tartari, and K.M. Lau, "Preliminary estimation of black carbon deposition from Nepal Climate observatory-pyramid data and its possible impact on snow albedo changes over Himalayan glaciers during the pre-monsoon season". *Atmospheric Chemistry and Physics Discussions* 10, 2010, pp. 1-38,
6. Tandong, Y.; Thompson, L.G. and Mosbrugger, V *Report of the 1st Third Pole Environment Workshop*" Beijing, 2009, Chinappa.20.
7. Immerzeel, W.M.; P.H. Ludovicus van Beek and Bierkens, M.F.P., "Climate Change will Affect the Asian Water Towers", *Science, V. 328*.11 June, 2010.
8. Edwards, S *et. al*., "The waters of the third Pole: sources of Threat, sources of Survival". *UCL Hazard Research Centre*, University College, London and Humanitarian Futures Programme, Kings College, 2010, London.
9. Ramanathan, V and Y, Xu, "The Copenhagen Accord for limiting global warming: criteria, constraints, and available avenues". *PNAS, V. 107* no. 18, 2010, pp. 8055-8062.

10. Qiu, J., The third pole, 454, Nature, 2008, pp. 393-396.
11. Lau, W.K.M.; M.K. Kim, K.M. Kim, "Asian summer monsoon anomalies induced by aerosol direct forcing: the role of Tibetan Plateau, *Climate Dynamics*, V. 26, 2006, pp. 855-864.
12. Ginsburg, G and Mathos, M, "*Communist China and Tibet*". The Hague: Martinul Nijhoff, 1964.
13. Arpi, C.,"Himalayan rivers: Geopolitics and Strategic Perspectives. in; Water: *Culture, Politics and Management*", India International Centre, New Delhi, 2010, India.
14. Mullik, B.N., "*My Years with Nehru—The Chinese Betrayal*", Allied Publishers, 1971, New Delhi.
15. Hasnain, S.I., "Tibetan Plateau needs international treaty for its protection" The *Economic Times*, New Delhi, December 16, 2010, p. 2.
16. Hasnain, S.I., "River Runs Through it", *Times of India*, New Delhi, August 17, 2010, edit page.
17. China Launches key construction along water diversion project", *People's daily* (online), June 5, 2009. Beijing.
18. Tiwari, R, "India, China Renew Brahmaputra Pact", *Indian Express*, New Delhi (online) April 29, 2010.
19. Public International Law and Policy Group (PILPG), *Negotiating Dispute Resolution in the Indus River Basin*, May 2011, p. 36.
20. Immerzeel, W.M.; P.H. Ludovicus van Beek and Bierkens, M.F.P., "Climate Change will Affect the Asian Water Towers", *Science,* V. 328.11 June, 2010.
21. USAID, Washington D.C., Changing glaciers and hydrology in Asia: addressing vulnerabilities to glacier melt impacts, 2010, p. 79.
22. UNEP, "Integrated assessment of black carbon and tropospheric ozone; summary for decision makers," 2011.

16

Himalayan Hydrology and Hydropolitics

Uttam Kumar Sinha

Introduction

Rivers are a prominent feature of the Himalaya mountain system, physically linking upstream and downstream users. The glaciers that are a source of these rivers are vulnerable to various exogenous impacts including global warming and growing levels of black carbon. Some of the findings/studies so far (not yet fully determined) indicate that the Himalayan glaciers are experiencing unprecedented rates of thinning and retreating.[1] Clearly, any change in the glacier profile will impact the flow-levels of the transboundary rivers and challenge inter-state water relations. Hydropolitics[2] in the region will potentially emerge as a critical issue. While undoubtedly the Himalayan hydrology offers ample opportunities for harnessing development benefits thus serving as a cornerstone for cooperation, it equally, given the competitiveness of the users (riparian states) and the uses of rivers (primarily in terms of consumptive utilisation[3]), can potentially trigger tension and strife. The Himalayan hydrology needs to be analysed, well in advance, as it will be one of the critical frontlines in the global approach to mitigate the impact of climate change on the water resources.

Quantitative Assessment of Water

The earth, referred to as the 'Blue Planet', has an instructive irony to its description. While water covers 73 per cent of the planet only 3 per cent is freshwater, of which 2 per cent is held in ice caps and glaciers. A mere 1 per cent in form of lakes, ponds, rivers, streams, swamps, marshes and bogs, is non-frozen, salt-free and accessible for human consumption. There are only three major parts of the globe covered with ice and snow throughout the year—the Arctic, the Antarctic and the Himalayan mountain system. The Himalaya, increasingly referred to as the 'Third Pole' and 'Asia's water tower', constitutes the largest non-polar ice mass with over 46,000 glaciers.[4] The vast quantity of water held in frozen reserve of the Himalaya greatly matters in sizing the future water challenges in Asia. With population increase and corresponding consumption patterns, it is projected that by 2030 the demand for water will be 40 per cent more than at present and 50 per cent higher in the most rapidly developing countries that include India and China.[5] The Himalayan watershed has about 4 per cent of world's annual renewable water resources that flow through several river basins.[6] Almost 90 per cent of water in the watershed region is consumed by the agriculture sector as compared to the world's average of 70 per cent. Except for Nepal and Bhutan, the per capita water availability of the Himalayan-basin countries is lower than the world average. Given the uneven endowment of water resources in the region, the issues and challenges are large, diverse and complex.

Profile of the Himalaya Mountain System

The Himalaya separates the Indian subcontinent from the Tibetan plateau. The mountain system, as the Himalaya is referred to, includes the Karakoram, the Hindu Kush and the Pamir Knot. Though geologically the Tibetan plateau is distinct from the Himalaya mountains, geologists often regard all the rivers, including those originating from Tibet, collectively as the "circum-Himalayan rivers".[7] ICIMOD describes the Himalayan mountain system as the Hindu-Kush Himalaya (HKH), which extends some 3,500 km from Afghanistan in the west to Myanmar in the east and includes the Tibetan plateau.[8]

Almost 1.5 billion people (18 per cent of the world population) live in the watersheds of the Himalaya mountain system. Ten major Asian river systems—the Amu Darya, Indus (including the Sutlej), Ganga, Brahmaputra, Irrawady, Salween, Mekong, Yangtse, Yellow and Tarim have their sources, with varying degree, in the Himalayan glaciers contributing to almost 70 per cent of water resources. Overall an estimated 2.5 billion people "directly and indirectly benefit from the food-water-energy provided by the river basin in the HKH."[9] The combined drainage basin of the river basin is about 8.6 million square km, equivalent to the size of Brazil.[10] The effects of global warming will be felt through the changes in the hydrological cycle. An effective adaptation policy cannot be delinked from the way water resources are managed and used.

A new thrust towards understanding the water-climate link was evident at the recent COP 16 meeting in Cancun in 2010,. The Green Group of six countries—Cape Verde, Costa Rica, Iceland, Singapore, Slovenia and UAE—stressed the need for water management and climate adaptation. Unchecked climate variation can cause unprecedented challenges to the waterways of Afghanistan, Bangladesh, Bhutan, China, India, Myanmar, Nepal, Pakistan and the countries in the Mekong basin.

Rivers originating from the Himalaya system will directly impact riparian relations. This can be analysed from three perspectives. First, majority of the rivers are trans-boundary in nature originating from, flowing through and draining into territorially defined boundaries. Second, there is no clear definitional demarcation as to whether such a resource is exclusively a 'public good' (defined as non rival and non-excludable) or a 'private good' (defined as rival and excludable)[11]. Third, planning any water resource utilisation policy will have to take into account the assessment of the impact of climate change in terms of seasonal flow and extreme events. River uses are deeply subjective in terms of where, what and how they are being used. New modalities of a water sharing agreement will have to factor in the shifting hydrograph.

This paper will examine the driving forces behind water demand; the political obstacles that stand in way of river cooperation and the behaviour of riparians—in the backdrop of the changing Himalayan hydrology. The hydrology and the hydropolitics in *realpolitik* framework suggest that the preciousness of water often translates into 'possessiveness' and at times 'resource aggressiveness'. 'Water insecurity' can be expressed in terms of availability, reliability and quality, while 'water security' can be described as insecurity arising from control of headwaters. The insecurity-security dynamics can be explained through upper-lower riparian relations and further examined through the claims of 'absolute territoral sovereignty' versus 'absolute territorial integrity'. With no binding watercourse treaty, the Himalayan hydrology assumes great significance.

Himalayan Glacier Profile

Many studies using remotely sensed images have identified and analysed specific glacierised region.[12] The mid-latitude, high-altitude Himalayan glaciers contain one of the largest reservoirs of snow and ice outside the Polar regions but with varied characteristics. Studies have confirmed that there is no "stereotypical" Himalayan glacier.[13] While the glaciers in the

Karakoram region of northwestern Himalaya are mostly stagnating, those in the western, central and eastern Himalaya are mostly retreating. Of these, the western glaciers have shown the highest rate of retreat.[14] Some studies, particularly the ICIMOD study (2007), have calculated the percentage of glacial melt contributing to the river flow of the 10 river system originating in the HKH. In South Asia, the Indus system has the highest glacial melt percentage (44.8), which is the highest of all the river system. The Brahamaputra and Ganges estimates are 12.3 and 9.1 per cent respectively. Others, for example, the Tarim has 40.2 per cent while the Irrawaddy has the smallest contribution. The Mekong, Salween, Yangtze and Yellow have 6.6, 8.8, 18.5 and 1.3 per cent respectively.[15] Glacial ice-melt from the HKH provides the Asian river systems with critical seasonal flows of water.

Scientific observations point that 90 per cent of the Himalayan glaciers melting is directly caused by black carbon soot and other industrial processes. Other studies point to the presence of debris such as pebbles and rocks as being an additional factor. The impact will gradually shrink glaciers resulting in the decrease of water runoff in the long-term. Clearly seasonal shifts in stream flow will create serious downstream problems. The timing of the glacial melt more than the percentage of its contribution to the annual water flow is important. If the melt is significantly higher during the monsoon then the floods will be intense and widespread. During the 'shoulder season', the period before and after the monsoon, the melt is crucial to maintain the steady supply of river flow. For the Kabul, Indus and the Ganges, the melt during the 'shoulder season' accounts for over 70 per cent of the flows. In the short-term earlier water runoff from glaciers when combined with seasonal rains can result in flood conditions. Many recent studies on the overall glaciers retreat and additional melt focus on water dammed or 'glacier lakes' that have the potential of generating dangerous outburst flooding.

In the coming years, rising concerns about the effects of climate change will take greater precedence over any physical changes associated with climate change. First, as the population grows, competition for food, energy and water will correspondingly increase. Increasing demand for food grains will claim larger areas for cropland and greater volumes of irrigation water. Second, given the risks attached to climate change, food-energy-water will be subject to various stresses and strains.[16] Perceptions of a rapidly changing ecosystem may prompt nations to take several measures, many of them unilateral, to secure resources and territorial sovereignty. Any willingness to engage in greater river basin cooperation will depend on a number of factors such as: the behaviour of competing countries, the economic viability, and other interests that states are reluctant to either compromise or concede. Since disputes over water are inevitable because of the changes, described above, understanding the processes of resolution and framing new mechanisms and approaches becomes a necessity.

Hydropolitics and Riparian Relations

Varied interpretations of the use of river water have resulted in the differing claims of riparian countries.[17] Upper riparian nations essentially base their claims on 'absolute territorial sovereignty', i.e., the right to use rivers unilaterally regardless of lower riparian concerns. The lower riparians, on the other hand, claim 'absolute territorial integrity' of rivers, stressing that upper riparian actions should not affect the water flowing downstream. Both the claims are extremely incompatible. There are, however, accepted legal norms of 'equitable utilisation', 'no-harm rule' and 'restricted sovereignty' that riparian states work through, and frame negotiations and treaties accordingly to overcome such differing positions.[18] But more often than not, these norms are rendered meaningless. It is almost a vague notion that nations are entitled to a "reasonable share of water".[19] Given that there is no legal-binding international treaty on water sharing, riparian relations will largely be influenced by the prevailing political dynamics and strategic considerations.

Overview

China and India, given the distances that the Himalayan rivers traverse the respective territories, are critical players in the hydropolitics of the region. India is simultaneously an upper, middle and lower riparian. China's hydrological position, on the other hand, is that of complete upper riparian supremacy. India's middle riparian position, increases its dependency (and is thus water insecure) on the headwaters of the rivers such as Indus, Sutlej and Brahmaputra which originate in the Tibetan plateau. China is equally water insecure but its insecurity relates to the disproportionate availability or uneven distribution of waters within its territory, the majority of which is in the south (Tibet Autonomous Region) with the north and west excessively water stressed. In terms of per capita water availability, China ranks among the world lowest. Water pollution is also a big worry with a vast majority of lakes and many of its largest rivers unsuitable for human use. The territorial source of the Himalayan rivers makes China far more water secure than India. It would be probably fair to say that China is the world's most independent riparian country.[20] This hydrological position gives it enormous latitude in shaping its larger political equations with its riparian neighbours. India, on the other hand, given its middle riparian position and its longstanding commitment to bilateral river treaties has to assiduously balance the anxiety and concerns of its lower riparians (Pakistan and Bangladesh) without compromising its own water requirements. China, in contrast, has no bilateral riparian treaties[21] and therefore is not bound by any water utilisation agreement with its neighbours. In fact, China was one of the three countries[22] that did not approve the 1997 UN Convention on the Law of the Non-Navigational Uses of International Waterways. On the Mekong River basin, China is only a dialogue partner in the Mekong River Commission (MRC), which was formed by the concerns of four lower-riparian countries—Cambodia, Laos, Vietnam and Thailand in 1995. Though China's non-binding participation in the Mekong basin has increased, it is unthinkable that it will join MRC as an active member.

For the Himalayan basin-states other than China, being water-dependent from external sources is a hydrological reality. The water dependency and the prevailing politics shape fears and perceptions. For example, the Mekong lower riparian countries will remain suspicious of China's upstream hydroelectricity projects.[23] Pakistan for example (which is heavily dependent on the sources of rivers outside its boundary), perceives India and not China as an upper riparian aggressor. This of course greatly relates to the grievances that Pakistan has over the Indus Water Treaty with India. Similarly with Bangladesh, the lowest riparian in the Ganga-Brahmaputra-Meghna basin, water becomes a political and emotional driver. China as the upper riparian player would like the water debate in Pakistan and Bangladesh to be directed and contested with India, without highlighting its own hydroelectricity plans either on the Indus or the Yarlung Tsangpo (Brahmaputra in India).

While China has no formal water sharing arrangements with its neighbouring countries, India has several treaties like the 1960 Indus Water Treaty with Pakistan and the 1996 Ganga Treaty with Bangladesh. With Nepal and Bhutan, treaties have been signed to share the benefits of water. Water treaties commit India to a dialogue-based water sharing approach and diplomatically become an important part of its neighbourhood policy. China, in contrast, would tend to take a strategic view of the water it commands and given its hydrological position factor water as a tool, leverage and a bargaining instrument in framing its regional policies. A snapshot of the riparian dynamics in the Himalayan watershed suggests that while there is considerable lack of trust on water issues between states, there is greater possibility of drawing India into regional water debate and breaking political deadlocks through sensible water sharing arrangements and resource development than with China.

China's thirst can Make Others Thirsty

China suffers from an annual shortage of 40 billion cubic metres of water and is expected to face 25 per cent supply gap for projected water demand by 2030.[24] Two-thirds of its cities already have difficulty accessing water. China has also one of the world's most uneven distributions of water. The region south of the Yangtze river, which accounts for roughly 36 per cent of Chinese territory, has 81 per cent of the country's water resources. The territories north of the Yangtze, where much of water is required, make up the remaining 64 per cent but have a meagre 19 per cent of water capacity. More than anything else, the water shortage becomes an impediment to China's goal of meeting food production and challenges its claims of self-sufficiency in food grains. Water deficiency, prolonged drought and below average rainfall in the nine provinces (including the six major wheat producing provinces) of the northern plains have resulted in a surge of grain imports not experienced in the last 15 years.[25] Reduced self-sufficiency is dangerous for a country with a population as large as China and therefore its search for water becomes an unending imperative. Electricity is equally crucial in China's economic development. With a GDP growth rate of 8-10 per cent a year, China's energy requirements are projected to increase by 150 per cent by 2020.[26] Its ever-increasing demand for energy requires it to look for various sources. While resource rich in coal and a net importer of oil, both of which are climate unfriendly, China is compelled to develop hydroelectricity as a clean and renewable source of energy. China has already half of the world's large dams including the Three Gorges. Its exploitable hydropower is estimated to be around 378GW with an annual power supply of 1.92 trillion KWh.[27] Huge hydroelectricity projects for energy and water diversion schemes for food sufficiency are strategic measures to China's growth path. The Tibetan Autonomous Region (TAR) in southwest China, which serves as the headwaters of many of Asia's largest rivers including the Yellow, Yangtze, Mekong, Salween, Brahmaputra and Indus is of existential importance. Without Tibet, China's hydrological supremacy would be overturned from being entirely water-independent to water-dependent. Mao Zedong long time ago realised the water resource value of Tibet when he remarked, "The south has a lot of water, the north little…If possible, it is okay to lend a little water." In fact Beijing's total control over Tibet in effect is the 'total' control over the water resources. 'Absolute territorial sovereignty' over the waters gives an added strategic sharpness to China's regional approach and its increasing 'hydro-egoism' dictate riparian relations with its downstream countries. The classic lines, "He who holds Tibet dominates the Himalayan piedmont; he who dominates the Himalayan piedmont, threatens the Indian subcontinent; and he who threatens the Indian subcontinent may well have all of South-East Asia within his reach, and all of Asia,"[28]

China's Dams and Diversion

China's dams and water diversions are important elements of its growth. Its increasingly aggressive hydro-behaviour, as stated earlier, is intended to secure its massive water requirements in its northern plains. Though not intended but as an accompaniment, the control over such a valuable natural resource gives Beijing enormous strategic latitude with its neighbours. On the one hand, water resource can be an effective bargaining tool when dealing with countries like India with whom it has testy political relations. On the other hand, hydro-diplomacy through partnership on dam construction and infrastructural development helps it to widen and deepen its influence on other downstream countries. China's water approach driven by its need and economic rationality also fits into, what many Chinese scholars observer 'non-combative aggressiveness'.

In the Mekong-basin, for example, while China is building eight dams on its territory, Chinese companies have also started constructing two hydroelectricity dams in Cambodia, one of the downstream basin countries. Lower riparian countries like Cambodia have much to lose by being too critical of China's dam construction, and would rather receive billions of dollars from China for the country's infrastructure. China's expertise and knowledge of hydrology and dam construction helps to enhance its power and influence and importantly weaken lower riparian coalition. Not surprisingly, its hydro-diplomacy is dam-centric with the underlying message that these engineering feats are essential to the livelihood of millions of people. China's uncompromising position on dams and diversion will only result in a dam spree in the lower riparian countries. The State Peace and Development Council (SPDC) of Myanmar and the Thai government are pushing ahead quickly and secretively with plans for a series of five giant dams on the Salween/Thanlwin River and its tributaries.

Likewise with Pakistan, China's strategic partnership has a strong component of dam building and infrastructural development. In August 2009, the two countries signed a MoU for the construction of the 7,000 MW Bunji dam on the confluence of the Indus and Gilgit in the Pakistan-Occupied-Kashmir (in Gilgit-Baltistan region),[29] which was soon followed by an agreement to be a major backer of the 4,500 MW Diamer-Bhasa dam[30] on the Indus. Many such 'build-operate-own-transfer' (BOOT) projects are planned in the future. Pakistan's east-west hydrological vis-à-vis India and Afghanistan has given ample scope for China to strengthen its ties with Islamabad and negate the remotest possibility of India and Pakistan uniting as lower riparians, since both the Indus and Sutlej flow from the Tibetan Autonomous Region (TAR). China has built its own storages and dams on the headwaters of Indus in the Ngari Prefecture of Tibet.[31] Chinese increasing presence and activity in the POK area irks India but it is a fact that Beijing will remain increasingly engaged with Islamabad. Pakistan considers its dependency on China less threatening than its water dependency on India and would willingly ally with China to strategically counter India.

Clearly for China its hydrological supremacy is an effective bargaining tool and potential weapon. China, accustomed to brinkmanship, will maintain for a large part a strategic silence over its river diversion plans, to keep downstream states nervous and worried. And with no legally binding international treaty on water-sharing, there is nothing to stop China from manipulating river flows and increasing downstream dependency.

The dam partnership between two of the world's biggest dam builders (China and Pakistan) has its undisputed ecological and societal consequences. Of the various stakeholders of water uses hydro-engineers are a dominant force in China. Many of its top leaders are hydraulic engineers and technocrats including Hu Jintao and Wen Jiabao.[32]

Yarlung Tsangpo/Brahmaputra[33]

China's Approach

Post the Chinese public announcement of the successful damming of the Yarlung Tsangpo on November 12, 2010,[34] there exist no free flowing rivers in Asia. The damming was a foregone conclusion. Yan Zhiyong, general manager of China Hydropower Engineering Consulting, has been quoted as saying: "By about 2020 most of China's hydropower projects outside of Tibet will have been completed, and the industry's focus will shift to the Jinsha, Lancang, the upper reaches of the Nujiang river and the Yarlung."[35] Unlike its earlier phases, this one will be far more contentious and controversial given the nervousness among the riparian neighbours.

The 510 MW Zangmu hydroelectricity power station being built on the middle reaches of the Yarlung between the counties of Sangri and Gyaca is equivalent to the entire existing hydropower generating capacity in Tibet.[36] Studies carried out by the Chinese Academy of Sciences' put the Yarlung hydropower potential second only to the Yangtze but the highest in terms of power-generating potential per unit of length. The site which has mild winters means that the work will continue unabated throughout the year. The hydropower development is expected to lead to a development boom with increased activities in mining and rapid growth in road and railways. The Zangmu dam marks the beginning of a new era in Chinese engineering and construction capability.

It was only in the last decade that the Chinese leadership focused its attention on river development plan in Tibet. Earlier it was low-key and mostly about Tibetan self-sufficiency. But with new studies revealing the larger hydropower potential of 114 gigawatts on the Yarlung basin including a possibility of 38-gigawatt hydropower facility at the Great Bend in Medog county (equal in power to the Three Gorges)[37], the Chinese leaders have now a water mine to exploit.[38] The Chinese hydro-obsession as seen in the Three Gorges will drive mega projects on the Yarlung. It has always been the Chinese way of development—a maximalist approach to resource development. Small-to-medium hydro power stations do not excite the Chinese leaders. Many 40-50 MW projects on the Yarlung were not constructed because of river-silt issue and displacement concerns. Now that the hydropower estimates are phenomenal, these same considerations have become inconsequential. The hydropower plants on the Yarlung come at a time when other infrastructural development has taken place or will soon be completed. By 2012 the electricity grid in Tibet will be connected to the national grid through Qinghai. The Qinghai-Tibet railway has now made it possible to move materials for building large dams on the Yarlung. The Chinese now have a clear policy direction on the Yarlung and with energy supply becoming ever more pressured, the time and condition is ripe to develop the rich hydropower resources of Tibet along with a no-holds-bar approach to taking total and complete political control of Tibet.

From an ecological perspective what is worrying is that Tibet's ecology is extremely vulnerable and the drive towards developing hydropower without an integrated approach will be dangerous. It is increasingly clear that rivers are not merely for provisions but they also have ecological functions. One need only look at the Yangtze and the Yellow rivers, both unfit for human use, to understand how important it is to follow the laws of nature regarding rivers. How far will the construction of dams take a holistic view to ensure that risks are assessed and harm to people and the natural ecosystem minimised depends on the dominant thinking in the leadership structure in China. Currently the dominant thought pays little attention to sustainability of the water resources.

India's Approach

India as the downstream riparian cannot legally question Chinese upstream projects. There is an argument in India that Chinese development on the rivers, particularly on the Yarlung, is of little consequence and blown out of proportion by the media and an anti-China constituency who perpetually see a threat in whatever China does. The IDSA's study on India's water security challenges, however, takes a view that water is a strategic issue for China and a better understanding of Chinese intentions is required to frame India's water policy. India needs to articulate its water concerns convincingly through the existing international rules of 'no harm' even if others like the 'equitable utilisation' and 'limited sovereignty' are contested and difficult to agree upon. For example, withholding data on the flows of the rivers or plans to build storages or dams or projects that divert water come under the 'no harm rules'. India's own water sharing structures with its neighbours—Pakistan and Bangladesh—are quite exceptional and take 'no harm rules' into consideration. Yet India has been defensive about China's development on the Tibetan rivers. More importantly, India is not one-river downstream with China. This makes India's riparian relations with China exceptional and critical. India is multi-river dependent with the Brahmaputra in the east and the Indus and the Sutlej in the west. The Ganga which originates in India has nine tributaries joining it from Nepal, three of which Karnali, Gandaki and Kosi arise in Tibet. Some figures indicate that about 354 BCM of water flows into India from the Tibet of which 131 BCM is accounted for in the Brahmaputra (the low-end figure is 78.10 BCM). It is important for India to draw China into a dialogue on water issues and consistently raise hydrological concerns at bilateral meetings. But to make the water dialogue effective and meaningful, India would have to unlock its ignorance regarding the implications of the reduced flow into its territory. Much more research and study of the geography, the hydrology and the topography is required to determine the changing flow and its potential impact on cultivation, fishery and the ecosystem.

India's other strategic concern, unrelated to China, is its pitiable storage capacity. Water storage per person (in cubic metres) in India is about 200. The US has roughly 6000 and China almost 2500. India needs to enhance its storage capacity to a reasonable and sustainable level based on proper environmental impact assessment and not entering into a dam race with China on the Yarlung/Brahamaputra. This is important, as India needs to take into consideration the concerns of Bangladesh. India's counter diplomacy should be to isolate China as an unreasonable upper riparian at the regional and global level.

Question of Tibet

The Tibetan glacial-fed rivers raise difficult and challenging questions both from political and ecological perspectives. The glacial melting has cast a new spotlight on Tibet's water resources. It has become a key issue for civil society and powerful environmental groups like the International Union for Conservation of Nature (IUCN) to campaign for Tibet as a vulnerable area to be protected from rampant resource exploitation. In fact in 2003, an area of 7.1 million hectares in Yunnan province, where the upper reaches of the Yangtze, Lancang and Nujiang run parallel was declared a World Heritage in 2003. Interestingly, China has ratified the Convention on the Protection of World Cultural and Natural Heritage which was adopted by the UNESCO in 1972.[39] China is usually suspicious of such activism and considers it to be West-driven interference. However, this does not demerit the fact that China needs to incorporate ecological costs in its development of water resources. This will also have a widespread approval of the local Tibetans who consider the ecology a way of life and have always derided Chinese blatant mismanagement of the environment.

From a political perspective, Tibet's water resources raise some contentious questions. Should China alone be the stakeholder of the waters in Tibet? For one, Tibet is an autonomous region of China, politically not settled. Secondly, as explained above, there are crucial sustainability and ecological issues that affect 2 billion people downstream. With the on-going works on the Yarlung and Nujiang,[40] China has exploited all the rivers of the Tibetan Plateau that Mao had once so gleefully looked at and desired to possess. None of these developmental plans take into consideration the concerns of the lower riparian. Terming the water resources in Tibet as a global 'commons' would draw international attention and possibly prompt China into a water dialogue with downstream countries on ways to preserve and share the benefits of the waters of Tibet. China also has a strong environmental constituency with activists, scientists and journalists, who despite odds, are sensitising local people and authorities to ecological concerns. However, China's environment and growth have a disjointed relationship. The pace of its development, which is not to be compromised, fails to calculate the long-term ecological damages. Only an environmentally conscious regime in Beijing, can make conciliatory moves allowing for the broad-based basin management of the Tibetan rivers.

Conclusion

The Himalayan rivers are crucial to peace and prosperity in the heavily populated and highly water dependent South and South East Asia. China's upper riparian position and its enormous domestic requirement gives water a strategic character and as such accounts for the differing levels of trust and mistrust and cooperation and misperception. Stability in this wide Himalayan watershed region would also mean the stable supply of water. Suggesting that Tibetan rivers, geologically called the 'circum-Himalayan rivers', be taken to be 'global commons' raises contentious and controversial questions which are clearly anathema for the Chinese leadership. However, this merits attention given the large numbers who are dependent on the water resources from the Himalayan glaciers. In more practical terms, China's water development approach in Tibet is unilateral and is bound to affect downstream riparians both in terms of water flow and ecological consideration. A coalition of lower riparians oming together specifically to draw China into a water dialogue is conceivable in spite the fact that many of them have a strategic partnership with China. The Himalayan watershed region desperately requires a framework of 'water resources management' and 'hydrosolidarity' that includes multi-purpose beneficial utilisation of water resources with the active participation of the lower riparian basin states.

Notes

1. The IPCC in its 2007 Report had projected that Himalayan glaciers "...if the present rate continues, the likelihood of them disappearing by the year 2035 and perhaps sooner is very high." The IPCC in January 2010 admitted to the error in its projection. See IPCC 'Statement on the melting of the glaciers', http://www.ipcc.ch/pdf/presentations/himalaya-statement-20january2010.pdf
2. Hydropolitics can be defined in terms of the existing and future hydrology and its interplay with the political reality.
3. Rivers have many uses. Some are 'consumptive' in nature and some 'non-consumptive'. The non-consumptive uses, such as navigation and hydroelectricity generation are less problem-generating than the consumptive uses such as drinking water and water for irrigation.
4. Of the 46,000 glaciers only about 600 are being monitored. http://www.sciencenews.org/view/generic/id/55455/title/Science_%2B_the_Public__IPCC_admits_Himalayan_glacier_error
5. Mckinsey Report, 'Charting our Water Future', November 2009. http://www.mckinsey.com/App_Media/Reports/Water/Charting_Our_Water_Future_Exec%20Summary_001.pdf

6. Freshwater Under Threat: South Asia, UNEP Report, 2008
7. "Geochemistry of the suspended sediments of circum-Himalayan rivers and weathering budgets over the last 50 years". See abstract http://adsabs.harvard.edu/abs/2003EAEJA....13617G. ICIMOD
8. www.icimod.org/?page=43
9. The Waters of the Third Pole: Sources of Threat, Sources of Survival, A Report prepared by UCL Hazaed Research Centre, China Dialogue, Humanitarian Futures Programme, Kings College, London, p. 7. http://www.chinadialogue.net/UserFiles/File/third_pole_full_report.pdf
10. *Ibid.*
11. Jaroslav Tir and John T Ackerman, "Politics of Formalised River Cooperation", *Journal of Peace Research*, vol. 46, no. 5, September 2009, p. 623. There is, however, a general view to perceive water resources as 'collective goods' or 'common pool resources'.
12. http://www.treehugger.com/files/2010/08/usgs-confirms-himalayan-glaciers-melting-climate-change-to-blame.php#
13. Study carried out by the Department of Geography, UC Santa Barbara. See Hindustan Times Report January 27, 2011. Also see, http://www.sciencedaily.com/releases/2011/01/110124162708.htm
14. *Ibid.*
15. See N. 9, p. 8.
16. As noted by John Beddington, UK Chief Scientist on March 18, 2009. See http://www.guardian.co.uk/science/2009/mar/18/perfect-storm-john-beddington-energy-food-climate
17. Riparian nations, are those nations "across which or along which a river flows, have legal rights to use the water of river", Law, International Water. http://www.waterencyclopedia.com/La-Mi/Law-International-Water.html
18. The rule of 'equitable utilisation', is based on the concept that an international drainage basin is a coherent legal and managerial unit, and embodies a theory of restricted sovereignty under which each nation recognizes the right of all riparian nations to use water from a common source and the obligation to manage their uses so as not to interfere unreasonably with like uses in other riparian nations.
19. While this is enshrined in the UN Convention on the Non-Navigational Uses of International Watercourses (1997), it is not binding because the Convention is not in force with only 16 countries have ratified it. To come into force it requires 35 countries. The UNGA approved the UN Convention by 104-3 in 1997. There is also the "no-harm rule" in the convention, which states riparian nations to take all "appropriate measures" to prevent causing harm to other watercourse nations.
20. 'Independent' in terms of rivers originating in its territory and not being 'dependent' on the headwaters from other countries.

21. Except for being a dialogue partner in the Mekong River Commission (MRC). The Lancang River begins in the northeastern side of Tanggula mountains in Qinhai province, flows through Tibet into Yunnan province, and is called Mekong River when it flows out in Mengla county in Yunnan province. Mekong runs all the way through Myanmar, Lao People's Democratic Republic, Thailand, Cambodia, and Vietnam and empties into the South China Sea near Ho Chi Minh in Vietnam. Lancang-Mekong River is 4,880.3 kilometers long; the section of Lancang River in China is 2,161.1 km long. So far, 25 major projects have been listed: 14 projects inside China, 4 inside Laos, 4 along the borders of Laos and Thailand and Laos and Cambodia and 3 projects inside Cambodia. These dams are estimated to produce 70 per cent of China's current electricity needs.
22. The other two were Turkey and Burundi.
23. Likewise, Kazakhstan and Russia will be concerned over China's diversion of the Irtysh and Ili rivers.
24. N. 5.
25. See *People's Daily* online, http://english.peopledaily.com.cn/90001/90776/90882/7275037.html
26. Gal Luft, "Fueling the Dragon: China's race into the oil market", http://www.iags.org/china.htm
27. See, http://www.our-energy.com/china_hydropower_as_the_right_solution.html
28. See George Ginsburg and Michiel Mathos, *Communist China and Tibet: The First Dozen Years*, Martinus Nijhoff, 1964.
29. *Daily Times*, August 24, 2009.
30. The Diamer-Bhasa is South Asia's costliest dam project. In spite of having signed a MOU with China International water and Electricity Corporation for the construction, international financing have bneen difficult. In January 2011, US following the World Bank has refused to fund big projects. See, http://damsandalternatives.blogspot.com/2011/01/bhasha-dam-project-in-dire-straits-us.html
31. Various news report captured Alice Albinia's book, *Empires of the Indus*, London: John Murray, 2008. The author describes a dam being built on the upper reaches of the Indus in Tibet. Reports suggest that the Senge-Tsangpo Hydropower station on the Indus has a capacity of 6,400 MW.
32. Hu Jintao was a hydraulic engineer having worked on the construction of Liujiaxia Hydroelectricity station in the 1960s. Wen Jiabao professional background is in geology and engineering.
33. The Yarlung originates in the Jima Yangzong glacier near Mount Kailash in Tibet and enters into Arunachal as Dihang. It then flows southwest through the Assam valley as Brahmaputra and south through Bangladesh as the Jamuna. There it merges with the Ganges and then splits into two the Padma and the Meghna to form a vast delta, the Sunderbans.

34. He Haining and Jiang Yannan, "A new era for Tibet's rivers", Chinadialogue, http://www.chinadialogue.net/article/show/single/en/4055-A-new-era-for-Tibet-s-rivers
35. *Ibid.*
36. *Ibid.* Zangmu will be the fourth in a row of five on the Sangri to Gyaca stretch of the river. Several well-known Chinese hydropower firms have already made their way into Tibet.
37. Chinese research and study have always delved into the possibility of building the world's largest hydropower station in Tibet. For example, Chen Chuanyou's article, "Could the World's biggest hydropower station be built in Tibet?" in *Guangming Daily* in 1998. In 2002, Chen published another paper in Engineering Sciences looking at the positive impact of the hydro development in the Great Bend in south-east Asian region.
38. See Chinese 11th Five Year Plan (2006-2010), http://www.gov.cn/english/special/115y_index.htm
39. See, http://whc.unesco.org/en/list/1083
40. Earlier, the National Development and Reform Commission had approved of the construction of 13 dams on the Nujiang to generate 21,000 MW of power that would have been bigger than the Three Gorges Project. With Premier Win Jiabao's intervention (February 2004), the massive projects were suspended. Now preparation work for Liuku dam has restarted. See, http://www.businessgreen.com/bg/news/2019684/china-start-controversial-nujiang-hydro-power-project

Section Five

CLOSER HOME

17

US Policy towards South Asia

Sumit Ganguly

As the second decade of the twenty-first century dawns, South Asia looms large in US foreign policy priorities. The reasons thereof are apparent. Unlike during much of the Cold War and beyond, the US now has critical, if not vital, economic, strategic and political interests in the region. None of these are likely to draw to a close in the foreseeable future. This chapter will provide a brief overview of the recent US involvement in the region, it will discuss the significance of its on-going commitments and it will explore the prospects and limits of the administration's goals in South Asia.

Toward a Strategic Partnership?

Even prior to assuming office, some of the positions that President Obama had taken while a junior senator from Illinois and subsequently as a presidential candidate had caused concern in New Delhi. Specifically, policymakers were distressed at his hostility toward the US-India civilian nuclear agreement when it was under discussion in the Senate. Subsequently, they were troubled because of his apparent opposition to the outsourcing of employment to India.

Their fears were reinforced following his assumption of the presidency. Unlike the Bush administration which had accorded it much significance during its second term in office, Obama's administration evinced no particular interest in India.[1] Instead, in the realm of foreign policy, it wanted to improve relations with the Muslim world in general, to wind down the war in Iraq, to boost the flagging Arab-Israeli peace process and begin a wider dialogue with the People's Republic of China (PRC).

To the extent that the administration was concerned about South Asia it appeared focused on working with Pakistan to ensure that it could bring about some degree of stability in Afghanistan and ensure that the Al Qaeda and the Taliban did not manage to regroup in the area.[2] Also, to the dismay of many of India's policymakers, key administration officials suggested that the Kashmir dispute would have to be addressed to ensure that Pakistan can concentrate its efforts on the Afghanistan-Pakistan border, the home of many anti-American militants.[3] Not surprisingly, India reacted with considerable alacrity and the administration backed away from its attempt to link the two issues. However, to India's continuing dismay some administration officials still sought to link the two questions.

Worse still during President Obama's first state visit to the PRC the concluding joint communiqué included a phrase that caused much distress in New Delhi. In this statement the US appeared to be inviting the PRC to serve as a mediator to resolve Indo-Pakistani differences. Given the PRC's long-standing security relationship with Pakistan and its on-going differences with India, such a suggestion was understandably viewed with much concern in New Delhi.[4]

Doubts about the significance that the Obama administration attached to India persisted even until the eve of the president's state visit to India in November 2011 despite the obvious personal bonhomie exhibited by the two leaders during Prime Minister Manmohan Singh's visit to the United States in November 2009.[5] As Indian political commentators, however, underscored, this personal cordiality did not translate into policy significance.

Policy differences persisted at multilateral, regional and bilateral levels. At a multilateral level, the two sides remained far apart on the vexed question of how best to tackle global climate change. These differences came to the fore at the Copenhagen Summit where India, much to the dismay of President Obama, made common cause with the People's Republic of China.[6] Regionally also India and the US remained at odds on how best to deal with the behaviour of two authoritarian regimes, Iran and Burma (Myanmar). In the first instance, India, despite some domestic opposition, proved willing to censure Iran at the International Atomic Energy Agency (IAEA). However, it made it showed its reluctance to impose unilateral sanctions on Iran and firmly opposed the use of force to end Iran's pursuit of a clandestine nuclear weapons programme. Similarly, despite its past displeasure with the authoritarian military regime in Burma, it chose to work with the junta on strictly pragmatic grounds.[7] Also, thanks to Pakistani pressure, the US did not wish to see an expanded Indian role and presence in Afghanistan especially in the security realm.[8]

Finally, the bilateral relationship too had its share of differences.

Both Indian politicians and businessmen alike objected to the president's continued attacks on the outsourcing of jobs to India and to the dramatic increase in visa fees for Indian professionals working in the United States.[9]

At another level, India's handling of the requisite legislation that would enable foreign companies to invest in its nuclear industry fell prey to the exigencies of India's fractious parliamentary system. It did not help matters, of course, that the legislation was debated against the backdrop of a court judgment against several senior Union Carbide executives for their role in the Bhopal gas leak tragedy of 1985.[10] As a consequence, the legislation that was passed placed onerous requirements on not merely the manufacturer of a nuclear power plant but also on the suppliers.[11] Both US policymakers as well as potential nuclear suppliers raised serious questions about the conformity of India's new legislation with global standards. Until questions pertaining to the issue of supplier liability are better addressed it is far from clear whether key American firms would prove willing to invest in India's nuclear power sector.

Bilateral differences also remained on how best to deal with Pakistan and its continuing complicity with a host of indigenous terrorist organisations.[12] Finally, some tensions persisted on the extent and scope of US counterterrorism cooperation with India. This issue came to the fore over Indian demands for the possible extradition of David Coleman Headley, an American indicted for his role in the terrorist attack on Bombay (Mumbai) in November 2008.[13]

Finally, the Obama administration had wanted India to move with some dispatch to decide on the acquisition of 126 medium multi-role combat aircraft (MMRCA). Two major American aerospace firms were in the fray. Unfortunately, India's labyrinthine defence acquisition process, its dilatory decision-making procedures and its contentious domestic politics have hindered a swift decision. Obviously, a deal of this magnitude could significantly bolster the Indo-US security relationship.[14] In the end, however, India chose to short list the French Rafale and the European consortium, EADS' Eurofighter.[15]

The Indian decision to overlook the two US options, the F-16/Block 70 and the F/A18E/F came as a significant shock to the US policy establishment because President Obama had made a concerted pitch for the American aerospace firms to his Indian interlocutors during his state visit to India in November 2010. It is beyond the scope of this chapter to discuss the visit in any detail. Suffice it to say that despite much prior scepticism about the significance of the visit it was mostly successful. Most importantly, the president proffered limited but vital support for India's quest for a United Nations Security Council seat, he committed his administration to lift a raft of sanctions on the sale of dual-use technology to two Indian state-run entities, the Defence Research and Development Organisation (DRDO) and the Indian Space Research Organisation (ISRO). He also extended his support for India's inclusion in a host of multilateral export control organisations.

All these developments pleased his Indian hosts and even went to some length in muting if not wholly silencing some of his most vociferous critics in India. However, the two sides remained at odds on three issues of varying significance. First, many Indians took umbrage at the president's public hectoring of India's relationship with the military regime in Burma/Myanmar. Second, they could not find common ground on the matter of how best to deal with Iran's clandestine quest for nuclear weapons. Third and finally, they publicly disagreed on how best to deal with Pakistan and its continuing involvement with terror.[16]

Despite these disagreements, the visit laid to rest a host of lingering Indian concerns about the place that had been accorded to India in the Obama administration's foreign policy calculus. The task before the administration as well as policymakers in New Delhi is to decide on how best to build on the achievements of the presidential visit. There is little question, however, that the Indian decision on the MMRCA purchase—the anodyne comments of some key Indian political commentators to paper over the breach notwithstanding—will have some adverse consequences for Indo-US relations in the near term.[17]

Dealing with a Reluctant Ally

In the waning days of the second Bush administration, some key officials belatedly came to the realisation that General Musharraf's cooperation on its "war on terror" was considerably less than fulsome. As a consequence, the administration sought to exert greater pressure on his regime to address American concerns about the recrudescence of the Taliban. However, before the shift in American policy could have any effect, domestic political turmoil within Pakistan brought Musharraf's tenure to a close.[18]

The Obama administration within days of assuming office appointed the veteran US diplomat, Richard Holbrooke as the Special Representative to Afghanistan and Pakistan (SRAP). His task was to coordinate policy across the entire US government to enable the US to achieve its strategic goals in the region.[19]

Holbrooke quickly seized the bureaucratic initiative and managed to create a substantial professional staff who reported directly to him. This had the effect of marginalising other entities such as the state department and the Central Intelligence Agency analytical units which had responsibility for Afghanistan and Pakistan. He managed to create this enclave because of his stature within the foreign policy establishment and also because of his close ties with the secretary of state, Hilary Clinton. As a consequence, for the most part, he came to dominate the policy process and was able to forge ties with Congress to authorise substantial assistance to the region.[20]

The Obama administration sought to elicit Pakistan's cooperation through a policy shift under the aegis of the Kerry-Lugar legislation.[21] This act had a twofold goal. First, it sought to re-focus American assistance to Pakistan. To that end it was designed to bolster hitherto neglected areas of Pakistan's polity such as education and health care. Second, it also sought to reassure its policymakers that the US was prepared to make a long-term commitment. Accordingly, the legislation offered Pakistan $7.5 billion over a five-year span.

Nevertheless much of the Pakistani populace remained sceptical about the seriousness of the US commitment to the country over the long term. Simultaneously, the Pakistani military establishment made it well known that they cared little about the apparent intrusiveness of the Kerry-Lugar Act.[22] Subsequently, tensions have persisted over the American resort to the use of pilotless drones to strike inside Pakistani territory to destroy Taliban strongholds or target specific Taliban leaders.[23] Later US/ISAF helicopter strikes within Pakistan that led to the deaths of three Pakistani soldiers contributed to attacks against supply convoys to Afghanistan.[24]

At bottom US and Pakistani strategic interests in the region and in Afghanistan are not congruent and are likely to remain at odds.[25] Pakistan's powerful security establishment is loth to abandon its ties to a host of jihadi organisations despite the domestic costs that they have imposed on Pakistan. Not surprisingly, US policy toward Pakistan still remains fraught despite Vice-President Biden's attempts to assuage Pakistani concerns during a visit to Afghanistan and Pakistan in January 2011. The principal difference at this juncture revolves around the unwillingness of Pakistan's powerful military establishment to undertake meaningful military operations against a number of militant organisations which are located in Waziristan. It is unlikely that this fundamental difference in strategic priorities will be resolved anytime soon.[26]

In the Graveyard of Empires?

Even before assuming office, as the Democratic presidential candidate, Barack Obama, called on the Bush administration to redress what he believed was a significant shortfall of troop levels in Afghanistan and urged it to send two combat brigades to the country in July 2008.[27] He also argued, while on a trip to Afghanistan, that the US needed to press Pakistan harder to eliminate terrorist sanctuaries in Afghanistan.[28]

However, the more significant policy shift occurred after the completion of the Afghan Policy Review in March 2009 under the Obama administration. Bruce Reidel, a former, senior officer in the Central Intelligence Agency (CIA), chaired this review.

The review spelled out an explicit set of new goals the most important of which was to defeat, disrupt and dismantle Al Qaeda in Afghanistan and Pakistan. To that end the administration sought to move 17,000 troops from Iraq to Afghanistan and to add 4,000 additional troops to train Afghan forces. It also specified that the US wanted to create an Afghan military of 134,000 and a police force of 82,000 by 2011.[29]

The next most significant turning point in the president's policy toward Afghanistan came in December 2009. The reasons for his decision to dramatically escalate the American military presence in Afghanistan need careful examination.[30] The decision in considerable part had stemmed from an earlier request on the part of General McChrystal, the commander of the American forces in Afghanistan, that the troop commitment had to be augmented with as many as 80,000 new personnel to make serious headway against the Taliban and Al Qaeda.

In the wake of this request, the president held extensive and far-reaching discussions with his national security team including the vice president, Joseph Biden, the secretary of defence, Robert Gates, the national security adviser, General James Jones, the secretary of state, Hilary Clinton, General Mike Mullen, the chairman of the joint chiefs of staff and General David H. Petraeus, the then regional commander. After much discussion of the goals of this additional deployment as well as the likely financial burdens, the president and his advisers concluded that they would only deploy another 30,000 troops. Also, the president made it clear that he was willing to augment the troop presence rapidly but wanted a timetable for a relatively swift withdrawal rather than an open-ended commitment.

President Obama then went on to spell out this new strategy in a major speech that he delivered at West Point, the US military academy. In this speech he made it clear that while he was committing additional troops he also expected to start drawing down American forces starting July 2011. In this speech he also provided further details about the new strategy which called for securing major population centres. Simultaneously, he bluntly rebuffed those who wanted the US to withdraw from Afghanistan, who wanted to maintain the status quo and those who sought a drastic open-ended escalation with no timetable for withdrawal.[31]

As part of this new strategy, in March 2010 the administration chose to augment the civilian component of the American presence in Afghanistan. Recent reports suggest that this expansion of American civilian expertise has had a limited but salutary effect on state-building efforts in the country.[32]

A key turning point in the Afghan strategy came as a consequence of an idiosyncratic development, namely the replacement of General McChrystal with General Petraeus as the Afghan commander in June of 2010. President Obama chose to replace McChrystal because of an ill-advised interview that the general had given to *Rolling Stone* magazine in which he had made a number of disparaging remarks about both the president and his advisers. Upon assuming this new command, Petraeus spelled out a new set of guidelines to measure progress toward the administration's stated policy goals. They included, but were not confined to, an assessment of how many Afghan troops were trained, a focus on the number of operations against Taliban strongholds and the success of attacks by US special operations forces.[33]

In the wake of the adoption of this new strategy at a North Atlantic Treaty Organization (NATO) summit in Lisbon, an agreement was forged to continue operations until 2014. To that end the administration envisioned a plan to transfer security duties in particular areas of Afghanistan to the country's indigenous military forces over the course of the next two years.[34] However, the plans did visualise that a substantial foreign troop contingent would still remain in Afghanistan for purposes of training and mentoring the Afghan armed forces.

In December of 2010, the administration carried out a strategic review of the conduct of the war in Afghanistan. Among other matters, the review established that insurgents were taking refuge in Pakistan and returning to wreak havoc in Afghanistan. Simultaneously, despite much urging, the Pakistani military establishment was proving unwilling to go into North Waziristan, the likely base of a number of insurgent organisations. Given the military's reluctance to conduct military operations in the region the US stepped up drone strikes against selected targets. It also noted that the Afghan security forces had exceeded growth targets and that Special Forces night raids had reduced Taliban influence in both Kandahar and Helmand provinces.[35]

In January 2011, administration decided to deploy an additional 1,000 marines in Afghanistan in an attempt to solidify prior gains made in the southern part of the country. The bulk of these new forces will be deployed in Helmand province where American forces have made steady gains but the possibility of the Taliban re-grouping remains.[36]

Unfortunately, with Holbrooke's abrupt death in December 2010, the responsibility for the coordination of US policy toward Afghanistan and Pakistan remains unclear. In his absence, it is, however, reasonable to assume that a bureaucratic turf battle may well be under away as those who had routine responsibilities for both policy and analysis now seek to assert themselves.

Conclusions and Prescriptions

US policy toward South Asia faces three critical challenges. The first is age-old. How does the Obama administration induce the Pakistani military to dovetail its priorities with those of the United States in Afghanistan? The most immediate and compelling challenge, of course, remains the necessity of inducing the Pakistani military to sever their ties with the Afghan Taliban and to then end their sanctuaries within Pakistani territory.

As argued in this paper, attempts to cajole, persuade and bribe the military establishment into aligning its policies with those of the administration have produced very limited results. There is little or no reason to believe, that left to it itself, the military apparatus will change its orientation. Unfortunately, the administration appears to have convinced itself that it has few viable, alternative choices. It appears to have persuaded itself that it can ill-afford to exert any pressure on Pakistan for fear that Pakistan will threaten the critical supply lines to the International Security Assistance Force. (ISAF).

This argument, sadly, has the quality of a self-fulfilling prophecy. Unless the administration reminds the Pakistani military establishment that American largesse, especially military assistance, cannot be taken for granted, there is little reason to believe that the military establishment will alter its behaviour. It has managed to convince itself that Pakistan's geographic proximity to Afghanistan and the lack of viable American options in supplying Afghanistan all but ensure that the US has limited leverage over its behaviour.

The second challenge is closely related to the first. It will need to decide on the level of commitment, both military and civilian, that it expects to maintain in Afghanistan after the drawdown of US forces in July 2011. Obviously, the answer to this question depends in considerable measure on the state of affairs that prevail within Afghanistan. Despite significant domestic pressures it is unlikely that the administration would risk a precipitate withdrawal. Nevertheless, an open-ended commitment to Afghanistan remains both politically and financially untenable.

Third and finally, the administration must come to terms with the fact of Indo-Pakistan competition in Afghanistan. There is every likelihood that the Pakistani military establishment will boost its efforts to dislodge India as the US and the ISAF presence in Afghanistan starts to wind down. In the absence of the security umbrella that American forces and the ISAF have provided India's highly regarded aid programme, its future presence, let alone success, may well prove to be at much risk.

Apart from these issues the evolution of the Indo-US relationship will have four critical drivers. First, much will depend on India's ability to sustain its high growth rate and simultaneously expand its commercial ties with the United States. If the Indian market continues to expand and is accessible to the US, it can become an important bulwark in the stabilisation of the relationship. Second, the US relationship with the PRC will also, in considerable measure, shape the future of the Indo-US relationship.[37] The state of the US-PRC relationship is of such significance because the PRC is and will remain India's principal strategic concern for the foreseeable future. Third, despite the professed American policy of "de-hyphenation" of its ties to India and Pakistan, influential individuals in India's foreign policy establishment continue to harbour doubts about such a de-linkage. In this context, President Obama's attempt to prod India to improve relations with Pakistan without forthrightly addressing the Pakistani military establishment's unwillingness to eschew the use of terror as an instrument of state policy, did little to assuage their concerns. Fourth and in a related vein, India's policymakers must forge a working consensus on the scope and limits of the US-Indian "strategic partnership". Though various Indian officials have spoken warmly about this partnership they have not been able to spell out their minimal expectations of the US and have not been able to clarify what India may be able to bring to this partnership. In the absence of such clarity the central US relationship in South Asia may well reach a plateau.

Notes

1. Sumit Ganguly and S. Paul Kapur, "The Transformation of US-India Relations: An Explanation for the Rapprochement and Prospects for the Future," *Asian Survey*, 47:4, July/August 2007, 648-649.
2. David Ignatius, "A Three-Pronged Bet on 'Afpak'", *The Washington Post*, February 22, 2009.
3. Mark Sappenfield, "Is Kashmir key to Afghan Peace?" *The Christian Science Monitor*, November 21, 2008.
4. Sumit Ganguly, "America's Other Strategic Dialogue," *The Wall Street Journal*, May 30, 2010.

5. Edward Luce, "Singh Embraced by US as "Natural Ally," *Financial Times*, November 24, 2009.
6. Damian Carrington, John Vidal and David Adam, "China and India join Copenhagen accord," *The Guardian*, March 9, 2010.
7. Sumit Ganguly, "India's Machiavellian Turn", *Asian Wall Street Journal*, July 30-August 1, 2010.
8. Sumit Ganguly, "Let India Train the Afghan Army," *The Wall Street Journal*, February 16, 2010.
9. Indo-Asian News Service, "Obama Again Slams Outsourcing to India, Calls it Unfair Practice," *The Hindustan Times*, February 12, 2010.
10. Elana Schor, " Bhopal Disaster Casts Indirect Shadow Over Obama's Visit to India," *The New York Times*, October 29, 2010.
11. Press Trust of India, "Parliament adopts nuclear liability bill," *Daily News and Analysis*, August 30, 2010.
12. Times News Network, "Obama Says Talk to Pak, PM Says not Until Terror Machine Stops," *The Times of India*, November 8, 2010.
13. Economic Times Bureau, "Govt Not Ruling Out Headley Extradition," *Economic Times*, April 1, 2010.
14. Ashley J. Tellis, *Dogfight! India's Medium Multi-Role Combat Aircraft Decision* Carnegie Report (Washington, DC: Carnegie Endowment for International Peace, January 2011)
15. See Sumit Ganguly, "US aircraft dropped from multi-billion dollar US defense deal," *Opinion Asia*, April 30, 2011. Available at: http://opinionasia.com/node/887
16. Sumit Ganguly, "The Diwali Summit," *The Deccan Chronicle*, November 14, 2010.
17. James Lamont and Amy Kazmin, "India shuns US in $11 billion fighter deal," *The Financial Times*, April 28, 2011.
18. Saeed Shah, "Pervez Musharraf Resigns as President of Pakistan," *The Guardian*, August 18, 2008.
19. K. Alan Kronstadt, *Pakistan: Key Current Issues and Developments* (Washington, DC: Congressional Research Service, July 1, 2010)
20. I am deeply grateful to my Professor Devin Hagerty of the Department of Political Science of the University of Maryland at Baltimore County for a discussion of this subject.
21. For an overview and a critique of the Obama administration's strategy toward Pakistan see Paul Staniland, "Caught in the Muddle: America's Pakistan Strategy," *The Washington Quarterly*, Winter 2011, 133-148.
22. Ayesha Siddiqa, "Turf Wars: Why Pakistan's Military Hates America's New Aid Package," *Foreign Policy.com* Accessed on January 26, 2010. Available at: http://www.foreignpolicy.com/articles/2009/10/14/turf_wars
23. Saaed Shah and Ewen MacAskill, "Anger in Pakistan at US Plan to Expand Drone Attacks," *The Guardian*, March 19, 2009.

24. See Farhan Bokahri, "ISAF supplies at risk as Pakistan takes tough line," *Jane's Defence Weekly*, October 13, 2010.
25. See Sumit Ganguly and David P. Fidler, "The Definition of Insanity," *Foreign Policy*.com Accessed on January 26, 2011. Available at: http://www.foreignpolicy.com/articles/2011/01/12/the_definition_of_insanity
26. Eric Scmitt, "Pakistan's Failure to Hit Militant Sanctuary Has Positive Side for US," *The New York Times*, January 18, 2011.
27. For an analysis and a critique of the Bush administration's policies and military strategies in Afghanistan see Seth Jones, *In The Graveyard of Empires: America's War in Afghanistan* (New York: W.W. Norton and Company, 2009)
28. Candace Rondeaux and Dan Balz, "Obama, Karzai Share a 'Working Lunch," *The Washington Post*, July 21, 2008; also see Jeff Zeleny, "Troops in Afghanistan Need Help, Obama Says," *The New York Times*, July 14, 2008.
29. See "Remarks By The President On a New Strategy For Afghanistan and Pakistan," March 27, 2009. Accessed on January 19, 20111. Available at: http://www.whitehouse.gov/the-press-office/remarks-president-a-new-strategy-afghanistan-and-pakistan
30. See the detailed analysis of the discussions that led to the surge. Peter Baker, "How Obama Came to Plan for 'Surge' in Afghanistan," *The New York Times*, December 6, 2009.
31. See "Remarks by the President in Address to the Nation on the Way Forward in Afghanistan and Pakistan," Eisenhower Hall Theatre, United States Military Academy at West Point, West Point, New York, December 1, 2009. Accessed on January 19, 2011. Available at: http://www.whitehouse.gov/the-press-office/remarks=president-address-nation-way-forward-afghanistan-and-pakistan
32. Paul Miller, "How the War Can Still Be won," *Foreign Affairs*, January/February 2011, 90:1, 51-65.
33. Helene Cooper, David E. Sanger and Thom Shanker, "Once Wary, Obama Now Relies on Petraeus," *The New York Times*, September 16, 2010.
34. Peter Baker and Rod Nordland, "US Plan Envisions Path to Ending Afghan Combat," *The New York Times*, November 14, 2010.
35. Helene Cooper and David E. Sanger, "Obama Cites Afghan Gains as Report Says Exit Is on Track," *The New York Times*, December 16, 2010.
36. Elisabeth Bumiller, "More Than 1,000 Extra Marines To Be Deployed in Afghanistan," *The New York Times*, January 6, 2011.
37. Sumit Ganguly and David P. Fidler, "The Shadow of the Sino-US "new normal" falls on India," *The Times of India,* January23, 2010. Accessed on January 23, 2010: http://timesofindia.indiatimes.com/home/sunday-toi/all-that-matters/Shadow-of-Sino-US-new-normal-falls-on-India/articleshow/7345109.cms

18

Indo-Pak Relations and the Balance of Emotions

Prakash Menon

Introduction

Passions have been the prime movers of history. This truism is unlikely to change. On the contrary, in a world that is getting increasingly connected, passions are easily influenced by images and words that are accessible to a significant part of the global population. Consequently national passions can significantly influence the political decision making process at the highest level. In fact, the decision making process is a product of this constant debate between the country's strategic logic and its political goals and objectives tempered over time. Strategic logic in itself is an outcome of interaction between the plethora of ways and means available to a nation. Broadly speaking, the macro constructs of geography, economy, military and diplomacy determine the contours of a country's strategic logic in the overall political climate and context. Geography is immutable but its endowments in terms of space, distance, time, terrain and weather are becoming dynamic with changing technology and innovation. However, this must also contend with the changing character of geo-politics and strategy. In other words, the geographical or physical space must contend with the "geography of the imagination" of its leaders and people that gives it a meaning in a particular political context. For instance, the Cold War stand off between the rival nuclear powers, the US and the erstwhile USSR was a classic example of how fear created through distrust combined with geographical disposition influenced the strategic behaviour of these countries. Undoubtedly, this "imagination" or "passion" was driven by rival political ideologies and technological development.

This construct of "passion", "imagination" or "emotion" perhaps requires a deeper examination. Images of the burning of the Holy Koran in distant United States led to rioting and arson in several parts of the world including in India's Shia dominated district of Budgam in the state of Jammu and Kashmir. The power of images to stimulate public emotion or unrest has increased manifold in recent decades, and is a phenomenon that is now unfolding notably in the Arab world. The Tunisian people's Jasmine revolution has led to a wave of protests in Egypt, Yemen, Jordan and Algeria, and when these passions will afflict other parts of the world where people's emotions have been long suppressed, particularly in authoritarian states, can simply not be predicted. The role of Al Jazeera in fanning the flames of passion might be difficult to estimate but it would not be out of place to remark that this phenomenon may not have been possible without its ability to communicate powerful "imagery and vocabulary" across the Muslim world. More will surely follow, and the wave in due course might sweep several other parts of the globe including our immediate neighbourhood. It seems as if we are witnessing a "geography of emotions" across the Muslim world. These emotions are being derived from the prevalent dominant political and religious narratives internalised over several decades within the affected states and between them.

This paper intends to examine the geography of emotions in the sub-continental context. In the first section, I lay out the framework of discussion based on Dominique Moisi's thesis of three emotions—fear, hope and humiliation.[1] In the next section, the two constructs, namely, the balance of geo-political power and emotions, are contrasted in the sub-continental context. The following section discusses the strategic rivalry between the two countries namely India and Pakistan. The next section explicates the myriad challenges that India faces in re-setting this relationship, and the final section suggests a plausible way ahead.

The Geography of Emotions

Dominique Moisi's book, *The Geopolitics of Emotions*, focuses on the three primary streams of public or political emotion: fear, hope and humiliation. He chooses these three shades of emotion because of their linkages with the notion of how nations and people at large address the national challenges they face as well as how they relate to each other. The absence of public confidence begets fear and this in turn creates apprehensions and colours expectations for the future. Hope in contrast is a reflection of confidence; for people, today is better than yesterday, and tomorrow shall be even better. Humiliation, as Moisi describes, is the injured confidence of those who have lost all hope for the future; and this lack of hope is often construed as the fault of the others. When the contrast between one's idealised and glorious past and one's frustrating present is too great, humiliation prevails. Moisi summarises these three emotions at three levels: One would say that hope is "I want to do it, I can do it and I will do it". Humiliation is; "I can never do it" and may lead to "I might as well destroy you since I can never join you"; Fear is "Oh my god, the world has become such a dangerous place; how can I be protected from it?"

Generally speaking, emotions whether religious, national, ideological or personal, have always mattered. The emotion of visceral hatred that unfortunately pervades significant sections of the Indian and Pakistani society has perhaps been the underlying impediment that makes Indo-Pak peace prospects seemingly intractable. The "emotional jelly" of the two countries is easily amenable to political manipulation. The challenge for the leadership on both sides is to find the means to transcend the emotional barriers in order to subdue mutual hatred and animosity, and morph it into lasting peace and friendship. At yet another level, the rise of China and its political, military and economic growth has given rise to the emotion of fear in large parts of Asia and also the world. This emotion runs parallel with growing economic engagement that gives rise to hope for prosperity and development. The combination of fear and hope has potential for both stability and strife. But in essence the degree of stability or strife is produced by the balance between hope and fear. Generally speaking, in a perennially uncertain world, both hope and fear are required for survival. But a predominance of hope provides greater space for stability than when fear is the overriding emotion.

The changing constellation of forces in the contemporary world has the potential for generating tensions that if guided by fear or humiliation could result in avoidable outcomes. In a worst case, it could result in a large scale war among major military powers, a prospect that is often discounted by those whose do not take into account the beast that lurks within humankind, a fact that is embedded in recorded history. With power in the process of shifting from the West to Asia, it should be obvious that many of the tensions that may be generated could be geographically located to a large extent within the Asian hemisphere. The Asian continent is home to approximately 60 per cent of the world's population and a plethora of ancient civilizations that are presently on a comeback trail towards what they perceive as their historical greatness. Though this is definitely not an all encompassing description of the Asian resurgence, it is certainly true in terms of the overall demographic numbers which include the people of China, India and South East Asia. But Asia also harbours substantial numbers of people who nurse a deep sense of humiliation based on a sense of historical dispossession in the present and even more for the future; a future that is in utter contrast to an idealised, glorified past, and a future in which their own political, economic, social, cultural conditions are dictated by other more powerful nations.

This sense of humiliation is reflected in the emergence of terrorism a form of trans-national violence adopted by radical religious fundamentalists among other groups. In this scheme, Indo-Pak relations have no doubt been impacted by the regional balance of power politics through the active machinations of China and the United States. The relations are also undermined by the "balance of emotions" between the people of both countries. An examination of the balance of emotions within the context of balance of power is therefore warranted.

Balancing Power and Emotion

The politics of maintaining the balance of power between India and Pakistan has been the strategic logic of both China and the United States. This logic was further reinforced by both nations going nuclear. It is a balance that is likely to be continued by both China and the United States at least for the foreseeable future. The underlying *meta*-logic is to constrain India's power to within the subcontinent. It is the natural game of jockeying for power by the nation states aimed at preserving their predominance and national interests. One of the primary interests of both United States and China is the preventing India from emerging as a maritime power, since the regional geography has endowed India with a location from where it could in large measure dominate the Indian Ocean region. The Indian ocean region assumes geo-strategic importance for most major nations of the world, and especially China whose dependence on it for trade and oil is growing by the day, and which is been referred to by the Chinese strategists as the 'Malacca dilemma'.

Since the Malacca Straits are the main sea route between the Indian and Pacific Oceans, and the United States controls all the oceans of the world and has a fleet in each one of them, it would be wary of India as a maritime power primarily because it would not be in its larger interest to have another major maritime power when it already and significantly dominates the Indian Ocean. This is the game that nations play and it is not going to be different between India and the United States unless India agrees to be part of a United States strategic alliance—a thought that is anathema to Indian political thought and the national leadership. India's long standing tradition of non alignment prevents it from being in an alliance with any nation other than under the flag of the United Nations.

There is also the fear that it will drag India into America's quarrels in other parts of the world that would not be in India's long term interests. The idea of a 'strategic partnership' that is contextually derived fits in easier with India's long standing political culture.

At yet another level, state power too seems to be undergoing a definitive change. In his recent book, *The Future of Power*, Joseph Nye argues that our outlook on the utility of state power—hard and soft—needs a review.[2] In his view, the issue is no longer about power transition but that of power diffusion. Power transitions were relevant when states moved from one ideological construct or form of governance to another. But now, since state power stands considerably diluted into the hands of the people at large and even non-state actors, it is all about managing the diffusion of power. This diffusion of power in turn is nothing but the sheer display of public "emotion" and "passion" through a highly interconnected world of Facebook, Twitter and other electronic media, as has been witnessed in the Middle East over the past few weeks. Diffusion of power not only breeds but propels and drives the emotions of fear, hope and humiliation, and currently being witnessed in the Muslim world.

Geographically, humiliation is the contemporary reigning emotion in the Arab world, except for the tiny city states of the Gulf. The presence of an Arab emotion of humiliation unites, however poorly, the Arab world against the West. This sense of humiliation is not confined to the Arab world but has spread to Afghanistan, Pakistan, sections of Indian society, Bangladesh, Indonesia, Philippines and Malaysia. The spread has been through the morphing of a predominant religious ideology from its moderate and tolerant strain into the Wahhabi/Salafi mould. Afghanistan and Pakistan are both now in the grip of radical Islam and personify the fountainhead of terrorism in the world. The muted reaction to Salman Taseer's assassination provides a good indication of the radicalisation of the Pakistani masses. The sense of humiliation is underpinned by a complete loss of hope resulting in despair. This in turn has lead to the fostering of a desire for revenge and hence finds its ultimate abode in destruction. The dominant narrative being, "If I cannot do anything about people who are humiliating me, I can make them feel what suffering means". The killing of innocents aided no less by technological means is the manifestation of the dominant narrative of humiliation. This is manifested in the advent of the suicide bomber and the increasing numbers of suicide bombers over the last couple of years is an indication of the depth of humiliation experienced by the radical religious elements.

Humiliation: For Pakistan, humiliation is derived mostly from its relations with America and in a larger sense, the West. The dominant narrative that drives emotions in contemporary times is the notion that

Pakistan is suffering because it is fighting America's war and has endured immense suffering because of it. The humiliation is further deepened by to the fact that Pakistan is economically and ecologically devastated, and survives on American financial steroids. The Kerry-Lugar bill meant to ensure that American aid reaches its intended constituents was perceived as humiliating by the Pakistanis. The impunity with which American drones strike targets in Pakistan resulting in large scale casualties of civilians, albeit with the tacit concurrence of the Pakistani rulers including its military is in the popular perception an insult to their sovereignty and deepens the sense of humiliation at the popular level. This humiliation has already morphed into a sense of loss of hope in the future and there is now a significant section that believes "I might as well destroy you since I can at least have my revenge". The fact that the powerful elite of Pakistan send their children to the West for higher studies, and hope for eventual absorption into the Western societies, is a point that the people on the Pakistani streets note. There is therefore a growing gulf between the power elite of Pakistan and the common masses, and they see the powerful political, bureaucratic and business elite as American proxies. The saving grace for the power elites is the lack of a coherent leadership that could galvanise the popular sentiment. The lack of leadership could be attributed to the fragmentation of its polity along ethnic lines and the ability of the Pakistan army to keep the emergence of such a leadership in check. The main militant group, Tehrik-e-Taliban (TTP) has limited political clout except as a terrorist group presently fighting the Pakistani state. The other omnipresent emotional state is fear. Fear is not only driven by the present state of internal affairs in Pakistan, but it is the fear that has been embedded in the Pakistan psyche ever since its creation in 1947 with India being the cause and source of that fear.

Fear: Throughout its existence or rather because of the manner of its creation, fear of India wanting to see the demise of Pakistan has been a deeply internalised emotion in the minds of most Pakistanis. This of course is derived from its attempts to create an identity of its own which remains an unfinished agenda that has still some way to go. Pakistani identity has been sought to be constructed by trying to unify the different ethnicities of the Punjabis, Pashtuns, Seraikis, Baluchis, Sindhis and Mohajirs. Zia ul Haq started the process of attempting to unify these identities through Islam—a process that has reduced Pakistan to a radicalised Muslim state without national unity. Concurrently with Islam ran the other notion for buttressing this flailing unity—the Indian military threat—a threat that was fulfilled by the dismemberment of Pakistan in 1971. The fact that there is much more to that story is not of relevance here since we are dealing with the popular belief that animates the fear born of the perceived India threat.

Emotion and Strategic Rivalry

The Pakistan army has been the prime institution that has ensured that the idea that India poses the main existential threat to Pakistan remains the centre piece of Pakistan's strategic and political logic. It has of course also been the driver that has kept the Pakistan army in power either directly or behind the scenes. The proxy war by Pakistan in Kashmir has been seen as a cost effective tool to keep India tied down so as to neutralise the perceived Indian threat materialising in a manner that can overwhelm the Pakistani state. Fear—both perceived and created—of the Indian threat has also dovetailed with China's need to keep India as a continental power and the Americans not wanting India to emerge as an independent actor in the Indian Ocean. The fact that Pakistan today is able to garner the support of both is testimony to this impression. The attempt to balance Pakistan against India invites an Indian reaction that falls in the domain of military preparations. This, therefore, starts a cycle of preparations and counter preparations that draws greater finance into the defence budget of both countries.

All this at a time when there are hardly any resources left for the many fields that beg for attention in order to facilitate development and growth. This lack widens the socio-economic gap between the rich and the poor—a gap that coupled with the spread of radical religious ideology could threaten the very existence of the Pakistani state. Any major threat to the Pakistani state stemming from internal disorder can, from an Indian standpoint, make Pakistan fabricate the Indian threat as an instrument to galvanize all Pakistanis and sink their immediate differences and unite. This may involve even elements like the TTP who are at present fighting the Pakistani state. The threat could be fabricated through an Indian reaction post another terrorist attack on Indian soil like Mumbai in 2008. To understand the likely Indian reaction the Indian emotional state needs elaboration.

Significant sections of Indian society due to historical reasons rooted in the partition, share a visceral hatred with Pakistan. Ironically this visceral hatred does not manifest itself at the individual level. At the individual level, Indians and Pakistanis can exhibit a level of bonhomie that stems from the fact that they are the same people with a common cultural and historical heritage. But viewed through the nationalistic framework, the individual level bonhomie withers and the visceral hatred predominates. On the Indian side this hatred has been deepened by the firmly held belief that Pakistan has used and continues to use terrorists, as assets to try and destabilise the Indian state not only in Kashmir but also in the rest of the country. The Mumbai attack and the capture of Ajmal Kasab provided ample proof to the Indian psyche of Pakistan's continued attempts to use terrorism as a tool of statecraft. There is therefore a solidification of the historically derived sense of visceral hatred. But there is also a deep sense of India's honour being hurt and a call for revenge—a call that is echoed in the Indian street, an emotion that is not lost on the Indian polity. The call is widespread enough not to remain or be left unheeded. But democratic India has so far shown great strategic restraint and displayed political wisdom to ensure that it is not dragged into the political chaos that is Pakistan, and has sought to concentrate on its economy in order to address the enormous challenges in terms of growth and development. But this does not guarantee that the Indian reaction to the next terrorist attack emanating from Pakistan could not be different. But it shall certainly put to test the tension that will ensue between strategic logic and political compulsions. It has been India's stated policy that it desires a stable and friendly Pakistan and use of force for resolving Indo-Pak differences cannot facilitate the peace that is desired.

Of course there is no denying the fact that there are some sections of the Indian strategic community who believe that the only way to ultimately achieve the peace is to bring about the demise of Pakistan. This point of view is often quoted in Pakistan to prove the existence of the Indian threat. But this view has no takers within the Indian ruling establishment for very good reasons. The main reason being that a stable Pakistan alone can un-lock India's trade routes to West Asia and Central Asia which remain politically blocked by the political situation in Afghanistan and Pakistan. But perhaps more significant is the fear of the mass influx of the Pakistani populace into India that a collapsing Pakistan can bring about. The dictates of geography have ensured that the population has no other way to go than across the international boundary or the Line of Control towards the Indian state. This large scale movement or exodus will no doubt also be aided by the familial connections that span both countries. An over-populated India would soon find this influx straining communal ties that constitute a fault line that secular India has managed to keep in check even though it raises its ugly head once in a while—a fault line that Pakistan has tried to deepen by attacking religious targets in India. But the collapse of Pakistan is an unlikely scenario though its possibility can never be ruled out, not because of Pakistan but because of the greater game being played between the two pre-eminent economic and military powers of the time—United States and China.

India's Emotional Challenge

Pakistan like Afghanistan is situated at the cross roads of West, Central and South Asia. The United States would not like to cede control of this region to any other power especially China or Russia. If the United States were to withdraw from the region, the Chinese will fill the political space a it is well placed to do so. The Chinese would be as wary as the Americans to ensure that radical religious forces do not come to power in Pakistan. The already existing deep relations between the Pentagon and Pakistani military will ensure that the Pakistan military gets their full support to keep the state together even if it uses the harshest means to suppress dissent. The only game changer here will be the extent of radicalisation that has taken place amongst the lower ranks of the Pakistani military. This is a possibility since no army can be completely immune to winds that blow so strongly in the society they are rooted in; coupled with this is the institutionalised Islamisation of the Pakistan Army that had commenced in Zia's time.

There is also another fault line that could unravel the Pakistan army—the ethnic divide especially the Pashtun-Punjabi divide. Nearly 24 per cent of the Pakistani army is Pashtun and the rest is mainly Punjabi. It could be one of the reasons that General Kayani has shown a reluctance to launch operations in North Waziristan. Overall, Pakistan retains the ability to keep itself alive even though it may remain in a state described by M J Akbar in his book, *Tinderbox,* as a "toxic jelly state".[3] So though the populace of Pakistan is operating from the emotional state of humiliation and fear, their survival will be ensured by a centralised authority like the army that is likely to suppress any large scale dissent ruthlessly. However with the attendant danger that such suppression might cause the unravelling of its most powerful institution—the army. This situation will threaten the interests of every one and may finally involve an intervention by the international community especially since nuclear weapons are involved.

On the other hand, and despite its manifold shortcomings, the Indian state has made impressive economic progress though its inclusiveness remains a challenge. The dominant emotional narrative in India is one of hope, and this is not to discount the existence of several regional sentiments and geographic spaces where humiliation experienced through deprivation has manifested in extreme political violence—left wing extremism being a recent example. But in the larger picture, the Indian state and masses have consistently embraced hope, and their aspirations have grown phenomenally. The fact that in most parts of India even among the lowest economic classes, the parents are hell bent on getting a good education for their children is a sign of the hope that resides in their emotional state. But accompanying this sense of hope is the fear in the strategic community of the uncertain strategic behaviour of the People's Republic of China with respect to its continued nuclear and military assistance to Pakistan, the unresolved boundary issue, its persistent and conflicting claim on Arunachal Pradesh, its ambivalence over the state of Jammu and Kashmir, and the general tenor of its international behaviour witnessed in the last couple of years. Fear is sometimes also generated by the home grown religious radicalism in India, which from recent indications spans the Muslim-Hindu religious landscape.

The challenge for India is how to translate this prevailing sense of hope better relations with its neighbours, especially Pakistan and China. This has to be done simultaneously by guarding against fears that are aroused and experienced. The balancing of hope and fear within the existing global political context and particularly with Pakistan and China forms the Indian security challenge. How can this be possible with Pakistan is the natural line of investigation especially since this paper deals with Indo-Pak relations?

Way Ahead

The architecture of any improvement in Indo-Pak relations must have as its foundation the bonhomie generated by people of both countries at several levels. It is not possible to sustain a mutually beneficial relationship when it is based on a visceral hatred for each other's country. This visceral hatred although rooted in the history of the partition has been deepened especially in the last two decades. But this is not an immutable condition that cannot be reversed and is largely a political matter that can be changed provided the leadership on both sides agrees to make prepare the ground for it. Incidentally, there are constituencies on both sides that support this notion. In a recent Pew poll, despite 53 per cent saying that India is the major threat to the country, most Pakistanis want better relations with India.[4] Roughly seven-in-ten (72 per cent) say it is important for relations with India to improve and about three-quarters support an increased trade with India and continuation of talks between the two rivals. Notwithstanding the fact this may not be truely reflected on the streets of India and Pakistan, it gives at least a glimpse of the reconciliation possibilities in the future. The "geography of emotions" can be addressed at three levels.

- The first level for making the soil fertile for better relations is to increase interaction across the spectrum in both the societies.

The interaction and dialogue must be continuous and include officials, politicians, scholars, mediapersons, as well as military and cultural exchanges amongst others. This is only possible if travel between the two countries is facilitated through a visa regime that is much more relaxed than what it is at present. The idea is get to know each other better than we think we do and to identify mutual areas of interests that can be taken forward. Simultaneously, official level talks or dialogue need to be institutionalised and must be totally immunised against political events and upheavals including acts of wanton terrorism.

- The second level, which must grow out of the first argument, is to facilitate trade through open borders and which in turn could become a core area that can over time link the people on both sides. Educational institutions on both sides too should be accessible in large measure. Political level talks must graduate to identify measures that will build confidence and minimise the fear that we are a threat to each other especially in military terms. There must be mutual acceptance of the fact that as nuclear powers, there is always the possibility of a nuclear exchange during a conflict especially through inadvertence. Both countries therefore must come to an understanding that force as an instrument of statecraft to resolve disputes must be eschewed. We must eventually work towards a no war pact. But this cannot be possible unless the United States and China give up the balance of power game in the region.
- Therefore, at the third level, measures taken by United States and China to balance power in the region through extensive assistance given to build the conventional and nuclear military capability of Pakistan might require a serious re-think. Ironically, it has strengthened the Pakistan armed forces and has been the major factor in preventing the building and return of democratic institutions. The strengthening of the armed forces has facilitated the use of terror groups as strategic assets, which in turn has resulted in backlash. The balancing that the United States and China have to undertake must recognise the changing utility of power and the diffusion of power to entities that may not be amenable to their control. Both countries are in a position to succeed if they attempt to reconcile their positions in cooperation with one another. It will also ease the Afghan situation which must finally be left to the Afghan people to decide for themselves.

There is always a groundswell in all societies that want peace and dignified living. It is this groundswell that must be politically captured by the national leadership in the two countries. The fact that the Pakistani contingent at the Commonwealth Games opening ceremony at Delhi in 2010 received one of the biggest applauses is an indication of the possibilities. Indo-Pak relations cannot make much progress until the prevailing emotional balance is rectified. It will require statesmanship of an extremely high order on both sides. In this regard, Pakistan's traditional supporters also need to change track.

Conclusion

Fear, hope and humiliation as dominant forms of public emotion have played up in myriad ways on the Asian subcontinent. India and Pakistan are now apparently at the opposite ends of hope and humiliation respectively. Both countries, born out of the flawed circumstances of the time, have behaved differently over their last six decades of existence. While India has continuously cruised along the trajectory of "hope" which in turn has tempered its long standing political culture, style and temperament, the state of Pakistan on the other hand has consistently fumbled along with self conceived images of fear and humiliation. These images have been charged over time by the policies of the extra-regional powers namely the United States and China. Notwithstanding their past and present policies, it is clear that is sufficient negotiating space between the regional power play and the animosity between the two countries. India and Pakistan need to shed the acuity of the emotional baggage of the past, and construct new ways and means to build a new future for their people. This alone can bring lasting peace and stability in the region.

Notes

1. Moïsi, Dominique. *The Geopolitics of Emotion: How cultures of Fear, Humiliation, and Hope are Reshaping the World*. New York *et al*: Doubleday. 2009.
2. Joseph S. Nye, *The Future of Power*, Public Affairs, New York, 2011.
3. MJ Akbar, *Tinderbox: The Past and Future of Pakistan*, Harper Collins, New Delhi, 2011.
4. Pew Global Attitudes India Report DRAFT 10-18-10.docx

19

How is India Managing its Asian Challenge?

Arvind Gupta

Introduction

For India, several new security challenges have arisen due to the shifting of geo-political landscape in Asia. The rise of China is one such challenge. Terrorism, maritime security and instability in parts of South Asia are some other challenges. This paper summarizes the broad geo-political and security changes in Asia; looks at the challenges India has faced in the last twenty years as a result of these changes and examines the Indian approaches for managing these challenges. The main argument of the paper is that while India has sought to engage with Asia and has been able to convert these challenges into opportunities through a combination of political, economic, and foreign policy measures, its approach in managing them has been ad hoc and not based on a forward looking grand strategy. Given the rapidity of change, India finds itself constrained by a number of developments which come in the way of deepening its engagements with Asia. Of the several challenges India is facing, management of its relationship with China will be the foremost in the near future. The lack of diplomatic capacities is also hampering India's capabilities to engage with Asia more intensely. India needs to articulate for public and the world at large its own view of how it is going to manage its relation with the rapidly changing region.

The end of Cold War bipolarity has created integrative trends in Asia and raised issues about Asia's future. The centre of gravity of power is shifting from the West to the Asia-Pacific region.[1] Asia, with 58 per cent of the world's population, now constitutes about 35 per cent of global GDP and a third of global trade.[2] During 2000-2009, Asia's global merchandise imports grew by 6 per cent annually, and exports by 8 per cent annually as compared to North America's one per cent in both.[3] In Asia, major political economic and military restructuring has taken place since the end of the Cold War. The most visible manifestation of the change in Asia is the simultaneous rise of China and India. What consequences these developments hold for Asia's future is a question for intense debate. Kishore Mahbubani has argued that the West is in the decline and it should learn from Asia's experience and the "Asian way". Some Western scholars see only intense competition, conflict and tension in the intense geopolitical churning that is taking place in Asia. They see the huge diversity in Asia as a stumbling block for the evolution of common political, economic and security structures despite some progress in these areas.[4] In short, the question of Asia's future is unsettled. Most analysts agree that the Asian order is still evolving and that there will be many surprises along the way.[5]

India has also been deeply affected by the changes in Asia. For India several new security challenges have arisen due to the shifting geopolitical landscape in the region. The rise of China is one such challenge. Instability in a part of India's neighbourhood is another. India has taken a number of far reaching measures to adjust its foreign and security polices to meet these challenges. This paper summarises the broad geo-political and security trends in Asia; details the challenges India has faced in the last twenty years as a result of these changes and examines the Indian approaches in managing these challenges.

I. Geo-political Trends in Asia

There is vigorous debate on the changes in Asia since the end of the Cold War and where Asia is headed.[6] The following is a broad summary of the emerging political, economic and security trends in Asia. These developments have a direct bearing on India's security environment and its well being. The broad trends in Asia point to a major restructuring but whose contours are not yet clear.

- **The Rise of China:** The rise of China is arguably the single most important development in the Asian and global context. After a century of humiliation at the hands of colonial powers, China is rising politically, militarily and economically. China is one of the few countries (India is another) which has weathered the global economic crisis (2008-09) well. It has been argued by some academics that the Chinese economic revolution is entering its second phase in which China will move from just being a manufacturing hub of the world to becoming a global financial centre and the Yuan is likely to become a global currency soon[7]. The currency issue—the appreciation of Yuan—has become a major bone of contention between the US and China. Increasingly, China has positioned itself as an engine of economic growth in Asia and the world. It has signed a Free Trade Agreement with ASEAN thus linking up its economy with the major regional economies, which now have a stake in China's economy. In 2006, 20.4 per cent of Japan's global imports were from China and 14.3 per cent of its global exports went to China. The figures for South Korea are 15.7 and 21.3 per cent and those for the ASEAN are 11 and 8.8 per cent respectively.[8] The share of major regions and countries in China's trade for the first quarter of 2010 is given in Fig 1. Nearly 18 per cent of China's trade was with North East Asia followed by 10 per cent with the ASEAN and 4 per cent with South and West Asia.

- China is also challenging the US hegemony in Asia and globally. The Chinese defence budget is growing at a double-digit rate over the last several years. The Chinese Peoples Liberation Army (PLA), in particular, thinks that China cannot play its global or regional role unless it has a strong military. China's growing military strength is a cause of concern for many countries. China is no longer coy about its military might. In a year-end interview disseminated by the state-owned Xinhua news agency on December 29, 2010, the Chinese defence minister Liang Guanglie said:

  > Without the military of more than two million troops…China will be a weak country. A major aspect of its strong national power, I think, is a strong defence. Despite growing national power and global influence, the international situation remains complex, while more and more unstable factors are shadowing China's security. A war which involves the whole country is impossible at the moment, but there is a possibility that a mishap or accident might ignite regional conflicts. To enhance our comprehensive national power, (we have to ensure) our core interests are not hampered. We have to put the nation's sovereignty and security in the top position that is the common understanding of the whole Party (Communist Party of China) and all the people.[9]

- China's economic and military rise is accompanied by the rise of nationalistic feeling in the country. This has unnerved many of its neighbours like Japan, for instance.

- **Japan-China Rivalry:** Despite considerable economic interdependence, there is a growing rivalry between Japan and China. Anti-Japan feeling in China is very high. The unresolved Senkaku island dispute is a continuing source of tension which came to a head in 2010 when the Chinese arrested the crew of a Japanese vessel for ramming into a Chinese boat near the Senkaku Island. Japan is concerned about the possible weakening of the Japan-US alliance. Will Japan militarise in reaction to the rise of China and the nuclearisation of the Korean peninsula? How will China react to an increasingly militarised Japan?
- **China and Taiwan Relations:** China will not compromise on its one-China policy. Taiwan is getting increasingly integrated with China even though thousands of Chinese missiles are still deployed pointing towards Taiwan. Taiwan is a major issue of discord between China and the US but the latter is increasingly accommodating China's concerns and does not encourage independence sentiments in Taiwan. China was furious at the $6 billion US military aid to Taiwan announced in 2010.

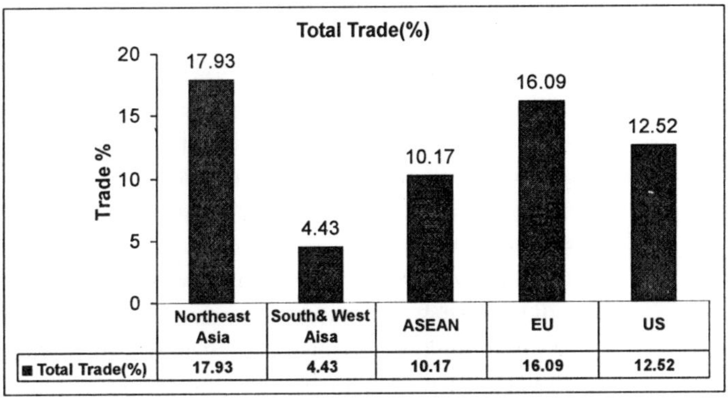

Figure 1: China's Total Trade (2010/01-04)

(*Source*: Ministry of Commerce, People's Republic of China.)

- **US Influence in Decline?:** The US has several key security alliances in Asia. It is seen by many countries as a provider of 'public good' in the shape of security. The war against terror and severe economic difficulties at home have raised the question whether the US power is in decline. In 2009 the US economy grew by only 1.3 per cent as compared to China's 9.2 percent. Are the current difficulties faced by the US temporary and will the US bounce back in a quick time? How will the US deal with a rising China? Will the weakening of the US dilute its alliance commitments? How will its alliance partners react to the decline of the US? Will the US and China be able to manage their rivalries in the future?

- **Tensions on the Korean Peninsula:** The Korean War during the fifties was a manifestation of the ideological struggle between capitalism and communism which under-pinned the Cold War for forty years. The Cold War has ended but the nuclearisation of North Korea has complicated the security situation and led to huge tensions on the Korean Peninsula. China continues to support the North Korean regime, fearing the massive influx of refugees in case of a North Korean collapses. China blames the US for the North Korea's intransigence. Six-Party Talks, the regional mechanism devised to deal with the Korean situation, have made little headway in resolving the problem. There have been clashes between North and South Korea. Will the two Koreas unite? How will Korean developments affect the regional security situation?
- **Japan-Korea Rivalry:** Although economically interdependent to a considerable extent, South Korea (Republic of Korea or ROK) and Japan have had an uneasy relationship. The former has not forgotten the past history of Japanese occupation. ROK has made significant economic progress and enjoys considerable military and economic power. It also has an alliance relationship with the US. The Japan-South Korea relations will be affected by the dynamics of the China-North Korea relationship as also by the issues between the two Koreas. Japan will be wary of a unified, nuclear Korea. North Korea has deliberately provoked Japan several times by firing missiles over the latter's air space. The future direction of the Japan-ROK relationship remains uncertain.
- **ASEAN as a "Driving Force" of Regional Integration:** The ten ASEAN states in South East Asia have sought to develop their relations with China while remaining wary about its military might and its claims over Paracel and Spratley islands in the South China Sea. Highly vulnerable to the surrounding geo-political eco-system with a history of instability, the ASEAN has constructed a web of relationships enmeshing China, India, US, Russia, Japan, South Korea, Australia and New Zealand in dense institutional arrangements. The latest such attempt is the ADMM (the ASEAN Defence Ministers Meeting), that consists of the defence ministers of ASEAN, China, US, Russia, India, ROK, Australia and New Zealand, for discussing common

security issues. They hope that the web of political and economic institutional relationships between them and the major powers will safeguard their security and stability. It is difficult to predict if such a web will be able to survive the rise of China.

- **The Af-Pak Region:** The Af-Pak region has emerged as a source of drugs, terrorism and instability that affects not only Asia but the entire world. The NATO and the US are desperately looking for an exit from Afghanistan even though the country is far from being stable. In Pakistan, which is already facing serious economic crisis, and is surviving on IMF loans and US munificence, Islamic radicalisation is taking root. Although a civilian government is in power, the army remains in the driving seat. The country is routinely rocked by sectarian clashes between the Shias and Sunnis. There is considerable concern about the growing dysfunction in the nuclear armed Pakistani state. The region is likely to remain unstable in the foreseeable future. The future of Afghanistan after Western forces leave appears to be that of Talibanisation and instability. Islamist radical forces could engulf the region and may spread outwards to the rest of the world.
- **Iran:** Iran's nuclear programme remains a top priority international security issue with the UNSC having passed several resolutions. Iran-Arab rivalry, Iran-Israel rivalry via the Hezbollah are likely to endure. Turbulence in the region will have damaging effect on the oil supplies which could impact the global oil market and the world economy.
- **Winds of Change in the Arab World:** The long standing Arab-Israel conflict which has been the cause of much international instability shows no sign of abating. The Obama administration has not been able to halt the construction of new Jewish settlements in occupied territory. The negotiations between the Israelis and the Palestinians have broken down. The prospects of a settlement are bleak and the prospects for conflict are high. In the mean time, following popular uprisings in Tunisia and Egypt, the Arab world is showing signs of ferment which could result in far reaching political, social and economic changes situation resulting in a restructuring of power equations that could affect the stability in the region.
- **Central Asia:** Central Asia is witnessing the "great game" being played out in a new context—that of competition for energy and mineral resources. China and Russia are deeply engaged in the region while the US, Iran and other countries have self

proclaimed interests there. The SCO seems to be gaining importance as a regional organisation but it may actually be masking the long term rivalry between Russia and China. The CAR countries are trying to define their identities but the regimes remain vulnerable, as has been seen in Kyrgyzstan. The region is experiencing some acute environmental problems like water shortages and the drying up of the Aral Sea. Islamic fundamentalist forces are also strong. China feels vulnerable due to the presence of large number of Uighurs in the various countries of the region. Central Asia looms large in the security calculus of the major Asian states.

- **Russia-Japan:** Although Russia and Japan have been trying to mend fences, the Kurile island issue remains unresolved. The Russian president's recent visit to the disputed islands has created tensions between the two countries.
- **Non-traditional Security Issues:** A variety of non-traditional security issues have emerged as major source of concern for Asian countries. These include terrorism, the safety of the sea lanes of communication, energy security, food security, illegal migration, financial instability, etc. How will the region deal with these issues?

From a realist point of view, stability in Asia is crucially dependent upon Sino-US relations. Will there be a concert of power between US and China in the shape of G-2 as suggested by Brzezinski or will there be a confrontation and the beginning of a new Cold War? The issue is far from settled with a range of differing views—both optimistic and pessimistic—being expressed. Hu Jintao's visit to the US in January 2010 attracted a lot of comment.[10] Taiwan continues to cast its shadow over their bilateral relationship. China is also conscious that the weakening of US-Japan security alliance may lead to increasing militarisation of Japan, to China's detriment. Thus, China's attitude towards US presence in Asia is nuanced. While it would like to see the US influence weaken, it is also conscious that the US remains the guarantor of regional stability to some extent.

The present trends indicate that while the economic integration is growing, Asia still lacks robust institutional structures to manage the fragile security situation. Due to the diversity of historical experiences, it is difficult to foresee the emergence of an Asian security community as in the case of Europe. The growing multilateralism in Asia is a positive sign but it remains a fragile experiment. Asian integration is vulnerable to a variety of plausible negative security shocks. A number of intractable disputes and issues continue to fester in different parts of Asia.

II. India's Growing Engagement with Asia

The end of the Cold War posed many serious challenges for India. With the disappearance of the Soviet Union, India's security environment changed drastically and to its disadvantage. The withdrawal of the Soviet forces led to the eventual rise of Pakistan backed Taliban regime in Afghanistan, which gave shelter to the Al Qaeda. The changed security environment led India to exercise its nuclear option. In 1998, it conducted nuclear tests and became a de facto nuclear weapon state. India's nuclear tests were followed by Pakistan's. Thus, the nuclear environment in Asia changed dramatically. India's very well being was jeopardised by the major economic changes in the world. However, India was able to convert these challenges into opportunities through a combination of policy measures which included inclusive economic growth, diversification of foreign policy, focussed diplomacy and participation in Asian regional institutions and building its military strength.

India's policy of nonalignment, pursued during the Cold War years with an intention of preserving independence in foreign and security policy, lost its erstwhile rationale as the Berlin Wall fell in 1989. The end of the Soviet Union in 1991 left India desperately looking for an entirely new foreign policy which was based on the hard geo-political realities of the rapidly changing world characterised by unipolarity.

India was in a deep economic crisis in 1991 as it came close to defaulting on its debt obligations. The economy was in shambles and there was intense political uncertainty at home. A former prime minister had been assassinated by the LTTE in 1991. India was also facing a proxy war in the shape of Pakistan sponsored groups infiltrating into Kashmir.

It was in this atmosphere of pessimism that India took the first tentative steps to revamp its economic and foreign polices. In 1991, India unveiled a slew of measures for economic liberalisation and opened up its economy to the world. The first steps were halting but telling. India decided to mainstream itself with the global economy. The economic reforms were not undertaken as a part of a grand strategy. They were more of a tactical adjustment in the face of a looming economic crisis. In hindsight, the 1991 economic reforms turned out to be arguably the most important decision by a government since India's independence in 1947.

The impact of economic liberalisation has been beyond anybody's expectations. India is now growing at 8-9 per cent annually, having emerged from a low economic growth trap. Economic growth picked up and the world began to see India as an investment destination. But the impact on foreign policy was even more remarkable. India began to pursue a bold economic agenda in its foreign policy. Sustained economic growth provided many foreign policy choices to the government as also the needed resources for revamping the defence forces. Economic diplomacy, aimed at wooing foreign investments and increasing trade, became a hallmark of Indian diplomacy in these years. This immediately led to improvement of relations with the Western countries from where most of the initial investment came. Western countries, describing India as an emerging economy, began to sense big economic opportunities in India. India joined the WTO, the ultimate symbol of free trade. India's image began to improve gradually. According to current economic forecasts India is seen as the third largest economy of the world by 2050.

Following the Indo-US civil nuclear cooperation agreement, India is now set to integrate with the global high technology trade. India has emerged as one of the leading suppliers of IT services in the world. India's pharmaceutical and biotech industries are also growing rapidly. Indian companies have a major presence in all parts of the world.

Asia is acquiring a higher profile in India's foreign trade. In 2009-10, Asian countries (East Asia, ASEAN, West Asia North America, North East Asia, South Asia, CARs) constituted nearly 55 per cent of India's global export and 63 per cent of India's global imports.[11] The rise of Asia's share in India's imports and exports is shown in Charts 1-4. Due to the heavy dependence on the import of crude oil, the West Asian region has the highest share in India's total trade with Asia. If the ASEAN, East Asian and North East Asian shares are combined (the regions covered by the 'look east' policy), the combined share of these regions is more than that of WANA (Charts 3 and 4). Charts 5 and 6 depict how the percentage share of Asia and North America in India's global export and import has changed over the last 50 years. The Asian share has been constantly rising indicating India's growing engagement with Asia.

Defence Cooperation: Another addition to India's foreign policy has been the intensification of defence cooperation with Asian countries. Maritime security issues—protection of the Sea Lanes of Communication, anti-piracy measures, terrorism, protection of exclusive economic zone—have assumed greater salience in India's security and foreign polices. As a result, in the last ten years India has stepped up defence exchanges with countries across Asia. This defence cooperation takes the form of defence agreements and defence MoUs, exchange of high level visits, joint exercises, training, defence production, cooperation in disaster management, joint patrolling in sensitive areas of the ocean, anti-piracy operations etc. Table 1 summarises the defence cooperation between India and the various Asian countries over the last eight years.[12] India has conducted joint exercises in the Persian Gulf as well as South China Sea. Indian ships regularly visit ports in the Indian Ocean and Pacific Ocean. Security and defence dialogues have now become a regular feature of India's defence diplomacy.[13] (See Table 1 in Annexure I)

Counter-Terrorism Cooperation: India's security cooperation in the area of counter-terrorism has grown over the years. India now has Counter Terrorism Joint Working Groups (CT-JWG) with 25 countries and two regional organisations.[14] The CT JWG helps to forge cooperation among security agencies, and also facilitates exchange of information and capacity building.

At the regional and bilateral level, India has taken several initiatives to revamp its goring policy.

Normalisation with China: One of the ways in which India has sought to address the Asian challenge is by trying to normalise and improve its relations with China. The rise of China, which is India's neighbour, with a long unresolved border, can prove to be a headache for India. China's influence in India's proximate neighbourhood is growing rapidly. The foremost challenge before India is to adjust to China's rise. In the late eighties and the nineties, India took some important steps for improving its relations with China, which had remained frozen between 1962—when the two countries fought a border war—and 1988, when Prime Minister Rajiv Gandhi visited China. In 1993 and 1996 respectively, India signed two key agreements with China which stabilised the unresolved border and set up a mechanism—a boundary working group—to deal with the border issue. This reduced the tensions between the two countries and set the stage for further growth of bilateral relations in the subsequent decade. In 2005, India and China signed a major agreement on political parameters and principles to resolve the boundary question. Both countries have been discussing the boundary issue at a senior level for a number of years now. China emerged as India's second largest trading partner in 2009-10. The intensity of political contact has increased. President Hu Jin Tao visited India in 2010, his second visit in four years. There is cooperation between India and China as well as competition. They will have to manage their complex relationship in a judicious way. India will have to watch carefully how US-China relationship evolves as this would have an impact on the Asian security environment.

Focus on South Asia: It is axiomatic that for India to continue to grow, its immediate neighbourhood must remain stable and peaceful. India's efforts towards that objective have produced mixed results. Several attempts have been made in the last two decades to improve and normalise relations with Pakistan but these have not succeeded. Since 1989, India has continuously faced militancy in Kashmir, sponsored by Pakistan. India and Pakistan became overtly nuclear in 1998. This further complicated the security situation for India. Pakistan used nuclear deterrence to launch a military conflict in Kargil in 1999 even as efforts were being made by the Vajpayee and Nawaz Sharif governments to normalise relations. The India-Pakistan relationship improved following the start of the peace process since 2004, however due to chronic instability in Pakistan since 2007 and the 26/11 terrorist attacks in Mumbai in 2008, the talks were suspended. Despite Pakistan's unwillingness to bring the perpetrators of the attack to justice, there has been a diplomatic engagement to restart the talks. However, it seems unlikely that Pakistan will shun its policy of using terrorism as an instrument of its foreign policy vis-à-vis India. India-Pak relations over the last two decades have been marked by mistrust and rivalry. Terrorism, nuclear weapons, external presence, Islamic radicalisation and several other factors have complicated this relationship for which an easy resolution is not possible. India will have to continue to find a way of dealing with a nuclear, unstable and hostile Pakistan.

Pakistan aside, India took some major steps to improve its relations with its neighbouring countries after the end of the Cold War. India signed a water sharing agreement with Bangladesh in 1996 thus bringing an end to a longstanding bitter dispute between the two neighbours. While the relationship with Bangladesh during the BNP government was not so good, the situation has changed dramatically since Sheikh Hasina came into power in 2009. In Nepal, India has tried to play a constructive role in ending the Maoist insurgency and to bring about a political solution. But, Nepali political parties and factions have not been able to arrive at a satisfactory agreement on the post insurgency political system in Nepal. India's options in Nepal seem to be limited due to the internal political dynamics. Instability in Nepal remains a matter of concern for India. However, India has succeeded in improving its relations with Bhutan, Sri Lanka and Maldives. The end of LTTE terrorism in Sri Lanka has created a situation for a further growth in Indo-Sri Lankan political and economic ties.

On the whole, South Asia remains a priority area for Indian foreign policy. China's growing influence in South Asia is an additional headache for India. India hopes that regional cooperation would strengthen under the new conditions and ease the tensions in the region. India's rising economy provides an opportunity for the neighbouring countries to benefit from an increasingly prosperous India. SAARC has been used as a vehicle to improve regional cooperation but with mixed success. Unfortunately,

India's trade with South Asian countries remains limited constituting a mere 4.5 per cent of India's global trade.

The Look East Policy: As result of geo-political changes in Asia, India has diversified its relations with the ASEAN countries. The Look East policy it launched in 1992 has borne fruit.

During the Cold War years, India had maintained a distance from ASEAN because it viewed these countries as being part of the US camp. However, the end of the Cold War wrought a change in India's outlook. It became a sectoral dialogue partner of ASEAN in 1992, a full dialogue partner in 1995, a summit partner in 2003 and a member of the East Asia Summit in 2005. It signed an FTA with ASEAN in 2010. It participated in BIMSTEC and Mekong Development Project. It has a series of defence cooperation agreements in the region. The integration of India's economy with ASEAN is growing apace. Following the adoption of the 'Look East Policy', India's engagement with ASEAN has grown rapidly. In 2009-10, exports to ASEAN constituted 10.2 per cent of India's global exports and imports from ASEAN constituted 9 per cent of India's global imports.

India is now set to expand its Look East Policy beyond the ASEAN. It is now has focussed on Japan, ROK. India and ROK have signed a Comprehensive Economic Partnership Agreement (CEPA) in 2009 India and Japan are discussing a similar agreement. With both countries, trade and investment relations are growing rapidly and so is the security dialogue. Indian navy is marking its presence in the Indian Ocean regularly. It also trains the personnel from Singapore. India has also focussed on Laos, Cambodia and Vietnam. Relations with Vietnam are taking on the shape of a strategic partnership.

China is watching India's foray into Asia with a wary eye. Yet, compared to China, India's engagement with Asia is still small.

India' Relations with West Asia: West Asia occupies an important role in India's foreign policy. India sources about 60 per cent of its oil imports from the region. Nearly 4.5 million Indians live and work in the Gulf counties and in West Asia. India's trade is dependent upon the security of the sea lanes of communications in the Gulf. Piracy in the Yemen-Somali waters affects India very seriously. India, with the second largest Muslim population in the world, has historical, cultural, religious and trade relations with the Gulf region.

Since the end of the Cold War, India has diversified its relations with West Asia and North Africa (WANA). WANA contributes 11 per cent to India's global exports and 28 per cent of India's global imports. (2009-10 figures, Department of Commerce)[15]. India has signed general strategic partnerships with the GCC. India has also built up some security relationships in the region with Indian naval ships being regular visitors to Gulf ports. Defence cooperation agreements have been signed with some countries. India has also sourced investment from the Gulf. India has traditionally maintained a close relationship with Iran but in the recent years the relationship has cooled off a bit due to Iran's unhappiness over India's vote at the IAEA against Iran on the nuclear question.

The success of the Look East Policy has compelled many people to urge the government to adopt a comprehensive Look West Policy to forge even closer ties with the Gulf countries. The attitude of the Islamic countries on the Kashmir issue has been a source of concern for India in its relations with the Gulf countries but that factor is losing its salience due to the prominence of economic factors. India's security depends on stability in the Gulf. India is vulnerable to Islamist radical trends emanating from the region. India has still to craft an overarching Look West Policy.

India and Central Asia: Although India took the lead in the nineties to recognise the newly independent central Asian countries and established diplomatic relations, the momentum was lost along the way due to many reasons. The instability in Afghanistan and Pakistan's intransigence ensured that India's relationship with the Central Asian could not develop at the desired pace despite the tremendous goodwill India has in the region. However, the situation may be changing. India has a small military presence in Tajikistan. Recently, India has joined the Turkmenistan-Afghanistan-Pakistan-India (TAPI) pipeline consortium which will help it to link with the hydrocarbon resources of Turkmenistan via Afghanistan and Pakistan. India has also signed some cooperation agreements with Kazakhstan on hydrocarbon exploration as well uranium ore supplies. India is likely to be admitted to the Shanghai Cooperation Organisation (SCO) when the regional grouping opens its gates for new members. Presently, India's relationship with the CARs remains at a low key due to a number of geopolitical reasons. The share of CARs in India's global exports and imports in 2009-10 was mere 0.15 per cent and 0.07 per cent respectively. This situation has not changed for the last one decade or so.

III. Conclusion

Two key challenges for Indian policymakers in the future will be to minimise the negative impact of instability in some parts of South Asia and managing its relationship with a growing and assertive China. India will have to build a robust framework of regional cooperation and devise mechanisms to minimise the instability in the South Asian region.

There is ample evidence that India's political and economic engagement with Asia is growing. But the question is whether it is sufficient to meet the challenges. Despite the growing engagement, India still lacks the capacity to engage with the region more deeply.

Since the end of the Cold War, India's foreign policy has undergone visible changes. But it appears that India's responses are based on expediency rather than on a well thought out long-term strategy. India has not yet come up with any public document on India's grand strategy, foreign policy orientation or a national security strategy.

India has sought to use a combination of economic and political measures to engage with Asia. India's economic rise and the strengthening of its hard power has improved India's statuie and created opportunities for greater engagement. This is reflected in the growing share of Asia in India's trade. However, given the rapidity of change, India finds itself constrained by a number of developments to deepen its engagement. Finally, India needs to articulate for public and the world at large its own views about the changes in the Asia and how it is going to manage its relationship with the rapidly changing region.

Chart 1

India's Imports from Asia

Chart 2

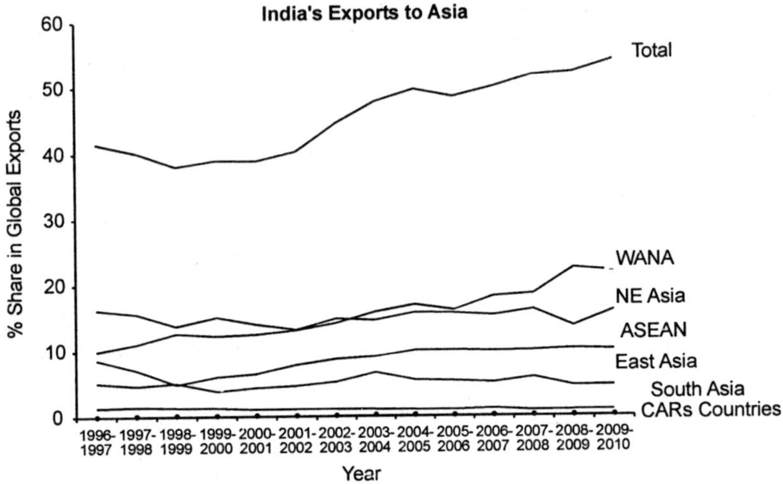

Source: Ministry of Commerce, Govt. of India (http://commerce.nic.in/eidb/ergn.asp) accessed on January 18, 2011.

Chart 3

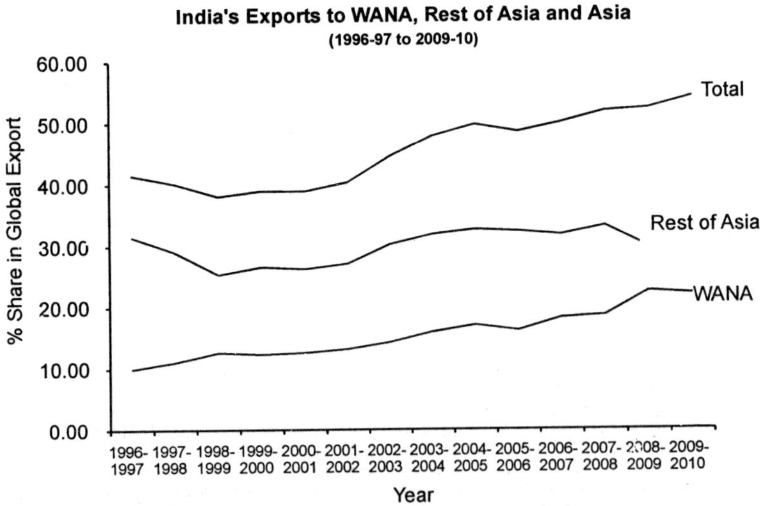

Chart 4

India's Imports from WANA, Rest of Asia and Asia

(1996-97 to 2009-10)

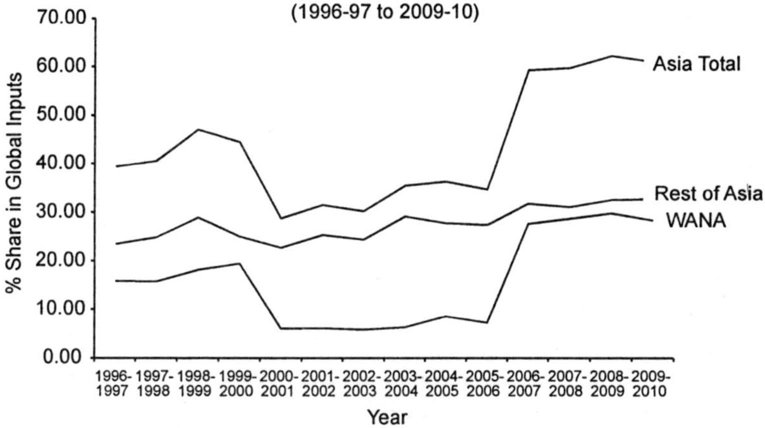

Source: Ministry of Commerce, Govt. of India (http://commerce.nic.in/eidb/ergn.asp) accessed on January 18, 2011.

Chart 5

Share in Global Exports

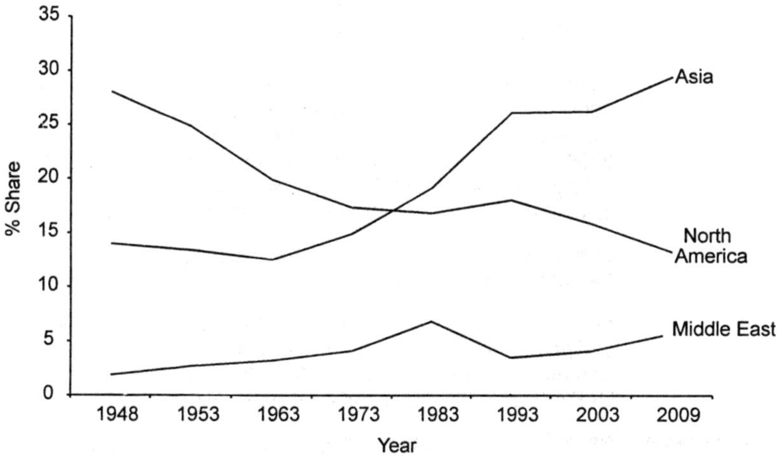

Source: World Trade Organisation, International Trade Statistics 2010, Table 1.6, Page 11, accessed on January 21, 2011.

Chart 6

Share in Global Imports

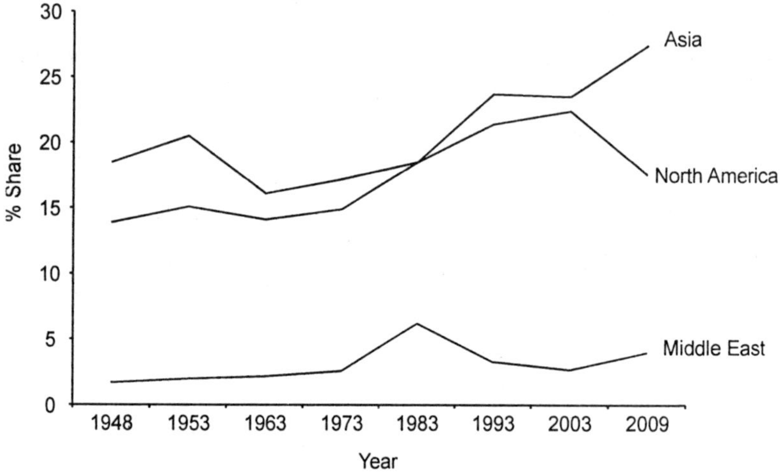

Source: World Trade Organisation, International Trade Statistics 2010, Table 1.7, Page 12, accessed on January 21, 2011.

Annexure-I

Table 1: India's Defence Cooperation

Oman	Joint Military Cooperation Committee (JMCC) meeting, (March 2009); Joint Air Exercise 'Eastern Bridge' (October 2009); India's Defence Secretary, Defence Minister visits (March 2009, March 2006); Oman's Minister of Defence visited India (February 2009); India purchased 03 Harbour tug boats from Oman (October 2009); Joint Maritime Exercise (March 2007, 2004).
Qatar	Defence cooperation agreement (November 2008); Indian Defence Secretary visit (June 2007).
UAE	First Training Squadron ships visit to the Persian Gulf (March 20 to April 10, 2009); Meeting of the Joint Defence Cooperation Committee (JDCC) (August 12-13, 2009, June 2008, April 2006); UAE's Chief of Air Force and Air Defence visits (2004).
Saudi Arabia	Naval Passage exercise (2009).
Egypt	Defence cooperation meeting (2009, June 2008, August 2006).
Iran	First Training Squadron visited Iran (2009); Passage exercises (2009); Commander of the Iranian Naval Forces visit (May 2004).
Israel	Naval exercises (2009, June 2006); Head of DSDE and Defence Export and Cooperation Department, Naval Chief visited India

	(July 2008, August 2007); Indian Defence Secretary visited Israel (November 2008); Joint Development Programme (July 2008); Joint Working Group meeting (November 2008, January 2007, December 2004).
Singapore	Joint Working Group meeting (April 2009); Defence MoU (November 2008); Defence Policy Dialogues (DPD) (May 2009, October 7-8, 2008); Training Squadron visited Singapore (January 2008, October-November 2006); DPSD-WG meeting (June 2007); Singapore, Minister of Defence visit (October 2007); Defence Minister visited Singapore (June 2007, June 2006); Joint navy exercise (March 2009, March 2008, 2007, February 2006); Air Exercise (January, June 2006, 2004); Army exercise (October 2006).
Indonesia	Naval Training (June 2009); Navy Joint exercise (March 2009, January 2008); JDCC meeting (June 2007); Chief of Indonesian Army visit (February 2007); Indian Defence Secretary visit (June 2007).
Malaysia	Malaysia-India Defence Cooperation Committee (MIDCOM) meeting from (January 2009); Defence Minister visited Malaysia (January 2008); Defence Minister, Deputy Prime Minister visited India (August 2007, June 2006); Air Force Training (December 2008, 2007); Joint defence cooperation meeting (December 2007, September 2006).
Philippines	Naval passage exercises (2009).
Thailand	Defence MoU (2007); Joint Working Group (JWG) meeting (August 2007); Joint exercises (2008, 2006).
Myanmar	Joint exercise (2008); Chief of naval staff visit (May 2007); Defence Cooperation meeting (September 2006).
Sri Lanka	Medical Team visit to Sri Lanka (March 2009); Sri Lankan defence secretary visit (May, September 2007); Naval Training (2008, October 2007).
Maldives	Naval exercise (December 2009); Defence Minister visit to India (February 2009, January 2007); Defence Cooperation meeting (February 2009, September 2008); Joint exercise (2006).
Vietnam	MoU signed (November 2009); Security Dialogues (October 2008, October 2006); Vice minister of defence visited India (February 2006); Defence Secretary visited Vietnam (October 2006).
Japan	Naval exercises (2009, April 2007, November 2006); Military to Military talk (February 2009); Indian Defence Minister, Prime Minister, chief of naval staff visits (November 2009, October 2008, August 2008); Japanese foreign minister visit (August 2008); Defence and Security agreement (May 2006); Defence dialogue (April 2007, May 2004).

Australia	Naval exercise (2009); Australian Defence Minister, chief of naval staff visits (April 2008, July 2007); Defence MoU (March 2006); Security dialogue (May 2009, February 2008).
Republic of Korea	The President, Defence Minister of Korea visits (January 2009, May 2007); Naval passage exercises (2009, December 2007); Defence MoU (March 2006).
Turkey	Naval passage exercises (2009, June 2006).
Kyrgyzstan	Defence MoUs (2008, 2006); Defence Minister's visit to India (November 2005).
Uzbekistan	Defence Cooperation agreement signed (April 2005).
Kazakhstan	Defence MoU (2006).
Russia	Naval exercises (January 2009, 2005); Russian Defence Minister, President Visited India (September 2008, December 2008, January 2007); Indian Prime Minister, Defence Minister visits (November, October 2007); India-Russia Inter Governmental Commission on Military Technical Cooperation (IRIGC-MTC) meeting (September 2009); MiG-29 contract signed (2008); Working Groups 'Military.Technical Cooperation' (MTC WG) meeting (August 2008); Joint Armies exercise (September 2007); Joint Air exercise (September 2007); Defence agreement (January 2007).
USA	Senior Technology Security Group (STSG) meeting May 2009; Defence Minister visits the US (September 2008); India purchased INS Jalashwa (L41) (June 2007); Defence MoU/Agreement (2007); Defence Procurement and Production Group (DPPG) and Military Cooperation Group (MCG) meeting (May, December 2007); Defence Policy Group meeting (January 2008, June 2004); Joint navy, army and air exercises (2006, 2004).
China	Defence Dialogues (January 2010, December 2008, November 2007); Joint Army training exercise (December 2007); Defence MoU (May 2006); Chiefs of Army, Naval Staff visited China (May 2007, April 2009); Chinese president, Defence Minister visited India (November 2006, March 2004); Indian delegation first time participated in PLA military exercise (September 2004).
Oman	Joint Military Cooperation Committee (JMCC) meeting, (March 2009); Joint Air Exercise 'Eastern Bridge' (October 2009); India's Defence Secretary, Defence Minister visits (March 2009, March 2006); Oman's Minister of Defence visited India (February 2009); India purchased 03 Harbour tug boats from Oman (October 2009); Joint Maritime Exercise (March 2007, 2004).
Qatar	Defence cooperation agreement (November 2008); Indian Defence Secretary visit (June 2007).
UAE	First Training Squadron ships visit to the Persian Gulf (March 20 to April 10, 2009); Meeting of the Joint Defence Cooperation

Committee (JDCC) (August 12-13, 2009, June 2008, April 2006); UAE's Chief of Air Force and Air Defence visits (2004).

Saudi Arabia Naval Passage exercise (2009).

Egypt Defence cooperation meeting (2009, June 2008, August 2006).

Iran First Training Squadron visited Iran (2009); Passage exercises (2009); Commander of the Iranian Naval Forces visit (May 2004).

Israel Naval exercises (2009, June 2006); Head of DSDE and Defence Export and Cooperation Department, Naval Chief visited India (July 2008, August 2007); Indian Defence Secretary visited Israel (November 2008); Joint Development Programme (July 2008); Joint Working Group meeting (November 2008, January 2007, December 2004).

Singapore Joint Working Group meeting (April 2009); Defence MoU (November 2008); Defence Policy Dialogues (DPD) (May 2009, October 7-8, 2008); Training Squadron visited Singapore (January 2008, October-November 2006); DPSD-WG meeting (June 2007); Singapore, Minister of Defence visit (October 2007); Defence Minister visited Singapore (June 2007, June 2006); Joint navy exercise (March 2009, March 2008, 2007, February 2006); Air Exercise (January, June 2006, 2004); Army exercise (October 2006).

Indonesia Naval Training (June 2009); Navy Joint exercise (March 2009, January 2008); JDCC meeting (June 2007); Chief of Indonesian Army visit (February 2007); Indian Defence Secretary visit (June 2007).

Malaysia Malaysia-India Defence Cooperation Committee (MIDCOM) meeting from (January 2009); Defence Minister visited Malaysia (January 2008); Defence Minister, Deputy Prime Minister visited India (August 2007, June 2006); Air Force Training (December 2008, 2007); Joint defence cooperation meeting (December 2007, September 2006).

Philippines Naval passage exercises (2009).

Thailand Defence MoU (2007); Joint Working Group (JWG) meeting (August 2007); Joint exercises (2008, 2006).

Myanmar Joint exercise (2008); Chief of naval staff visit (May 2007); Defence Cooperation meeting (September 2006).

Sri Lanka Medical Team visit to Sri Lanka (March 2009); Sri Lankan defence secretary visit (May, September 2007); Naval Training (2008, October 2007).

Maldives Naval exercise (December 2009); Defence Minister visit to India (February 2009, January 2007); Defence Cooperation meeting (February 2009, September 2008); Joint exercise (2006).

Vietnam MoU signed (November 2009); Security Dialogues (October 2008, October 2006); Vice minister of defence visited India (February 2006); Defence Secretary visited Vietnam (October 2006).

Japan	Naval exercises (2009, April 2007, November 2006); Military to Military talk (February 2009); Indian Defence Minister, Prime Minister, chief of naval staff visits (November 2009, October 2008, August 2008); Japanese foreign minister visit (August 2008); Defence and Security agreement (May 2006); Defence dialogue (April 2007, May 2004).
Australia	Naval exercise (2009); Australian Defence Minister, chief of naval staff visits (April 2008, July 2007); Defence MoU (March 2006); Security dialogue (May 2009, February 2008).
Republic of Korea	The President, Defence Minister of Korea visits (January 2009, May 2007); Naval passage exercises (2009, December 2007); Defence MoU (March 2006).
Turkey	Naval passage exercises (2009, June 2006).
Kyrgyzstan	Defence MoUs (2008, 2006); Defence Minister's visit to India (November 2005).
Uzbekistan	Defence Cooperation agreement signed (April 2005).
Kazakhstan	Defence MoU (2006).
Russia	Naval exercises (January 2009, 2005); Russian Defence Minister, President Visited India (September 2008, December 2008, January 2007); Indian Prime Minister, Defence Minister visits (November, October 2007); India-Russia Inter Governmental Commission on Military Technical Cooperation (IRIGC-MTC) meeting (September 2009); MiG-29 contract signed (2008); Working Groups 'Military.Technical Cooperation' (MTC WG) meeting (August 2008); Joint Armies exercise (September 2007); Joint Air exercise (September 2007); Defence agreement (January 2007).
USA	Senior Technology Security Group (STSG) meeting May 2009; Defence Minister visits the US (September 2008); India purchased INS Jalashwa (L41) (June 2007); Defence MoU/Agreement (2007); Defence Procurement and Production Group (DPPG) and Military Cooperation Group (MCG) meeting (May, December 2007); Defence Policy Group meeting (January 2008, June 2004); Joint navy, army and air exercises (2006, 2004).
China	Defence Dialogues (January 2010, December 2008, November 2007); Joint Army training exercise (December 2007); Defence MoU (May 2006); Chiefs of Army, Naval Staff visited China (May 2007, April 2009); Chinese president, Defence Minister visited India (November 2006, March 2004); Indian delegation first time participated in PLA military exercise (September 2004).

Sources: Ministry of Defence, Government of India, Annual Report 2002-2010, available on www.mod.nic.in (accessed on January 2011).

Notes

1. See Kishore Mahbubani, *The New Asian Hemisphere: the Irresistible Shift of Global Power to the East* Public Affairs, New York, 2008 for a sceptical view of Asia rise see Josuha Kurlantzich, "The Asian Country? Not Quite Yet", in *Current History,* Jan 2011, pp. 26-31.
2. David Shambaugh and Michael Yahuda, *International Relations of Asia*, Pentagon Press, New York, 2010, p. 6.
3. *WTO International Trade Statistics,* 2010, p. 8.
4. The rise of Asia has been vigorously debated in the academic literature. See, for instance, David Shambaugh and Michael Yahuda, *International Relations of Asia. op. cit.*, Also relevant are *David C. Kang,* "Hierarchy in Asian International Relations, 1300-1900," *Asian Security 1,* no. 1 (January 2005): 53-79; Aaron L. Friedberg,"Will Europe's Past Be Asian's Future?" *Survival* 42, no. 3 (Autumn 2000):147-159 and Amitav Acharya, "Theoretical Perspectives on International Relations in Asia" *in International Relations of Asia* by David Shambaugh and Michael Yahuda, *op. cit.* p. 57.
5. David Shambaugh and Michael Yahuda (ed.), *International Relations of Asia, op. cit.* Pentagon Press, New York, 2010, p. 6.
6. *Ibid......*
7. Talk on the "Second Global Shift and China's Rise in the 21st Century", by Prof. Victor Sit, Hongkong Baptist University, at the Institute for Defence Studies and Analyses, New Delhi on 29th December, 2010.
8. See table 6.1 and table 6.2 in Philips C. Saunders, "China's Role in Asia", in David David Shambaugh and Michael Yahuda (ed.), *International Relations of Asia,* Pentagon Press, New York, 2010, pp. 134-135.
9. "Military Must be Self-reliant: Minister" *People's Daily*, Dec. 29, 2010 http://english.peopledaily.com.cn/90001/90776/90786/7245014.html,
10. See Henry Kissinger, "Avoiding a China-US Cold War", *Washington Post*, Jan 13,.2011 available at http://www.washingtonpost.com/wp-dyn/content/article/2011/01/13/AR2011011304832_pf.html; Z. Brzezinski, "How to Stay friends with China", *New York Times*, Jan 03,.2011, http://www.nytimes.com/2011/01/03/opinion/03brzezinski.html; Michael J. Green, "Hu and the Communiqués", *The National Interest*, 18.01.2011, http://nationalinterest.org/commentary/hu-the-communiqu%C3%A9s-4734. Also see "Sino-US Joint Statement", January 19, 2011, issued at the conclusion of Hu Jintao's visit to the US.
11. "Export Import Data Bank", Department of Commerce, Govt. of India (website http://commerce.nic.in). In the terminology of the Indian Ministry of Commerce, East Asia constitutes 11 countries, namely, Australia, New Zealand and nine islands in the Pacific Ocean. ASEAN includes 10 countries, who are members of the regional grouping. The

WANA region includes six GCC countries. The North East Asia includes China, Korea RP, Japan, Hong Kong, Taiwan, Korea DP RP, Mongolia and Macao. South Asia includes Nepal, Pakistan IR, Sri Lanka DSR, Bangladesh PR, Bhutan, Afghanistan TIS and Maldives. CARs countries include Kazakhstan, Uzbekistan, Tajikistan, Turkmenistan and Kyrgyzstan.

12. The table has been compiled from the annual reports of the Indian Ministry of Defence.
13. India's has signed defence cooperation agreements/MoUs with Oman, Qatar, UAE, Saudi Arabia, Egypt, Israel, Singapore, Indonesia, Malaysia, Thailand, Myanmar, Maldives, Vietnam, Japan, Australia, Republic Of Korea, Kyrgyzstan, Uzbekistan, Kazakhstan, Russia, China and USA. It has conducted military, navy and air exercises with Oman, UAE, Saudi Arabia, Iran, Israel, Singapore, Indonesia, Malaysia, Thailand, Myanmar, Maldives, Japan, Australia, Republic Of Korea, Turkey, Russia, China and USA.
14. Indian Foreign Minister's reply to the Indian parliament. See http://indiacurrentaffairs.org/india-signs-agreements-to-counter-terrorism-with-25-countries/(accessed on 31.1.2011)
15. Department of Commerce (website http://commerce.nic.in/). (Accesed on 19.1.2011)

An Indian View on a Changing Asia*

Shivshankar Menon

Thank you for asking me to speak at the closing session of IDSA's 13th Asian Security Conference, "Towards a New Asian Order". You have a very impressive list of participants and speakers, and there would seem to be little that I can add to what has come before.

I must confess to some confusion when I first heard the title of your conference, "Towards a New Asian Order". I was not sure whether you meant a new world order that was Asian, or whether you meant a new order in Asia; whether you meant new Asian rules of the game and stakes in the international order, or a new ordering of the way things are done and managed in Asia. It took only a second's thought to make it clear that the answer was the second choice, especially as your conference is about Asian security. But the fact that the question even arose in my mind suggests how rapidly and how extensively world geopolitics has changed, and how the centre of geopolitical gravity has moved to Asia.

At the risk of repeating what may have already been said, I thought I would give you one Indian's view on how Asia is changing, on the new security challenges that result from these changes, and on the sort of Asian order that might meet these new challenges.

Changing Asia

There is no question that the rise of Asia, with China as the most prominent example, is the major geopolitical fact of our times. What Asia has achieved in the last three decades in terms of the rapid accretion of wealth to begin with, and now power, has never been done so fast, in history, by

---

* Based on the closing remarks delivered at the 13th Asian Security Conference, Institute for Defence Studies and Analyses, New Delhi, February 18, 2011.

any other region. Nor has any other power in history grown as rapidly as China in this period. The relative prominence and importance of Asia and China have been accelerated by the world economic crisis since 2008, and a crisis of confidence in the Western developed world. We Asians are reminded nearly every day that Asia is now pivotal to global politics, and urged to fulfil our responsibilities, (which I sometimes suspect is merely a polite way of asking us to do what others find more convenient than our following our own interests.)

But, stupendous as the Asian achievement is, we must remember that it has several unique characteristics.

- Asia has seen several powers rising simultaneously, and has yet to find a new internal equilibrium of its own. Unlike what occurred when Japan rose in the thirties, China is not the only rising Asian power today. Within Asia itself, on China's immediate periphery, other powers like South Korea, Vietnam, Indonesia and India are also developing rapidly. We already see in Asia classical responses to the rapid rise of new powers in terms of internal and external balancing. We are see increasing defence budgets throughout the region over the last two decades.*
- These growing Asian powers increasingly owe more and more of their economic growth to their links and interdependence. Several of them still have overwhelming domestic preoccupations and will therefore concentrate*
- on those rather than on external entanglements. It would be natural to expect that these powers would concentrate on creating external environments conducive to their continued economic growth and social transformation rather than focussing purely on the security aspects of their external environment. For quite some time to come Asia will see the development of powerful states whose citizenry as individuals will enjoy lower standards of living than those in the developed world. These states will therefore not replicate the behaviour of previous powers. The new equilibrium in Asia is likely to be as much a result of production chains and regional and global market integration as of purely security driven alliances or structures.
- Asia's geopolitics are complicated by the presence of several global and extra-regional powers who are now integral to Asian security in this age of the globalisation of economics, security and technology. Powers such as the USA, Russia and Japan are present and have long established interests of their own in Asia.

An Asian order which ignores their interests is unlikely to be stable.

- And lastly, Asia as a whole is and will remain dependent upon the rest of the world for its own continued growth and security, whether in terms of energy security, food security or, in the more conventional calculation of the sources and providers of security capabilities and technologies. In several respects that matter, (concepts, technology, security, energy etc), Asia is still and will remain a net global consumer for some time to come. This will remain true of the major Asian developing economies for the foreseeable future with younger populations, and the task of maintaining the growth momentum necessary for generating jobs for new entrants to the labour force,. (Historically speaking, this is very different from the situation before the mid-eighteenth century when Asia last enjoyed relative global pre-eminence. Then Asia depended on itself for its needs and markets. Again, the contrast with the Soviet Union's autonomy from others could not be greater. So expect different behaviour from Asian powers.)

In other words we are speaking of an Asia that will be largely developing, increasingly integrated within itself, and simultaneously dependent upon the broader global community and environment,—a powerful but poor Asia.

None of this is to say that the relative rise of Asia will not continue. I would expect it to continue over the medium term, though straight line extrapolations have seldom worked in history. This is just to say that the issues that we face in dealing with the changes in Asia are different from those that we see in historical analogies of rising powers in the past. (This is not 19th century Europe, amenable to balance of power solutions such as a "Concert of Asia".) For the present, Asia and the world have yet to work out a security architecture that accommodates the changes in Asia and the legitimate interests of all the countries concerned without reducing them to a zero sum game. Our preference is that the new architecture be open, inclusive, and flexible. But this still is a work in progress.

The Issues

What are the new challenges that this situation in Asia throws up for us? To my mind three are three:

1. Connectivity domains like maritime security, cyber space and outer space;
2. The increasing security divide within Asia;
3. The institutionalisation of security cooperation in Asia.

The first probably explains itself. Asia's pattern of growth and the fact that its r physical and economic security will be linked to that of the rest of the world make her connectivity critical. Hence the need to assure security of maritime communications, cyber space and outer space. Many of you know much more than I do about these domains, but it is no accident that these are precisely the domains where Asia is witnessing latent and explicit competition and contention, a build up of national offensive and defensive capabilities which in other areas would be called an arms race.

You might well ask why I also mentioned the security divide and a lack of institutions. Why not choose traditional security issues like border disputes or terrorism or inter-state conflict or nuclear proliferation? Because these problems, though very real and worrying, which exist in virulent forms in Asia, have not prevented the stupendous transformation that has resulted in the accumulation of power and wealth in large parts of Asia that we spoke of. In other words, so far Asia has successfully managed individual issues of nuclear balance and several conventional security threats, and even the proliferation of WMD and missile technologies. These will require continued and considerable effort, but increasingly these problems will occur in new forms.

Take piracy, for instance; one of the oldest of all conventional security threats. Asia was successful in tackling it around the Strait of Malacca but the international community have been much less so off the Horn of Africa. Why is this so? To my mind this is largely because there are parts of Asia which have not been transformed or changed. While we were able to act on both land and sea near the Strait of Malacca, we are unable to do so in the Horn. In those parts of Asia which are not developing as rapidly, traditional security issues now affect Asian security differently. There is a growing security deficit or divide within Asia.

To take another instance, there is an increasing danger of terrorism spreading from those parts of Asia, like Pakistan and Afghanistan, which have not been part of the Asian economic miracle. The security of nuclear materials and weapons in those same parts of Asia is another example of the sort of problems that we face. These are the problems I mean when I speak of the increasing security divide within Asia. There are parts of Asia which are falling behind, not just in economic terms but in terms of the normal security attributes of sovereignty that we take for granted. And the consequences for Asian security should concern all of us in Asia.

The New Asian Order

So what sort of new Asian order will meet the needs of these new challenges? Should we proceed in security as we did in the economy, building block by building block? Or do we move to collective security in Asia, something that each of the two super powers tried and failed to achieve during the Cold War?

I do not claim to have answers to these questions, but I can try to describe some of the desiderata of a new security order in Asia that should be part of our consideration of this problem.

- It should be inclusive. Given the diversity in Asia this is essential if the order is to work. And it must include all relevant powers, including those geographically external but intrinsic to Asia's security in practice and presence.
- It should be extensive, from Suez to the Pacific and including the entire Eurasian landmass. If not, it will not be possible to address the security consequences of the growing security divide within Asia.
- It should be plural. No one-size solution or simplistic prescription will work. We should learn from the failure of Cold War alliance systems in the area, and from the earlier Asian collective security proposals.
- Its institutions should be consultative, respecting the Asian cultural bias towards consensual solutions. Centring the institutions on ASEAN would be logical and practical, and the ASEAN Defence Ministers' plus Eight Meeting offers a potentially very useful way forward.

While these desiderata may sound utopian, they are suggested in the hope that we will be able to apply them to the issues we face, thus building habits of cooperation in security as we have already done in the economic sphere.

And if we do succeed in the years ahead to build such a security order in Asia and successfully handle the multiple security challenges that we face, I suppose the world order will indeed be "Asian" in character, and the question I started with would have been answered.

Contributors

A.K. Antony (President, IDSA)

Mr. A.K. Antony is India's Minister of Defence. An advocate by profession, Mr. Antony started his political career as a student activist of the Kerala Students Union. He has held many important political assignments: President, Kerala Pradesh Youth Congress Committee, President and General Secretary of Kerala Pradesh Congress Committee, General Secretary of All India Congress Committee and Member, Congress Working Committee. He was first elected to the Kerala Legislative Assembly in 1970 and was the chief minister of Kerala three times during 1977-78, 1995-96 and 2001-2004. He was also Leader of the Opposition in Kerala Legislative Assembly during 1996-2001. Mr. Anthony was Member Rajya Sabha twice from 1985-1991 and 1991-1995. He was union minister for civil supplies, consumer affairs and public distribution during 1994-1995. He was elected to the Rajya Sabha in May 2005 from Kerala. Mr. Antony edited a Malayalam weekly *Kalasala* between 1964-65 and later a Malayalam daily *Veekshanam* from 1978-1982.

Prof. William R. Thompson

William R. Thompson is a professor at the Political Science Department at Indiana University. He is also the managing editor of the International Studies Quarterly, the official journal published by the International Studies Association. Mr. Thompson graduated with a bachelor's degree in Economics and Political Science from the University of Washington, where he also went on to finish his Master's and Ph.D. He has been on the advisory board of the IU International Studies Programme between 2003 to 2006. He is currently a visiting faculty at GLOPEM (Global Politics and Euro-Mediterranean Relations) at the University of Catania, Italy. He has earned numerous honours and has been accredited with many distinctions for his contribution to the field of Political Science. Right from the N.D.F.L. Title VI Fellowship in 1969 to the more recent award for Excellence in World Society Research where he won first place at the World Society Foundation in Zurich, Switzerland. He also won in 2008, the First Round Prize, Second Round Prize, and was a Third Round Prize finalist at the 'Inequality Beyond Globalisation' World Society Focus Paper Competition, World Society.

Dr. Zhenjiang Zhang

Dr. Zhenjiang Zhang is currently an associate professor in Jinan University, Guangzhou, China. He is the vice dean of the Department of International Relations and the deputy director of the Institute of Southeast Asian Studies in Jinan University. Professor Zhang obtained his Ph.D from Nanjing University, China. He finished his post-doctoral research at Sun Yat-sen University in Guangzhou, China. His research interests cover history and theory of International Relations, American diplomatic history, East Asian International Relations, and ASEAN studies. He has published two monographs, *The Cold War and the Civil War: The Soviet-US Power Rivalry* and the *Origin of the Chinese Civil War,* 1944-1946 (2004) (in Chinese) and *From Sterling to Dollar: a Study on the International Economic Power Transition, 1933-1945* (2006) (in Chinese). He has published papers in key journals of international studies in China, which include *Shijie jingji yu zhengzhi* (World Economy and Politics), *Waijiao Pinglun* (Foreign Affiars Review)), *Dongnanya yanjiu* (Southeast Asian Studies), *Dangdai Yatai* (Contemporary Asia-Pacific), *Shixue yuekan* (History Monthly) and so on. His English published papers are *in American Journal of International Politics and Development Studies, Journal of Ritsumeikan Social Sciences and Humanities* and so on.

Dr. Niklas L.P. Swanström

Niklas L.P. Swanström is the director at Institute for Security and Development Policy (www.isdp.eu). His main areas of expertise are conflict prevention, conflict management and regional cooperation; Chinese foreign policy and security in Northeast Asia. He is the editor of the *China and Eurasia Quarterly*. Selected books: *Sino-Japanese Relations: The Need for Conflict Prevention and Management*; *Conflict Prevention and Conflict Management in Northeast Asia*; *Transnationell brottslighet: ett säkerhetshot*? (Transnational crime: a security threat); *Regional Cooperation and Conflict Management: Lessons from the Pacific Rim and Foreign Devils, Dictatorship or Institutional Control: China's Foreign Policy Towards Southeast Asia*. Dr. Swanström holds a Ph.D. in Peace and Conflict Studies from Uppsala University.

Vincent Wei-cheng Wang

Vincent Weicheng Wang is presently professor and Chair of the Department of Political Science at the University of Richmond. His teaching and academic experience spans across Schools from all over the world for over about two decades including National Chengchi University (Taipei, Taiwan) where he was a visiting Professor during the spring of 2010 and the El Colégio de México (Mexico City, Mexico) where he was a visiting Professor back in the summer of1991, apart from being associated with a host of other academic institutions. His areas of specialization include International Relations, International Political Economy, Comparative Political Economy (East Asia and Latin America), Asian politics and international relations, Taiwan, China, science and technology policy, industrial policy; Third-World development strategies, economic reform, and democratization.

Prof. Robert J. Art

Dr. Robert J. Art is a professor of International Relations at Brandeis University, where he teaches international relations and specialises in national security affairs and American foreign policy. He is also a senior advisor for the Security Studies Programme at the Massachusetts Institute of Technology, and director of MIT's Seminar XXI Programme. Professor Art received his Ph.D. from Harvard University in 1967. Professor Art is a former member of the secretary of defence's Long Range Planning Staff (1982), a former dean of the Graduate School of Arts and Sciences at Brandeis University, and has consulted for the Central Intelligence Agency. He is also a member of the editorial boards of the scholarly journals *International Security*, *Security Studies*, and *Political Science Quarterly*. Since 1982, he has co-edited, along with Robert Jervis and Stephen Walt, Cornell University's "*Series in Security Studies.*" Professor Art has lectured at numerous American and foreign universities and research institutes, including the U.S. National War College, the National War College (Beijing), the Fuhrungsakademie (Hamburg), the Institute for War Studies (King's College), the NATO School at Oberammergau, and the School of International Studies at Peking University. He has written various books including *The TFX Decision: McNamara and the Military* (1968); *A Grand Strategy for America* (2003)—a finalist for the Arthur B. Ross Award of the Council on Foreign Relations; and *America's Grand Strategy and World Politics* (2009).

Masako Ikegami

Dr. Masako Ikegami is Professor of political science at the Department of Political Science, Stockholm University, formerly Director (2001-08) of the Center for Pacific Asia Studies (CPAS) at the university, and Adjunct Professor at the School of Policy Studies, Kwansei-Gakuin University, Japan. She holds Doctor of Sociology from the University of Tokyo (1996), and Ph.D. in peace and conflict research from Uppsala University (1998). Her research ranges from empirical analysis of defence R&D and production, defence policy-making process, arms control & disarmament, to East Asian regional security and confidence building measures. She has published two monographs, *Military Technology and US-Japan Security Relations* (1998), *The Military-Industrial Complex: The Cases of Sweden and Japan* (Dartmouth 1992); and books chapters such as 'Japan: a latent but large supplier of dual-use technology' in H. Wulf (ed.) *Arms Industry Limited*, (Oxford University Press 1993), 'Japan' in R. Singh (ed.) *Arms Procurement Decision Making, Vol. 1*, (Oxford University Press 1998), 'China's grand strategy of 'peaceful rise': a prelude to a new Cold War?' in Hsiao & Lin (eds) *The Rise of China* (Routledge 2009); and articles, e.g. 'NATO and Japan: strengthening Asian stability', *NATO Review* (summer 2007), 'Asia in Global Security — an epicentre of new instability' (SIPRI), and 'Time for conflict prevention across the Taiwan Strait', *China Brief*, (March 2008). She gave lectures on East Asian security issues at eminent institutes and conferences in the United States, Europe and Asia, and was POSCO Fellow (2005) at the East-West Center, Honolulu, and a member of the International Institute for Strategic Studies (IISS), London, and the Pugwash Conferences on arms control & disarmament and nuclear non-proliferation (awarded Nobel Peace Prize, 1995).

Dr. Mumin Chen

Dr. Mumin Chen, is currently associate professor at the Graduate Institute of Political Science, National Chung Hsing University, Taichung, Taiwan. Chen earned his Ph.D. and M.A. degrees in International Studies from the Graduate School of International Studies, University of Denver, USA (2004 and 1997, respectively); and a B.A. in Political Science from National Taiwan University, Taiwan (1992). From 2002 to 2004, he worked as special assistant to Vice President Annette Lu of Taiwan in charge of speech drafting and foreign policy analysis. His previous academic experiences include: adjunct foreign-language lecturer, School of International Studies, Peking University (2001-2002); assistant professor, Graduate Institute of Political Science, National Changhua University of Education (2004-2008); visiting research fellow at East Asian Institute, National University of Singapore (2008); visiting scholar at Department of East Asian Studies, University of Delhi (2009). Chen's research focuses include: international security theory, East and South Asian security issues, Chinese foreign policy, and China-Taiwan relations.

Dr. Alexander Lukin

Alexander Lukin is the director of the Centre for East Asian and Shanghai Cooperation Organisation Studies at the Moscow State Institute of International Relations of Russia's ministry of foreign affairs. Alexander Lukin received his first degree from Moscow State Institute of International Relations in 1984, a doctorate in politics from Oxford University in 1997 and a doctorate in history from the Russian Diplomatic Academy in 2007. He worked at the Soviet foreign ministry, the Soviet embassy in the PRC, and the Institute of Oriental Studies of the Soviet Academy of Sciences. He holds the position of advisor to the governor of the Moscow region on foreign economic relations and serves on the editorial board of the *International Problems* journal published from Belgrade. In 2009 he was awarded by President Hu Jintao a medal for the "Outstanding Contribution to the Development of Sino-Russian Relations." The medal was awarded to 60 Russians to mark the 60th anniversary of the establishment of the diplomatic relations between the PRC and the USSR. He is also a member of the Russian National Committee of The Council for Security Cooperation in the Asia Pacific (CSCAP).

Dr. Arshin Adib-Moghaddam

Arshin Adib-Moghaddam is the author of *'The International Politics of the Persian Gulf: A cultural genealogy'* (Routledge, 2006, 2009), '*Iran in World Politics: The question of the Islamic Republic*' (Columbia University Press, 2008, 2010) and '*A Metahistory of the Clash of Civilisations: Us and them beyond Orientalism*' (Columbia/Hurst, 2011). Educated at the universities of Hamburg, The American (Washington DC) and Cambridge, he was the first Jarvis Doctorow Fellow in International Relations and Peace Studies at St. Edmund Hall and the department of politics and international relations, University of Oxford. Since 2007, Adib-Moghaddam is university lecturer in Comparative—and International Politics at SOAS, university of London. His writings have been translated into many languages and he is a frequent contributor to leading newspapers and TV channels around the world.

Dr. Paul Salem

Paul Salem is the director of the Carnegie Middle East Centre. Prior to joining Carnegie in 2006, Salem was the general director of the Fares Foundation and from 1989 to 1999 he founded and directed the Lebanese Centre for Policy Studies, Lebanon's leading public policy think tank. Salem was a member of the Lebanese National Commission for Electoral Law Reform, a blue ribbon commission tasked with revising Lebanon's electoral laws and proposing a new system. In 2002, Salem was a member of the Senior Review Committee for the UNDP Arab Human Development Report. He has also held various positions at the American

University in Beirut. He regularly seen on television, and gives radio interviews, and writes newspaper articles on political issues relating to the Arab world. Selected Publications: *Bitter Legacy: Ideology and Politics in the Arab World* (Syracuse University Press, 1994); *Conflict Resolution in the Arab World: Selected Essays,* editor, (American University Press, Beirut, 1997); *Handbook for Arab Parliamentarians Against Corruption,* editor, (Beirut and Kuwait: Arab Parliamentarians Against Corruption and Transparency, Lebanon, 2006); *Administrative Decentralization in Lebanon: Issues and Applications,* co-editor with Antoine Messara (Lebanese Centre for Policy Studies, 1996); *The First Elections in Lebanon After the War,* co-editor with Farid el-Khazen (Dar al-Nahar and Lebanese Center for Policy Studies, 1993)

Dr. Joseph Chinyong Liow

Dr. Joseph Liow is currently associate professor and associate dean at the S. Rajaratnam School of International Studies (RSIS), Nanyang Technological University, Singapore. He received his Ph.D from the London School of Economics, U.K. Joseph's research interests are in Muslim politics and societies in South East Asia, with a primary focus on Malaysia and southern Thailand, and the international politics of the Asia-Pacific region. He teaches foreign policy and security issues in South East Asia and Malaysia: state, society, and politics graduate seminars at the RSIS M.Sc. programme, as well as in various training courses at the defence ministry, civil service college, home team academic, and civil defence academy in Singapore, and has consulted for MNCs such as BHP Billiton, Chevron, and Shell. He also comments regularly on regional affairs in various local and international media. Joseph has authored or edited 11 books, including *Islam, Education, and Reform in Southern Thailand*: *Tradition and Transformation* (Singapore: ISEAS, 2009), *Piety and Politics: The Shifting Contcurs of Islamism in Contemporary Malaysia* (New York: Oxford University Press, 2009), and *Confronting Ghosts: Thailand's Shapeless Southern Insurgency* (Sydney: Lowy Institute, 2010). He is currently working on a comparative book project on insurgencies across Southeast Asia.

Tai Ming Cheung

Tai Ming Cheung is an associate research scientist at the University of California Institute on Global Conflict and Cooperation (IGCC) located at the University of California San Diego in La Jolla. He directs the Minerva program on Chinese security and technology, a multi-year academic research and training project funded by the U.S. Defense Department to explore China's technological potential. His responsibilities include managing the institute's track two program the Northeast Asia Cooperation Dialogue, which brings together senior foreign ministry and defense officials as well as academics from the United States, China, Japan, South Korea, North Korea, and Russia for informed discussions on regional security issues. Dr. Cheung is also an associate adjunct professor at UCSD's Graduate School of International Relations and Pacific Studies (IRPS) where he teaches courses on Asian security, Chinese security and technology, and Chinese politics. Dr. Cheung is a long-time analyst of Chinese and East Asian defense and national security affairs, especially defense economic, industrial and science and technological issues.

His latest book, *Fortifying China: The Struggle to Build a Modern Defense Economy*, was published by Cornell University Press in 2009. The book examines the economic, commercial and technological foundations of China's long-term defense modernization that examines the development of the defense industrial complex, the role and prospects for civilian-military integration, and the military dimensions of science and technology policies.

Prof. Xia Liping

Professor Xia Liping is dean and professor at the School of Political Science and International Relations at Tongji University in Shanghai. He is General-Secretary of Shanghai Institute for International Strategic Studies (SIISS), and vice president of the Shanghai Center for RimPac Strategic and International Studies (CPSIS). He is also Senior Guest Fellow of Institute of International Technology and Economics at the Centre for Development Studies under the PRC State Council. He is vice president of Shanghai Association of International Relations. He specialises in Asian security and nuclear non-proliferation and China's foreign strategy. He holds a Ph.D. in world history from the East China Normal University. He got the Master Degree of Law from the PLA Foreign Language University in 1991. He has published many books and papers. Three of his latest books are *Contemporary International System and Strategic Relations among Major Powers; China's Peaceful rise" and "Security and Arms Control in the Asia-Pacific Region.* He was Senior Fellow at the Atlantic Council in the United States from 1994-1995, visiting scholar at Monterrey Institute of International Studies in 1999, the Hong Kong University in 2002, the Stockholm University in 2005 and visiting scholar at Friedrich-Ebert-Stiftung EU Office in Brussels in April-May, 2009.

Prof. Yaacov Vertzberger

Professor Yaacov Vertzberger received his Ph.D. summa-cum-laude from the Hebrew University of Jerusalem, and is currently professor of international relations at the Hebrew University. He has also held visiting teaching and research positions at the East-West Centre, Honolulu; UN University, Tokyo; The Netherlands Institute for Advanced Study; The Graduate School of International Relations and Pacific Studies, UC San Diego; The Swedish Institute of International Affairs, Stockholm; The Rockefeller Bellagio Study and Conference Centre; The Institute for International Studies, UC Berkeley; McGill University, Montreal; and Stanford University. His areas of interest and research include theory of international relations; political psychology; international political economy; and foreign policy and security in South Asia. He has published extensively in these areas. These publications include books and monographs, articles that have appeared in the major journals as well as book chapters. Professor Vertzberger's current research focuses on (1) learning and change in national strategic policies, and (2) the architecture and practice of global disaster management. He is also a member of the editorial boards of *Political Psychology* and *International Studies Quarterly.*

Prof. Syed Iqbal Hasnain

Dr. Syed Iqbal Hasnain is a distinguished visiting fellow at the Stimson Centre. He currently serves as chair of the Glacier and Climate Change Commission established by the state government of Sikkim (India). He is a member of several high-level committees, including: the United Nations Environment Programme (UNEP) Committee on Global Assessment of Black Carbon and Troposphere Ozone and the International Mid-Term Review Committee commissioned by international donors on (ICIMOD) International Centre for Mountain and Research Development, Kathmandu, Nepal. Between 2002 and 2006, Hasnain served as a vice-chancellor (President) of the University of Calicut, Kerala, India. Previously, he held the post of professor of Glaciology, Jawaharlal Nehru University (India), where he introduced glaciology as a subject at the post-graduate research level. He has published research papers and books in numerous peer-reviewed journals such as the *Journal of Glaciology* and the *Journal of Hydrology*. Through lectures, television interviews, and articles featuring him in popular international magazines like *Nature* and leading Indian newspapers, he has long advocated the impact of long-lived carbon dioxide and short-lived climate forcers like black carbon, methane, and ozone on the Himalayan-Tibetan glaciers. *Time* magazine prominently featured professor Hasnain's work with his photograph in the December 24, 2009 article "Climate Change: The Tragedy of the Himalayas". Professor Hasnain was awarded Padma Shri in 2009 by the President of India for his contributions to advancing the science of glaciology in India.

Dr. Uttam Kumar Sinha

Dr. Uttam Kumar Sinha is research fellow at IDSA, New Delhi. He earlier worked for the national daily *The Pioneer* and wrote the weekly column 'Strategic Eye'. At IDSA, he serves on the editorial board of *Strategic Analysis* and is assistant editor of the *Strategic Digest*. He is a visiting faculty member of the Malviya Centre for Peace Research, Banaras Hindu University and visiting fellow to the International Peace Research Institute, Oslo. In 2008, he was awarded the Chevening 'Gurukul' Scholarship for Leadership to the London School of Economics and Political Science. Presently, he is member and co-chair of the Near East and South Asia Regional Network on Water Security, National Defence University, Washington. His research areas include climate change and resource issues with particular attention on transboundary rivers. He has coordinated and co-authored two IDSA Reports: *India and Water Security: External Dynamics* (2010), and *Security Implications of Climate Change for India* (2009).

Prof. Sumit Ganguly

Professor Sumit Ganguly holds the Rabindranath Tagore Chair in Indian Cultures and Civilizations at Indiana University in Bloomington. He has previously been on the faculty of James Madison College of Michigan State University, Hunter College of the City University of New York and the University of Texas at Austin. He has also taught at Columbia University in New York City. His research and writing focused on South Asia has been supported by grants from the Asia Foundation, the Ford Foundation, the Carnegie Corporation of New York and the W. Alton Jones Foundation. He serves on the editorial boards of *Asian Affairs*, *Asian Survey*, *Current History*, the *Journal of Strategic Studies* and *Security Studies*. He is also the founding editor of both the *India Review* and *Asian Security*, two referred journals published by Taylor and Francis, London. His most recent books are '*Fearful Symmetry: India and Pakistan Under the Shadow of Nuclear Weapons*' (co-authored with

Devin Hagerty) jointly published by Oxford University Press (New Delhi) and the University of Washington Press (Seattle) and '*More Than Words: U.S.-India Strategic Cooperation Into the Twenty-First Century*' (co-edited with Brian Shoup and Andrew Scobell) published by Routledge, London. He is currently at work on a book, *India Since 1980*, under contract with Cambridge University Press, New York.

Lieutenant General Prakash Menon, PVSM, AVSM, VSM, PhD

Lieutenant General Prakash Menon, PVSM, AVSM, VSM, PhD has been commandant of the prestigious National Defence College, New Delhi. In 39 years of service, the general has earned the reputation of a "Soldier Scholar". He has been an outstanding commander in the field. His command at the battalion, brigade and divisional levels in the counter insurgency environment of Jammu and Kashmir earned him two military awards for distinguished service. As a company commander in 1989, he was awarded for his performance during active operations in Siachen. He is one of the rare generals to have been awarded a Ph.D for his thesis on "*Nuclear Deterrence and Limited War in the Indo Pak context*" by the Madras University. He has been nominated by the cabinet as member of the expert group for the establishment of the Indian National Defence University (INDU). He is currently Deputy National Security Adviser.

Dr. Arvind Gupta

Dr. Arvind Gupta is an officer of the Indian Foreign Service, (batch of 1979). He presently holds the Lal Bahadur Shastri Chair in Strategic and Defence Studies at the Institute for Defence Studies and Analyses. Prior to joining IDSA, Dr. Gupta was Joint Secretary at the National Security Council Secretariat from 1999 to 2008. During his tenure at the NSCS, he dealt with a wide spectrum of national security issues. Dr. Arvind Gupta has wide-ranging diplomatic experience gained while working in Indian missions abroad. He has handled a number of assignments in the ministry of external affairs in different capacities. His current interests include: the impact of globalisation on India's security; the security, foreign policy and diplomatic challenges before India; energy security, climate change, institutional reform, India's neighbours, internal security, and technology and security. He has several books, articles and papers to his credit.

Mr. Shivshankar Menon

Shivshankar Menon is National Security Advisor to the prime minister of India since January 23, 2010. Menon was previously foreign secretary of India from October 2006 to August 2009. He has served as ambassador and high commissioner of India to Israel (1995-7), Sri Lanka (1997-2000), China (2000-2003) and Pakistan (2003-2006). He was also a member of the Atomic Energy Commission in 2008-9. A career diplomat, he has served in China (three times), in Japan, and in Austria and in the Mission to the IAEA and UN. He was seconded to the Department of Atomic Energy in the early eighties. His professional experience included India's relations with her neighbours, atomic energy and disarmament issues; and India's relations with the major powers. Menon speaks Chinese and some German. He has an MA in ancient history from Delhi University. He is married to Mohini, and has two children.

Index